ALLE ZEIT WACH 1842

Gertz I. Likhtenshtein

Chemical Physics of Redox Metalloenzyme Catalysis

Translated by A. Beknazarov

With 84 Figures

Springer-Verlag
Berlin Heidelberg New York
London Paris Tokyo

Prof. Dr. GERTZ I. LIKHTENSHTEIN
Institute of Chemical Physics
Academy of Sciences of USSR
142432, Chernogolovka, Moscow Region
USSR

Title of the Russian edition:
Polynuclear Redox Metalloenzymes
published by Nauka, Moscow

ISBN-13:978-3-642-73102-0 e-ISBN-13:978-3-642-73100-6
DOI: 10.1007/978-3-642-73100-6

Library of Congress Cataloging-in-Publication Data. Likhtenshteĭn, G.I. (Gerts Il'ich), 1934– Chemical physics of redox metalloenzyme catalysis. Translation from Russian. Includes index. 1. Metalloenzymes. 2. Oxidation-reduction reaction. I. Title. QP601.7.L55 1988 574.19'25. 87-28877

Softcover reprint of the hardcover 1st edition 1988

Typesetting: K+V Fotosatz GmbH, D-6124 Beerfelden

2131/3130-543210

Dedicated to the memory
of my beloved son
DMITRI (1965–1984)

Preface

"Chemical physics is a science of the physical foundations of chemical transformations" (N. N. Semenov). The main objective of chemical physics is to disclose the detailed mechanism of chemical reactions and to learn to control these processes. The physico-chemical approach hinges upon the extensive application of methods of molecular physics and chemical kinetics. Based originally on simple gas-phase processes, chemical physics gradually extended its sphere of interest to liquid-phase reactions and to processes taking place in solids, including polymers. At present, we witness the fact that the ideas and methods of this science penetrate deeper and deeper into modern molecular biology, including enzyme catalysis.

This monograph treats, from the standpoint of modern chemical physics, the principal general and individual features of the structure and mechanism of action of various classes of oxidation-reduction (redox) metalloenzymes.

There are several reasons for which this branch of science attracts the attention of specialists in various fields – from biologists and those working in medicine to chemists and theoretical physicists.

First of all, is the enormous biological and biochemical importance of processes catalyzed by metalloenzymes. These include biological oxidation with oxygen, oxidative and photo-phosphorylation, atmospheric nitrogen fixation, assimilation of nitrates and sulphites, phototransport of electrons, hydrogen evolution and hydrogenation and photo-oxidation of water, which is far from a complete list of such processes.

Second, metalloenzymes and metal carriers are attractive objects for the solution of the general problems of enzyme catalysis. The high selectivity and catalytic activity of metalloenzymes provoke the deep interest of chemists and physical chemists who try to produce fundamentally new effective catalysts and industrial processes that will simulate model biological processes.

And, finally, the area of investigation of metalloenzymes is a branch of science where a number of scientific areas merge, such as biochemistry, molecular biology, molecular biophysics, bioorganic and bioinorganic chemistry and the chemistry and physics of coordination compounds. It may be expected that the methodological and theoretical armamentarium of modern chemical physics will be especially useful at the junction of various sciences.

This monograph consists of three parts. Part I (Chaps. 1–5) offers general information on metalloenzymes, theoretical and experimental approaches to their investigation and also the foundations of chemical transformation theory. Part II (Chaps. 6–9) is devoted to the analysis of basic physical findings on the structure and mechanism of action of various metalloenzyme classes. Part III (Chaps. 10 and 11) considers the energy, conformational and molecular-dynamic aspects of enzyme catalysis, and also the mechanisms of elementary acts at active sites of metalloenzymes and metal carriers, including multi-electron mechanisms.

The author has the pleasure to express his gratitude to Academician N. N. Semenov, Corresponding Member of the USSR Academy of Sciences A. E. Shilov and Prof. A. P. Purmal, who drew the author's attention to the problem of metalloenzyme catalysis and, in particular, to the biological fixation of molecular nitrogen, the oxidation of paraffins and photosynthesis, and helped and encouraged the author throughout all the stages of writing this book.

The work carried on by the author over the years and the results obtained have, to a considerable extent, been made possible due to the participation of the Laboratory of Kinetics of Enzyme Catalysis of the Institute of Chemical Physics of the USSR Academy of Sciences in a thorough investigation of biological processes and their chemical models carried out at the Department of the Institute of Chemical Physics under the guidance of A. E. Shilov. Many of the problems were set up and solved due to constant cooperation, exchange of information and fruitful discussions between chemists and biochemists. For example, there were absolutely no data on the central enzyme of biological fixation of molecular nitrogen in the mid-1960s and the understanding of the basic principles of the structure and mechanism of action of the enzyme in the late 1970s is the inspiring result of the efforts of several biochemical and chemical laboratories in the USSR, the USA, and England.

The author is very grateful to his colleagues, R. I. Gvozdev, L. A. Syrtsova, L. A. Levchenko, A. P. Sadkov, A. V. Kulikov, A. I. Kotelnikov, V. V. Kochetkov, O. V. Belonogova, V. P. Fogel, V. M. Mekler, S. A. Marakushev, V. R. Bogatyrenko, I. Z. Mitsova, T. S. Pivovarova and E. S. Cherepanova, who have shared the bad times and good times over the many years of investigation of metalloenzymes, nitrogen fixation and photosynthesis, and have made substantial theoretical and experimental contributions to this field.

The author is deeply indepted to T. V. Maltseva, V. V. Kochetkov, T. G. Yukhanova, S. I. Druzhinin and V. S. Belova for their constant help in the preparation of the manuscript and also to A. F. Shestakov and A. V. Kulikov for valuable comments on some of the chapters. Finally, the author is grateful to L. V. Vorobiev for help in the preparation of the bibliography.

Moscow, January 1988 G. I. LIKHTENSHTEIN

Contents

PART I

PART II

PART III

Abbreviations

ADTC	Azidodithio carbonate
ATP	Adenosine triphosphate
Bchl	Bacteriochlorophyll
BLC	Bovin liver catalase
Bph	Bacteriopheophytin
BSA	Bovin serum albumin
CCR	Cytochrome c peroxidase
CD	Circular dichroism
C-side	Cytoplasmic side of the membrane
EITC	Eosin isothiocyanate
ELDOR	Electron-electron double resonance
ENDOR	Electron-nuclear double resonance
EPR	Electron paramagnetic resonance
ESL	Electron-scattering labels
EXAFS	Extended X-ray absorption fine structure
FAS	Flavin adenine dinucleotide
FMN	Flavin mononucleotide
Hb	Hemoglobin
HFS	Hyperfine structure
Hgn	Hydrogenase
HiPIP	High-potential iron proteins
HRP	Horseradish peroxidase
HSA	Human Serum album
ICAO-MO	Linear combination atomic orbitals-molecular orbitals
IHP	Iosine hexaphosphate
IHS	Isotropic hyperfine structure
IR	Infrared spectroscopy
LEFE	Linear electric field effect
Lh	Leg hemoglobin
LL	Luminescent labels
Mb	Myoglobin
MCD	Magnetic circular dichroism
MHS	Magnetic hyperfine structure
MO	Molecular orbitals
MOx	Methane monooxygenase

MOR	Magnetic optical rotation
M-side	Matrix side of the membrane
MUA	Mercaptoundecanoic acid
NADH	Nicotinamide adenine dinucleotide
NGR	Nuclear gamma resonance
NME	Nuclear magnetic modulation effect
NMR	Nuclear magnetic resonance
NR	Nitroxide radical
OR	Optical rotation
PCMB	Parachlormercury benzoate
PES	Photoelectron spectroscopy
PM	Paramagnetic atoms of the metal
PMR	Proton magnetic resonance
PTA	Phosphotungstic acid
RC	Reaction centres
RI	Relaxing ion
RRS	Resonance raman scattering
SA	Serum albumins
SH	Sulphhydryl group
XPS	X-ray photoelectron spectroscopy

PART I

CHAPTER 1

General Information on Metalloenzymes and Metal Carriers

1.1 The Discovery of Metalloenzymes

The first scientist to point to the important role of transition metals in the functioning of biological systems was apparently Berthrand [1]. He also coined the term "laccase" which is used at present as the name of one of the copper-containing enzymes. Though Berthrand erroneously believed that the active component of the enzyme is manganese, his supposition concerning the participation of a transition metal was prophetic. The presence of copper in oxidation-reduction (redox) enzymes was detected in the late 1930s [2]. The first copper-containing enzyme was isolated in a pure state in the 1950s [3]. Thus, nearly 60 years elapsed from the discovery of the function of the enzyme to its preparation in a pure state; 40 years of this period were spent in establishing the chemical nature of the metal.

Other enzymes, however, have had a shorter and more intensive history.

For example, for the central enzyme of the microsomal system of oxidation of heme-containing cytochrome P-450, the corresponding stages are as follows: the detection of the function in 1955 [4]; the discovery of the pigment in 1964 [5]; the isolation in a pure state in 1974 [6, 7].

The iron-molybdenum enzyme xanthine-oxidase, isolated from cow's milk, was purified in 1939 [8] and crystallized in the 1950s [9]. Only a few years passed from the time of detection of the EPR signal of an iron-sulphur protein of the ferredoxin type to the isolation of this carrier [10, 11]. In mid-1965 the iron-containing protein rubredoxin was obtained [12].

Very instructive and somewhat dramatic is the history of the discovery of nitrogenase, which is the central enzyme in the fixation of molecules nitrogen present in nitrogen-fixing microorganisms.

Over 200 years elapsed from the time when the Swedish druggist, Carl Wilhelm Scheele, and the Scotch botanist, Daniel Rutherford, discovered independently a colourless and odourless gas in the atmosphere. The chemical inertness of the gas was so sharply pronounced that Lavoisier later named the gas "azote" because it supported neither life nor combustion.

The chemists were so convinced of the inertness of molecular nitrogen that they paid no attention to the remarkable discoveries of biologists at the turn of the century, which demonstrated the biological fixation of nitrogen by microorganisms.

As early as 1838 in a communication published in the Bulletin of the French Academy of Science, Boussingault pointet to the possibility of biological fixation of atmospheric nitrogen by the legumes of beans [13]. Later, the work of the Russian scientist Vinogorodsky, titled "Concerning the assimilation of free atmospheric nitrogen by microbes", was published [14]. This work marked the beginning of the long path of the scientist and his pupils whch eventually led to the complete proof of the reduction of N_2 into ammonia due to the hydrogen present in carbohydrates of the medium [15].

The enormous demands for bound nitrogen, the presence of an ocean of molecular nitrogen (its content in the atmosphere is estimated to be equal to 4×10^{15} t) and the reality of the process of its fixation under mild conditions by microorganisms could be challenging enough to biochemists and chemists. The biochemists were challenged to isolate from microorganisms the active principle responsible for the fixation, to study its structure and to disclose the chemical mechanism of its action, and the chemists to make nitrogen enter into chemical catalytic reactions under mild conditions.

However, many decades passed before scientists took this challenge seriously and achieved the first successes.

The problem of the preparation of the individual components of microorganism fragments that are capable of fixing molecular nitrogen, N_2, was first formulated by Bakh in the 1930s [16].

Bakh presumed that the process of conversion of inert molecular nitrogen to ammonia or other combined forms could also be effected by the action of a special enzyme or a group of enzymes, which could be isolated, purified and studied by physico-chemical methods. This intuitive supposition was completely confirmed after nearly 40 years, during which period many unsuccessful attempts were made to isolate the enzymatic system of nitrogen fixation and to make if function outside microorganisms. The first failures caused scepticism and disappointment among many biochemists. Theories appeared, according to which the problem put forward by Bakh was unrealistic because the nitrogen-fixation system can function only in the composition of whole, intact cells undamaged by external effects.

The causes of the failures were clarified at a later time. It turned out that the experimentally, highly formidable taks of isolation of a given enzyme from microorganisms, containing at least 2000 various proteins and a large number of other compounds was, in the case of nitrogenase (such was the name of the nitrogen-fixing enzyme), even more complicated. The enzyme proved to be extremely sensitive to traces of oxygen and some metals and for it to function, the reducing forms of a number of compounds and a source of chemical energy, adenosine triphosphate (ATP), were required. The partly purified nitrogenase was first obtained only in 1966 by the well-known American biochemists Bullen and Mortenson [17, 18], who took into account the above mentioned factors and carried out the process in oxygen-free (anaerobic) media in the presence of the appropriate reducing agents and ATP. A few years more elapsed before the scientists of various countries developed the procedure for a more thorough purification of the enzyme.

The analysis carried out confirmed the hypothesis advanced by Bertels and based on the microbiological data obtained in the 1930s, indicating that molybde-

Table 1.1. Principal metalloenzymes and metal carriers

Trivial name	Systematic name	Transition metal, prosthetic group	Molecular weights thousands of daltons	Reference
Non-heme iron-sulphur carriers				
Tetranuclear centres		4 Fe, 4 S		
Ferredoxins				
Clostridium		8 Fe	6	[28]
Chlorobium		8 Fe	6	[28]
Chromatium		8 Fe	10	[28]
Rhodospirillum rubrum		8 Fe	13	[28]
Azotobacter vinelandii		8 Fe	15	[28]
Desulfovibric bacillus		4 Fe	6	[28]
High-potential iron proteins				
Chromatium		4 Fe	9.65	[28]
Trinuclear centres				
Ferredoxins		3 Fe, 3 S		[29]
Desulfovibrio gigas		3 Fe	6	[30]
Azotobacter vinelandii		3 Fe	6	[30]
Binuclear centres				
Ferredoxins		2 Fe, 2 S		
Spinach		2 Fe	10.6	[28]
Microcystis		2 Fe	10.3	[28]
Scenedesmus		2 Fe	10.6	[28]
Azotobacter vinelandii		2 Fe	21	[28]
Pseudomonas putida		2 Fe	12.5	[28]
Escherichia coli		2 Fe	12.6	[28]
Pir adrenodoxin		2 Fe	30	[28]
Mononuclear centres		1 Fe 4 S		
Rubredoxins				
Clostridium		1 Fe	6	[28]
Desulfovibrio gigas		2 Fe	7.9	[31]
Non-heme enzymes and proteins				
Hydrogenases	H_2; ferredoxin Oxidoreductase			
Clostridium pasteurianum		4 Fe	60	[33]
Desulfovibrio vulgaris		12 Fe	59.5	[34]
Pyruvate dehydrogenase	Pyruvate: oxidoreductase	6 Fe, thiamine	240	[36]
Succinate dehydrogenase	Succinate: coenzyme oxidoreductase	8 Fe, FAD	97	[36]
NADH dehydrogenase	NADH: coenzyme oxidoreductase	16 Fe, FMN	220	[37]
4-Methoxybenzoate O-Dimethyl-mono-oxidase	Substrate: O_2 oxidoreductase	4 Fe	120	[36, 37]
Trimethylamine dehydrogenase		4 Fe, lumazine	147	[36, 37]

Table 1.1 (continued)

Trivial name	Systematic name	Transition metal, prosthetic group	Molecular weights thousands of daltons	Reference
Adenylylsulphate reductase	AMP, sulphate: acceptor oxidoreductase	14 Fe, FAD	220	[36, 37]
Sulphite reductase	H_2S: $NADP^+$ oxidoreductase	16 Fe, 4 FAD 4 FMN	670	[36, 37]
Dihydroorotate oxidase	L-5,6-Dihydro-orotate: O_2 oxidoreductase	4 Fe, 4 FAD 7 FMN	110	[36, 37]
Fumarate reductase *E. coli*	NADH: fumarate oxidoreductase	9 Fe, FAD	110	[39]
Ferritin		Up to 4000 Fe	500	[40]
Transferrin		2 Fe	66 – 60	[41]
Hemerythrin		2 Fe	14	[42]
		2 Fe	35	[502]
Heme carriers				
Cytochrome *a*, mitochondria		1 heme		[43]
Cytochrome a_3, mitochondria		1 heme a_3	25	[43]
Cytochrome b_5, microsome		1 heme b	25	[44]
Cytochrome *c*, tuna		1 heme c	12.4	[45]
Cytochrome c_3, *Desulfovibrio desulfuricans*		4 hemes c	13	[364]
Cytochrome c_{552}, *D. salexigenes*		6 hemes c	70	[365]
Heme enzymes				
Catalase, bovine liver	H_2O_2: H_2O_2-oxidoreductase	4 hemes	230	[46]
Peroxidase, horseradish	Donor: H_2O_2 oxidoreductase	1 heme	40	[47]
Cytochrome *c*, peroxidase, yeast	Ferricytochrome *c*: H_2O_2 oxodoreductase	1 heme	31	[48]
Iodide peroxidase, hog thyroid	Iodide: H_2O_2 oxidoreductase	1 heme	64	[49]
Flavoprotein-dependent monooxidase, microsome	RH, reduced flavoprotein: O_2 oxidoreductase	1 heme	49	[51]
Camphor-5-monooxidase *Pseudomonas putida*	Camphor, reduced ferredoxin: O_2 oxidoreductase	1 heme	45	[52]
Alkane-1-monooxidase, yeast	Alkane, reduced rubredoxin: O_2 oxidoreductase	1 heme	50	[53]

Table 1.1 (continued)

Trivial name	Systematic name	Transition metal, prosthetic group	Molecular weights thousands of daltons	Reference
Steroid 11β-monooxygenase, hog thyroid	Steroid, reducing ferredoxin: O_2 oxidoreductase	1 heme	49	[54]
Nitrite reductase, *P. aeruginosa*	Ferredoxin: nitrite oxidoreductase	2 hemes c	120	[50]
Nitrite reductase, spinach	Ferredoxin: nitrite oxidoreductase	4 Fe, siroheme FAD	63	[831]
Sulphite reductase, *M. barkeli*	H_2S: $NADP^+$ oxidoreductase	1 siroheme F_4S_4	23	[55]
Fe-Mo enyzmes				
Nitrogenase		Donor: N_2 oxidoreductase		[56]
Mo-Fe protein of nitrogenase from:				
C. pasteurianum		2 Mo, 24 Fe	220	
A. vinelandii		2.54 Mo, 34–38 Fe	216	
K. pneumoniae		2.01 Mo, 32.5 Fe	218	
Fe protein				
C. pasteurianum		4.05 Fe	56	
A. vinelandii		3.45 Fe	64	
K. pneumoniae		4 Fe	62	
Mo-Fe protein complex				
Complex of Mo-Fe protein and Fe-protein from *A. vinelandii*		2 Mo, 30 Fe	276.5	[57]
Nitrate reductase	NADPH: nitrate oxidoreductase	Mo, 8 Fe	140	[58]
Xanthine dehydrogenase	Xanthine: NAD^+ oxidoreductase	2 Mo, 8 Fe, 2 FAD	280	[59]
Xanthine oxidase	Xanthine: O_2 oxidoreductase	2 Mo, 8 Fe, 2 FAD	280	[38, 60]
Sulphite oxidase	Sulphite: O_2 oxidoreductase	Mo, siroheme	11.5	[60]
Formate dehydrogenase	Formate: NAD oxidoreductase	Mo, 14 Fe, 1 heme		[60]
Sulphite dehydrogenase	Sulphite: ferricytochrome *c* oxidoreductase	Mo, 20 Fe	12	[60]
Cu-containing enzymes				
Cytochrome *c* oxidase	Ferricytochrome *c*: O_2 ox idoreductase	2 Cu, 2 hemes	210	[61]
Ascorbate oxidase	Ascorbate: O_2 oxidoreductase	8 Cu	136	[62]

Table 1.1 (continued)

Trivial name	Systematic name	Transition metal, prosthetic group	Molecular weights thousands of daltons	Reference
Ceruloplasmin	Ferro: O_2 oxidoreductase	8 Cu	142	[62]
Laccase	Donor: O_2 oxidoreductase	6–4 Cu	125	[35, 62]
Tyrosinase	O-diphenol: O_2 oxidoreductase	2 Cu	126	[63]
Dopamine-β-mono-oxygenase	3.4-Dihydroxyphenylethylamine ascorbate: O_2 oxidoreductase	8 Cu	290	[64]
Peroxide dismutase	Peroxide: peroxide oxidoreductase	2 Cu, 2 Zn	32	[62]
Monoaminooxidase	Amine: O_2 oxidoreductase	3 Cu	250	[62]
Galactooxidase	D-Galactose O_2	1 Cu	42	[62]
Cu-containing carriers				
Plastocyanin		2 Cu	21	[62]
Azurin		1 Cu	14	[62]
Ni-containing enzymes				
Hydrogenase	H_2: ferredoxin oxidoreductase			
D. Salexigens		Ni, 12–15 Fe, Se	62	[69]
D. gigas		Ni, 11 Fe	89	[88]

num is required for the fixation of molecular nitrogen [19]. Another component of the enzyme was found to be non-heme iron-sulphur centres [17, 18].

At present over 100 metalloenzymes are known (Table 1.1). The most thoroughly studied enzymes are: some iron-containing proteins (hemoglobin, myoglobin, cytochromes, catalase, peroxidase, plant and bacterial ferredoxins), the iron-molybdenum enzyme xanthine oxidase, copper-containing carrier proteins, and also ceruloplasmin and laccase. A number of laboratories in the USSR, USA, England and FRG are intensively engaged in the study of nitrogenase.

1.2 The Main Biological Role of Metalloenzymes

In accordance with their basic chemical ability to lose and gain electrons readily and to change their redox state upon interaction with various substrates, metalloenzymes and carriers play the key role in processes that support life activity by supplying the required substances and energy.

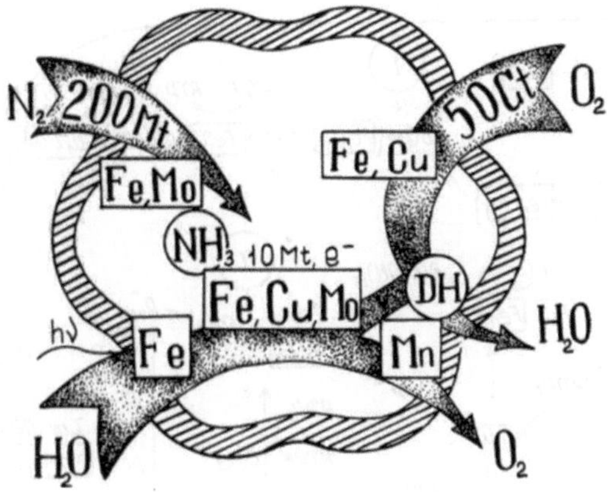

Fig. 1.1. The basic multi-tonnage annual production of metalloenzymes

Figure 1.1 presents the basic processes in which the class of enzymes under consideration participate. They are, first of all, the large-scale processes of molecular nitrogen reduction to ammonia (approximately 200 million tons per annum), the oxidation of substrates by oxygen and the photo-oxidation of water to O_2 (several billion tons) [20].

This most important function is performed by enzymes due to their remarkable ability to activate small, chemically inert molecules, such as N_2, O_2, H_2O. The scope of the processes of intracellular electron transport via iron- and copper-containing proteins is also surprising (Fig. 1.2).

Polyatomic metalloenzymes and, in particular, those containing iron and copper, are involved in the oxidative phosphorylation chain, as a result of which several hundred million tons of phosphate assume and "active" form in the composition of ATP molecules.

Molybdenum, iron and copper enzymes play an important role in the occurrence of processes which are not so efficient as far as the yield is concerned, but are nevertheless vital to life. Mention may be made here of the reduction of nitrates and sulphates and the oxidation of purines and pyrimidines; and example is the oxidation of xanthine to uric acid (iron and molybdenum). Other examples include the oxidation of ascorbic acid (copper), the hydroxylation of saturated hydrocarbons and various aromatic compounds (iron), the oxidation of substrates by hydrogen peroxidase (iron), the metabolism of molecular hydrogen (iron, nickel) and the transport and storage of iron and copper. These processes will be considered briefly in the corresponding chapters. For a more detailed study of the various problems of bioinorganic chemistry, the reader is referred to the literature [21–26].

There is no doubt that as new data on the structure, mechanism of action and regulation and physiological role of metalloenzymes are accumulated, such important practical fields as medicinal and agricultural biochemistry, agrochemistry, pharmacology and model chemical catalysis will have a firmer scientific foundation.

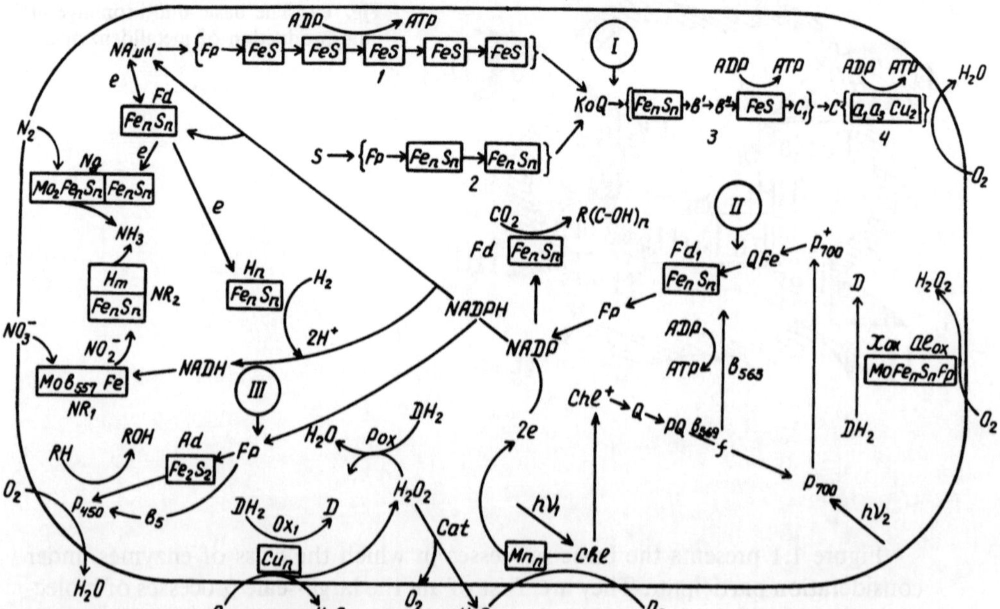

Fig. 1.2. Chemical transformations catalyzed by metalloenzymes and metal electron carriers. The abbreviations used for polyatomic enzymes and carriers are as follows: *Ng* nitrogenase; NR_1 nitrate reductase; NR_2 nitrite reductase; *Hn* hydrogenase; *Fd* ferredoxin; *Fp* flavoprotein; *Hm* heme; *Ad* adrenodoxin; *Ox* oxidases; *Xox* and *Alox* xanthine and aldehyde oxidases; *Cat* catalase; *Pox* peroxidase; *I* mitochondrial oxidation and phosphorylation system; *II* photosystems; *III* hydroxylation system

1.3 Classification and Nomenclature of Metalloenzymes and Metal Carriers. Mechanisms of Enzyme Action

According to the classification recommended by the Enzyme Commission of the International Union of Biochemistry in 1971, all enzymes are divided into six major classes [26, 27]. The oxidation-reduction (redox) function of metalloenzymes allows them to be classified as class 1: oxidoreductases.

According to the systematic nomenclature, the names of enzymes are formed by using the following formula: "donor": acceptor oxidoreductase". In the names of oxidoreductases using NAD or NADP, the cofactor is always indicated as acceptor (except for group 1.6). If the acceptor is unknown, parentheses are used; an example is: succinate: (acceptor) oxidoreductase. When the donor is unknown, no systematic name is recommended.

The second figure denotes the subclass, indicating the specific group subjected to oxidation; the third figure indicates the group of donors that comprise the subclass. The fourth figure is the serial number of enzymes in a particular subclass. The number 99, which follows the second figure, implies the presence of the yet unidentified substrates in a given class.

Trivial nomenclature is also widely used. Dehydrogenating enzymes are classified as "dehydrogenases" or "transhydrogenases". If the hydrogen transfer has not been proven, the term "reductase" is used. The term "oxidase" is referred to in cases where the acceptor is O_2, and the term "oxygenase" is employed for cases where oxygen is incorporated into the substrate. In the latter case there are two variants: "monooxygenase", if one oxygen atom is incorporated into the substrate, and the other joins the water molecule; and "dioxygenase" if both oxygen atoms are incorporated into the substrate.

For enzymes that catalyse the reactions of hydroxylation of substrates and that have not been given a systematic name, the notation "substrate n-hydroxylase" is used, where n indicates the position in the substrate molecule into which the hydroxyl group is introduced. For this type of enzyme the name "monooxygenase" with a mixed function" is also used.

Table 1.1 gives examples of names of metalloenzymes in accordance with the classification of enzymes together with the frequently used trivial names.

Formally, the classification of oxidoreductases also extends to cover many metal carriers, for example, non-heme iron- and sulphur-containing proteins. However, perhaps because of the absence of a strict substrate specificity, most authors prefer using their trivial names with an indication of their biological source. For instance, metalloproteins containing one iron atom per molecule of protein are called rubredoxins, those with two iron atoms per molecule of protein are plant ferredoxins and metalloproteins containing four or eight iron atoms are known as bacterial ferredoxins.

According to the new recommendations of the Enzyme Commission of the International Union of Biochemists, the following classification of non-heme iron-sulphur proteins is proposed: ferredoxins, high-potential iron proteins (HiPIP), rubredoxins and conjugated iron-sulphur proteins with an indication of the origin [26, 27]. In the case of rubredoxins, the source and function of a given preparation must be indicated.

The term cytochrome implies a hemoprotein, whose function consists in the transfer of reducing equivalents associated with the reversible change of the oxidation states of the prosthetic group.

Four main groups of cytochromes are distinguished:

1. Cytochromes *a* in which the prosthetic group is a heme with a formyl side chain.
2. Cytochromes *b* contain a heme without a formyl group (protoheme) which is not covalently bound to the protein.
3. Cytochromes *c* with covalently bound heme groups.
4. Cytochromes *d* with a partially hydrogenated tetrapyrrole ring.

Trivial names are retained for metalloenzymes or their complexes, for which it has not been determined with which substrates they directly interact.

The authors of reviews and monographs often classify enzymes according to the type of transition metal and the manner in which the metal interacts with the protein moiety of the enzyme: heme proteins, iron-sulphur proteins and copper- or molybdenum-containing enzymes. Mixed names are also encountered: iron-molybdenum enzymes, iron-sulphur carriers, iron-molybdenum-flavin proteins, etc.

For convenience, in a treatment of the structural and catalytic properties of polyatomic (polynuclar) metalloenzymes, it is considered expedient to classify

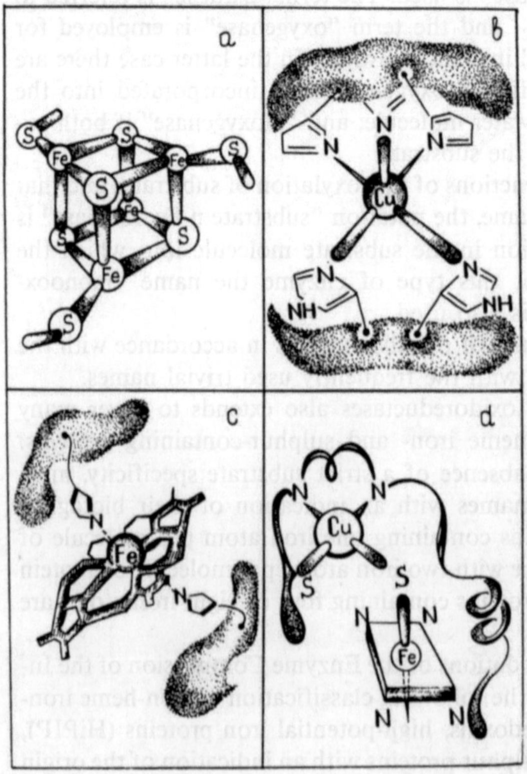

Fig. 1.3 a–d. Active sites of metalloenzymes and electron carriers: **a** ferredoxin [28]; **b** superoxide dismutase [67]; **c** myoglobin [66]; **d** cytochrome *c* oxidase [68]

them according to the character of their interaction and metal atom distribution in the polyatomic assembly. Accordingly, we may distinguish [24] the following types of active sites of enzymes and carriers (see Fig. 1.3):

Type I. A strong interaction between transition-metal atoms with the delocalization of electrons. In functional and structural respects the groups of atoms act as entities; if the atoms move away from one another, the catalytic properties are completely destroyed; the interatomic distance is equal to 0.3 – 0.4 nm. An example of this type is provided by ferredoxins and, possibly, by certain groups of copper atoms in copper-containing enzymes.

Type II. A sufficiently strong interaction between liganded metal atoms (centres). In its main function the complex is an entity which an interatomic distance of 0.3 – 0.6 nm. But during the functioning a separation of the sites may occur, e.g. on different subunits or polypeptide blocks. With one of the sites being moved away, the auxiliary function may be retained. There is ground for believing (see Chaps. 7 and 11 for more details) that this type includes the diatomic molybdenum site in nitrogenase and, possibly, some atomic groups in copper-containing enzymes, and the heme a_3-Cu^{2+} system in the cytochrome *c* oxidase.

Type III. A relatively weak interaction between the sites upon practically complete localization of electrons at the site. The functional-chemical interaction, i.e. the transfer of electrons, is effected at a distance, e.g., by a "tunnel" mechanism (the distance between the sites is 0.6 – 1.5 nm). The system may be uncoupled practically without loss of the specific function of each individual site. There are numerous examples of this type, the electron-transport system in the biological oxidation taking place in mitochondria and microsomes; the photoelectron-transport system; oxidative and photophosphorylation systems; a number of enzymes are associations of type III (nitrogenase, xanthine oxidase, etc.) composed of sites of types I and II.

Type IV. The spatially separated and chemically inert sites, which are, however, connected functionally via, e.g. an allosteric conformational link. An illustration of this type is hemoglobin composed fo four subunits, each of which contains a heme group.

In the literature, use is often made of the term "cluster" for polyatomic assemblies. In the chemical literature, a cluster is a complex in which the metal atoms are linked by covalent bonds. Here, we shall use the term in a broader sense, i.e. associations of interacting metals, which are structural and functional entities (polynuclear system of the first and second types).

In this monograph, attention is mainly focussed on the first three types of active sites (centres) of individual enzymes.

Complex enzymatic membrane systems of the mitochondrial, microsomal and photosynthetic membrane types are considered only briefly in connection with the problem of the generality of electron transfer or oxygen activation mechanisms.

As will be shown in subsequent chapters, the oxidation reduction (redox) reactions catalyzed by metalloenzymes, especially by polynuclear metalloenzymes, proceed by a large variety of mechanisms, which almost always involve several steps. For a concise and unambiguous description of these mechanisms we make use of the following notations: the figures 1, 2, 3, 4, etc., denote the multiplicity of the mechanism, i.e. the number of electrons or hydrogen atoms transferred in a given elementary step. The order of these figures indicates the sequence of electron transfer from the reducing to the oxidizing agent. The direction of the transfer is indicated by arrows; the figures enclosed in parentheses denote the number of steps of a given type required for the process to be completed. Brackets are used to designate the steps that proceed intramolecularly on the enzyme. The general scheme may be represented as follows:

$$
\begin{array}{l} \text{Reducing} \\ \text{agent} \\ n_{red}(m_0) \end{array}
\rightarrow
\left[
\begin{array}{l} \text{Primary} \\ \text{electron} \\ \text{acceptor} \\ n_A(m_1) \end{array}
\rightarrow
\begin{array}{l} \text{Primary} \\ \text{carrier} \\ n_i(m_2) \end{array}
\rightarrow
\begin{array}{l} \text{Electron} \\ \text{donor} \\ n_D(m_i) \end{array}
\right]
\rightarrow
\begin{array}{l} \text{Oxidizing} \\ \text{agent} \\ n_{ox}(m_{i+1}) \end{array} ,
$$

where n_{red}, n_A, n_D and n_{ox} are respectively, the numbers of electrons transferred in a given step; m_0, m_1, m_2 and m_i indicate the number of times the step is repeated.

For example, the reduction of a certain one-electron acceptor A_1 (e.g. cytochrome *c* or ferricyanide) by the wellknown two-electron reducing agent nicotinamide adenine dinucleotide (NADH) via a quinone mediator, e.g. flavin mononucleotide (FMN), and a one-electron carrier X may be illustrated by the following scheme:

$$\begin{array}{lllll} \text{NADH} & \rightarrow [\text{FMN} & \rightarrow \text{X}] & & \rightarrow A_1 \\ 2 & \rightarrow [2 & & \rightarrow 1\,(2)] & \rightarrow 1\ . \end{array} \tag{1.1}$$

The FMN-X system serves as some kind of device which switches the reaction over from the two-electron to the one-electron mechanism [24].

In this particular simple case, because of the obvious coincidence of the multiplicity of the electron transport from NADH to FMN and from X to A_1 and also because of the total number of electrons being retained in the linear chain, scheme (1.1) may be written in a simpler manner:

$$[2 \rightarrow 1]\ .$$

A further example is the multi-electron reduction of molecular oxygen to H_2O_2 by way of hydrogen transport by a one-electron donor D_1. Multi-electron processes become necessary if the formation of a one-electron intermediate product of the superoxide radical O_2^- is energetically unfavourable. As will be shown in Chapter 9, such a case occurs in the mitochondrial electron transport system (cytochrome *c* oxidase). Two variants of the mechanism of two-electron reduction of O_2 to the peroxide derivative are possible. In the first variant, two internal one-electron carriers are arranged successively and the scheme of the mechanism is written as:

$$1 \rightarrow [1\,(2) \rightarrow 1\,(2) \rightarrow 2]$$

or, more simply, as

$$[1 \rightarrow 1 \rightarrow 2]\ .$$

The parallel arrangement of the carriers is described by the following simplified scheme:

$$\begin{bmatrix} 1 \searrow & \\ & 2 \\ 1 \nearrow & \end{bmatrix}\ .$$

For complex, multi-component systems the notation suggested here allows one to describe reaction mechanisms in a concise and graphic manner.

CHAPTER 2

Physical Methods of Investigation of Metalloenzymes

The solution of the complicated problems of the structure and action mechanism of such complex systems as metalloenzymes is possible only by using an all-round physico-chemical approach. This approach calls for the consistent application of a wide spectrum of methods, from the "simplest" chemical analyses and biochemical procedures of protein isolation to the utilization of quantum-mechanical calculations and the latest physical methods, including the entire range of spectroscopic techniques, from gamma resonance to radiowave spectroscopy.

The most precise and unequivocal method of investigating enzymes and their active sites is the X-ray diffraction analysis. At present, this method has led to the establishment of the structure of many metalloenzymes and metal carriers. There is no doubt that the number of structures studied by X-ray diffraction will continuously increase. However, a long period of time is required to determine the structure of a large number of the most interesting enzymes, whose molecular weight exceeds 100000 daltons, and also of non-crystallizable enzymes, e.g. membrane enzymes. At the same time, as has been shown for a number of metalloenzymes (Chaps. 6–9), physico-chemical methods used jointly allow us to obtain sufficiently complete information on the electronic structure of a transition-metal atom and its ligand environment in the active site. X-ray diffraction analysis does not cover such important structural and dynamic properties as the charge density distribution, the energy of electron shells, the intramolecular dynamics of electrons and nuclei, conformational transitions, the chemical and physical transformations of molecules in biological systems and in solutions. Therefore, the construction of an X-ray diffraction structural model of the enzyme is only a prerequisite for further detailed investigations.

In this chapter, we shall briefly consider the principles of physico-chemical methods that have found wide application in the study of the structure of the active sites of metalloenzymes. Major emphasis will be placed on spectral methods and also on methods based on the analysis of physical properties that are sensitive to interatomic interactions.

2.1 Physical Methods. General Principles

Two effects are observed in the interaction of radiation with matter: the scattering and absorption of radiation.

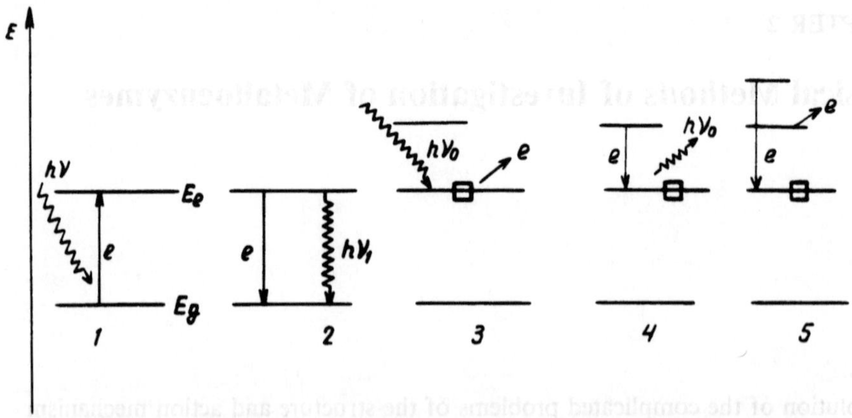

Fig. 2.1. The principles of the primary and secondary processes upon absorption of electromagnetic radiation by a quantum system. The *solid arrows* indicate the direction of electron transfer (*e* is an electron); the *wavy arrows* indicate the absorption or emission of quanta; the *squares* represents an electron vacancy: *1* absorption of radiation; *2* luminescence; *3* photoelectron or X-ray photoelectron effect; *4* X-ray fluorescence; *5* Auger effect

The efficiency of the scattering of electromagnetic radiation without change in frequency (Rayleigh scattering), gamma rays, X-rays and light by molecular systems is determined, to a considerable extent, by the electron density of the constituent atoms, by the manner in which they are distributed in the molecule and the degree of ordering of molecules in space [71]. Based on the principle of scattering recording are such most important methods as X-ray diffraction analysis and small- and large-angle X-ray scattering techniques. An analogous principle forms the basis of electron microscopy and electron diffraction, which use the scattering of fast electron beams on molecular structures. Neutron diffraction has recently found application; it is based on the recording of the scattering of neutrons by atomic nuclei [72, 73].

Of paramount importance in the study of metalloenzymes and model transition-metal complexes are spectral methods associated with the absorption of electromagnetic radiation. A key role in this method is played by the quantum properties of molecular systems and the discrete nature of the energy levels in nuclei, atoms and molecules (Fig. 2.1).

Table 2.1 shows the tentative arrangement of discrete energy levels in transition-metal atoms. According to the energy differences, these levels are conveniently divided into the following groups: nuclear levels, inner electron shells, outer electron shells (mainly d-orbitals or orbitals formed with their participation), and levels that split in a magnetic field (so-called Zeeman electron levels) for electrons and nuclei (the latter property is essential for systems that possess an uncompensated electron or nuclear spin).

The transitions between levels with different energies ΔE involve the absorption or emission of a quantum of electromagnetic energy with frequency ν. Here, the following resonance condition is satisfied:

$$\Delta E = h\nu \quad (2.1)$$

Table 2.1. The energy and time characteristics of spectral methods

Transition energy E, eV	Frequency of method, Hz	Lifetime of excited state τ^*, φ	Range of variation of parameters measured in various complexes		Method
			E, eV	With respect to transition energy	
2.4×10^{-7} to 1.2×10^{-6}	6×10^{7} to 3×10^{8}	10^{-5} to 10^{-2}	10^{-12} to 10^{-9}	10^{-6} to 10^{-3}	NMR spectroscopy
3.6×10^{-5} to 5.4×10^{-4}	10^{10} to 1.5×10^{11}	10^{-12} to 10^{0}	10^{-9} to 10^{-5}	10^{-4} to 1	EPR spectroscopy
2.4×10^{-2} to 4.8×10^{-1}	6×10^{12} to 1.2×10^{14}	10^{-14} to 10^{-12}	10^{-6} to 10^{-1}	10^{-4} to 10^{-1}	Infrared and Raman spectroscopy
1.6 to 6.2	1.5×10^{15} to 3.7×10^{15}	10^{-7} to 10^{-10}	10^{-5} to 1	10^{-5} to 10^{-1}	Electron spectroscopy, circular dichroism, optical rotatory dispersion, magnetic circular dichroism, magnetic optical rotatory dispersion, fluorescence
10^{3} to 6×10^{3}	3.6×10^{17} to 2×10^{18}	10^{-17} to 10^{-18}	0.1 to 10^{+3}	10^{-4} to 1	X-ray photoelectron spectroscopy; X-ray absorption
10^{4} to 10^{5}	3.6×10^{18} to 3.6×10^{19}	10^{-7} to 10^{-8}	10^{-8} to 10^{-6}	10^{-12}	Nuclear gamma resonance

Table 2.2. The effect of spin state on the parameters of spectral methods

Transition energy, eV	Nature of spectrum change with variation of the spin state	Range of variation of energy level with variation of spin state, eV	Method
0.25×10^{-6} to 1.2×10^{-6}	Line shift, variation of the line shapes	10^{-12} to 10^{-9}	Nuclear magnetic resonance
3.6×10^{-5} to 5.4×10^{-4}	Line shift, variation of the lines shapes	$10^{-9}-10^{-5}$	Electron paramagnetic resonance
$\sim 10^3$	Line splitting, appearance of satellite lines	$\sim 10^1$	X-ray photoelectron spectroscopy
10^4 to 10^5	Line splitting, variation of line shapes	10^{-7} to 10^{-6}	Nuclear gamma resonance

Tables 2.1 and 2.2 show the scope of energy differences between levels corresponding to the frequency of transitions between them, the spectroscopic method used and the approximate range of variation of the spectral parameters with change in the spin state of the d-shell of a transition metal. Since the specialists in various areas usually employ different units for energy measurements, we give the conversion factors for different systems: 1 kcal = $0.34975\ cm^{-1}$ = 4.1840 J = 4.3361×10^{-5} eV = 1.0485×10^4 MHz = 0.503217 kelvin K = 4.1840×10^7 erg.

An important parameter of the spectral methods is the lifetime of the excited state, τ^*, which is connected, in the case of a single spectral line, with the level width (δE) and, accordingly, with the line width ($\Delta\nu$) by the Heisenberg uncertainty principle:

$$\delta E\, \tau^* \sim h \ . \tag{2.2}$$

As the value of τ^* decreases the line is broadened (*homogeneous broadening*). The shape and width of the spectral line are influenced by a number of factors. This is primarily the presence of additional interactions which cause so-called fine and hyperfine splittings of the energy levels and the heterogeneous nature of the environmental effect on the position and splitting of the levels. The influence of these factors results in an *inhomogeneous broadening* of the lines.

A number of physical methods are based on the study of secondary phenomena that accompany the absorption of radiation [74, 75] (Fig. 2.1). The emission of quanta from excited levels leads to the appearance of luminescence. The measurement of the kinetic energy of an electron "knocked out" from the orbit upon absorption of a quantum provides information on the position of the one-electron levels of the system (photoelectron and X-ray photoelectron spectroscopy). The filling of the electron vacancies formed upon ejection of electrons from the inner shells of the atom upon X-ray absorption causes X-radiation. The scattering of this radiation by the surrounding atoms may serve as a source of information about their nature and mutual arrangement. This information is obtained by analyzing the fine structure of the X-ray absorption spectrum (EXAFS, extended X-ray absorption fine structure). The wavelength and intensity of secondary emission and also the properties of the secondary electron released upon vacancy filling may be associated with the properties of the valence electrons of the atom (X-ray fluorescence, Auger spectroscopy).

Among other methods of investigation of transition-metal complexes containing unpaired electrons, special mention should be made of the method of measuring magnetic moments from the extent of their interaction with an external magnetic field [76].

2.2 Mössbauer Spectroscopy

The nuclear gamma resonance is based on the resonance absorption of gamma quanta, which results in a transition of the nuclei of some elements (Mössbauer atoms, e.g. [^{57}Fe] and [^{119}Sn]) to an excited state. The absorption is effected in accordance with the general resonance law expressed by Eq. (2.1). The basic condition for the absorption of gamma quanta is the small magnitude of the recoil momentum upon absorption and emission of a quantum [77–80].

Mössbauer discovered that this condition is realized if a Mössbauer nucleus can impart a greater part of its momentum to the lattice. If the bond with the lattice is weak or the environment has considerable mobility, the recoil energy is high and the resonance condition is violated. The difference in resonance energy between the gamma-quantum source and absorber is compensated by their relative motion. The dependence of the absorption intensity on the velocity of motion is known as the nuclear gamma resonance spectrum or the Mössbauer spectrum (Fig. 2.2). Obviously, the nuclei of the source and absorber must have nearly identical energy levels. These conditions are realized, in particular, for a pair of [^{57}Fe] (absorber) and [^{57}Co] (source).

The quantitative relationship between the resonance absorption intensity (f′), the square of the mean amplitude of nuclear displacement in the direction of recoil, $\langle x\rangle^2$, and the wavelength of the gamma quantum, λ, is expressed in the simplest case of the harmonic approximation by the following relation:

$$f' = \exp(-4\pi^2\langle x\rangle^2/\lambda^2) \ . \qquad (2.3)$$

Another important parameter, which determines the possibility of observing the Mössbauer effect, is the natural line width (Γ), which is connected with the lifetime of the nucleus in the excited state (τ_n^*) by the relation:

$$\Gamma \sim h/\tau_n^* \ . \qquad (2.4)$$

For the isotope [^{57}Fe] the value of τ_n^* is equal to 1.4×10^{-7} s.

Apart from the above mentioned absorption intensity and line width, the main experimentally measured parameters are: the isomer (or chemical) shift (δ), quadrupole splitting ($\Delta_Q E$), the splitting in a magnetic field (ΔH_M), the line shape and the fine structure of the lines (Fig. 2.2).

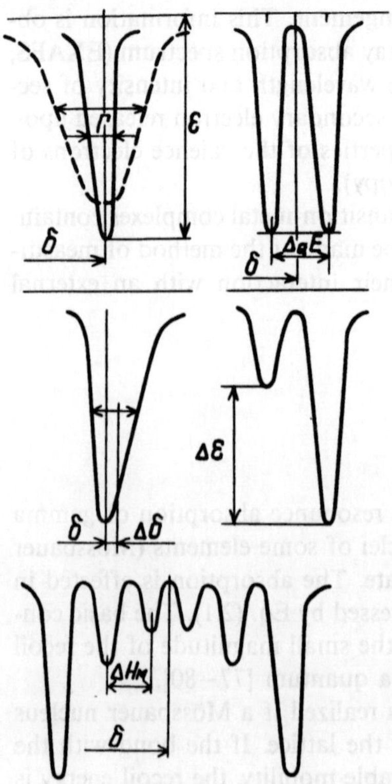

Fig. 2.2. Types of Mössbauer spectra and the spectrum parameters associated with the structure of complexes [77–80]

The first two parameters are connected with the effect of the electronic environment of the nucleus. The magnitude of the chemical shift is determined by the following relation:

$$\delta = a_\gamma(|\Psi_{(0)1}|^2 - |\Psi_{(0)2}|^2) \ , \qquad (2.5)$$

where a_γ includes the nuclear parameters, which are practically constant for a given nucleus, and the second factor characterizes the difference in total electron density between the source and absorber nuclei. Since the electron density at the nucleus has a finite value only for s-electrons (the Fermi or contact interaction), the effect of the population of the other orbitals is indirect; it may, for example, be realized through hybridization for p-electrons and through shielding for d-electrons. The quadrupole splitting (Δ_QE) in the NGR (nuclear gamma resonance) spectra arises due to the asymmetry of the interaction of the electric quadupole moment of the nucleus with an inhomogeneous electric field.

In those cases where the nucleus is in an internal or external magnetic field, e.g. if the Mössbauer atom is paramagnetic, a magnetic hyperfine structure (MHS) may appear in the spectrum. The [^{57}Fe] nucleus in the ground state has a magnetic moment with a magnetic quantum number of I = 1/2, which may have two orientations in the magnetic field: one along the magnetic field and the other against it. The magnetic moment of the excited nucleus (I* = 3/2) may take four possible orientations: along the field, against the field and at two various angles to the field. Differently oriented magnetic moments can give different contributions to the energy of interaction, which is why the nuclear gamma resonance spectrum may consist of six components.

The magnitude of internal magnetic fields in paramagnetic compounds reaches 1 – 10 T, and therefore in a number of cases a splitting in the absence of a magnetic field is observed. In other cases, the application of a field of high strength can reveal its hyperfine structure. According to theory, the relative line intensities depend on the angle θ between the directions of the applied magnetic field and the propagation of gamma quanta. For isotropic polycrystalline samples the line intensity ratios must be (3:2:1:2:3).

A set of the Mössbauer spectrum parameters, which depend on the electronic structure and symmetry of a transition-metal complex, may be used to establish the valence and other properties of the central atom. For example, if the value for sodium nitroprusside is used as the zero standard, the value of the isomer shift (δ) in mm s^{-1} for ferric compounds (the d^5-state) lies in the range of $\delta = 0 - 0.4$ for covalent low-spin complexes and in the range of 0.4 – 0.9 for ionic high spin complexes. The magnitude of quadrupole splitting $\Delta_Q E$ varies respectively, within the range of 0 – 8 mm s^{-1} for covalent complexes and 0 – 1.0 mm s^{-1} for ionic complexes. For divalent iron (the d^6-state) in high-spin compounds, the value of δ ranges from 1 to 1.8 mm s^{-1}, which corresponds to an additional shielding of electrons when one more electron is placed on the d-orbital. Low-spin complexes are characterized by lower values of δ (0 – 0.4 mm s^1). The value of $\Delta_Q E$ for these complexes varies in the range of 0.2 mm s^{-1} [78].

Of primary importance for the solution of the problem of the interacton between transition-metal atoms is the analysis of the magnetic hyperfine structure of the spectra and the effect of magnetic fields on it. It is evident that the pairing of electrons on the orbits in low-spin complexes and the appearance of anti-ferromagnetic interactions between the atoms must result in an averaging of the magnetic fields for the time τ_n^* and in the disappearance of the magnetic hyperfine structure.

An interesting additional procedure, which aids in revealing the hyperfine structure, may be the superposition of a weak magnetic field of the order of tens of mT. This effect, which is associated with fine electron-nuclear interactions, may lead to the appearance of a magnetic hyperfine structure in a case where the spectral structure is "blurred" because of the presence of fluctuating weak fields before the field is applied [80].

Among the other promising applications of Mössbauer spectroscopy, mention should be made of the use of synchrotron radiations for the study of Rayleigh scattering, three-dimensional analysis and protein dynamics [70].

2.3 X-Ray Absorption. X-Ray Photoelectron and Photoelectron Spectroscopy

The principles of the methods based on the measurement of X-ray absorption intensity and the study of secondary processes are presented in Fig. 2.1. The absorption of a hard X-ray quantum of frequency ν_0 is associated with an electron transfer from the inner shell of the atom to the high-lying free levels of the discrete spectrum or with photoionization. The transfer energy, which determines the position of the band, depends on the nature of the atom and its environment. The filling of the vacancy by electrons from the high-lying shells is accompanied by secondary X-radiation with frequency $\nu < \nu_0$ and also by the ejection of a secondary electron (the Auger effect).

Since these secondary phenomena are associated with transitions from the outer valence shells, the secondary X-ray emission and Auger spectra may contain unique information on the energy and properties of systems, which is essential for the understanding of their reactivity [74, 75, 81].

X-ray photoelectron spectroscopy (XPS) and photoelectron spectroscopy (PES) rely on the phenomenon of the photoelectron effect, i.e. the radiation energy is consumed to overcome the binding energy of the electron (E_b) situated at the inner shells of atoms, for the work function of the material of the spectrometer (φ_1) in cases where the photocurrent is measured by the spectrometer, and to impart a kinetic energy (E_k) to the electron. The measurement of E_k at a known radiation frequency ν makes it possible to determine the binding energy from the equation:

$$E_b = h\nu - E_k - \varphi_1 \; . \tag{2.6}$$

The value of the quantum energy lies within the range of 1 keV and the line width ranges from 0.1 to 10 eV.

The high sensitivity of the method, the possibility of investigating the surface layers of very complex materials and, what is very important, the relationship between E_b and the electronic structure and magnetic properties of atoms make XPS and PES highly effective and promising tools for the investigation of metalloenzymes [82, 83].

Thus, the magnitude of the chemical shift E_B in closely related paramagnetic complexes varies approximately linearly with the effective change q, which characterizes the oxidation state of the atom, and with the electronegativity of the neighbouring atoms and is proportional to the atomic number. For example, for iron complexes with sulphur ligands the following relations are satisfied [84]:

$$E = \alpha_E + \beta_E q \; , \tag{2.7}$$

where $\alpha_E = 162.2$ and 52.2 eV respectively, for the 2p-level of sulphur and the 3p-level or iron; the values of β_E for these levels are 4 and 2 eV per unit charge respectively. Relation (2.7) may be used to establish the oxidation state of iron in biological systems.

An important property of the spectra is line splitting in paramagnetic complexes, which enables one to analyze their magnetic structure. It should be pointed out that the method has the shortest "observation time" ($\tau = 10^{-15}$ to 10^{-16} s). It is therefore possible to observe practically instantaneously magnetic states which cannot be recorded by "slower" methods, such as magnetic susceptibility measurements, EPR (electron paramagnetic resonance) or NMR (nuclear magnetic resonance).

The splitting of the spectra from s electrons is associated with the interaction of these electrons with unpaired electrons residing on the closely spaced energy levels. Such a distribution, which is known as the multiplet distribution, is observed for 2s lines in the case of O_2 and NO, and also in compounds with unpaired 3d and 4f electrons.

Another consequence of the presence of unpaired electrons in an atom is the appearance of additional satellite lines for p-electrons as a result of the shake-up phenomenon. The physical significance of this phenomenon is that part of the photoelectrons impart some of the energy to an unpaired electron residing on a higher lying orbit, which must lead, according to Eq. (2.6), to the appearance of a new line with a shifted E_b. According to the theory [85], the probability of this effect is approximately proportional to the number of unpaired electrons in the d-shell. Therefore, the intensity of satellite lines is associated with the total spin of the atom.

The possibility of such an "instantaneous" observation of the magnetic state of the system allows one to distinguish between the covalent binding of electrons and the weaker anti-ferromagnetic binding. It is evident that this capability of the method is very important for the solution of the problem of the nearest neighbours of a given atom and the degree of electron sharing in an ensemble.

2.4 Extended X-Ray Absorption Fine Structure (EXAFS)

The development of the proton synchrotron technique has recently made it possible to produce continous high-power radiation in the X-ray region of the spectrum [86–88]. This has increased to a large extent the efficiency of all the methods associated with the absorption or scattering of radiation, including the X-ray method. It will suffice to say that the exposure time in the new technique is 10^2 to 10^3 times shorter than in the case of conventional radiation sources.

Especially promising for the problem of interest to us, namely the problem of the relative arrangement of atoms in complexes, is the application of synchrotron radiation for obtaining X-ray absorption spectra in the region of K-absorption (the EXAFS method). The fine structure observable in this region of the spectrum is associated with the number, nature and arrangement of the atoms present in the environment surrounding the transition-metal atom under study.

Figure 2.3 explains the physical significance of the phenomena that occur during the interaction of an X-radiation photon with the central atom and its environment. The absorption of a photon with the ejection of an electron is accompanied by the appearance of photoelectron emission from the absorbing atom. To the outgoing photoelectron wave are added the waves arising from its reflection from the adjacent atoms. If the amplitudes of the outgoing and reflected waves are added at the central atom, a maximum appears in the absorption spectrum; if they are subtracted, the spectrum shows a minimum (Fig. 2.3). A remarkable property of the spectra is that they contain information not only on the first sphere of the surrounding atoms but also the subsequent ones. The dependence of the absorption amplitude $\chi_{(k)}$ at a given value of the wave vector k on the number of atoms in the sphere, N_i, and their distance from the central atom (r) is specified by the equation:

$$\chi_{(k)} = \sum \frac{N_i |f_{(k)}|}{k r_i^2} \exp(-\sigma_i^2 k^2) \sin[2 k r_i + \alpha_{(k)}] \ , \tag{2.8}$$

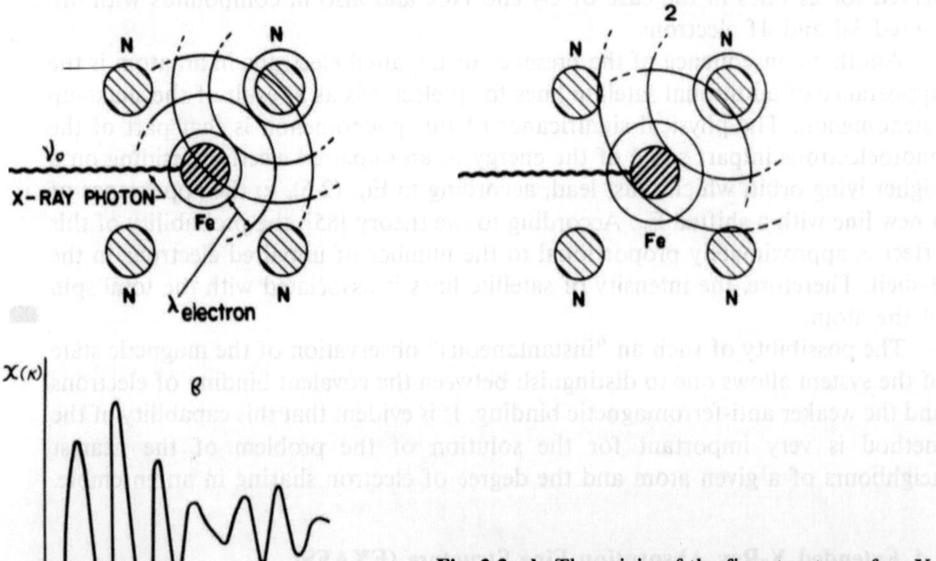

Fig. 2.3 a, b. The origin of the fine structure of an X-ray absorption band (**a**) and the spectrum of a molybdenum-containing complex of nitrogenase (**b**) [86, 88]: **a** amplitude of the outgoing and scattered waves add up (the maximum in the absorption band) (*1*) and the amplitudes substract (the minimum in the absorption band) (*2*); *Fe* and *N* are the central atom and the atoms of the ligand respectively; ν_0 is the frequency of X-radiation; **b** EXAFS of the Mo atom of nitrogenase

where $f_{(k)}$ and $a_{(k)}$ are the parameters dependent on the nature of atoms and radiation energy; σ_i is the dispersion of r_i.

The experimental determination of these parameters for simple compounds will permit the solution of the most complicated task of elucidating the structure of the active sites of metalloenzymes and metal carriers. It is also possible to solve the inverse problem: to calculate these parameters from X-ray spectrum data with the structure being known. For cases where the above mentioned parameters are known beforehand, the method allows one to establish the relative differences in the distance r_i.

2.5 Ultraviolet and Visible Spectroscopy

2.5.1 Absorption Spectra

The position of the absorption line is determined by differences in the electronic energy levels, between which transitions occur [89, 90]. When a quantum of radiation is absorbed, only one electron is transferred. According to the Franck-Con-

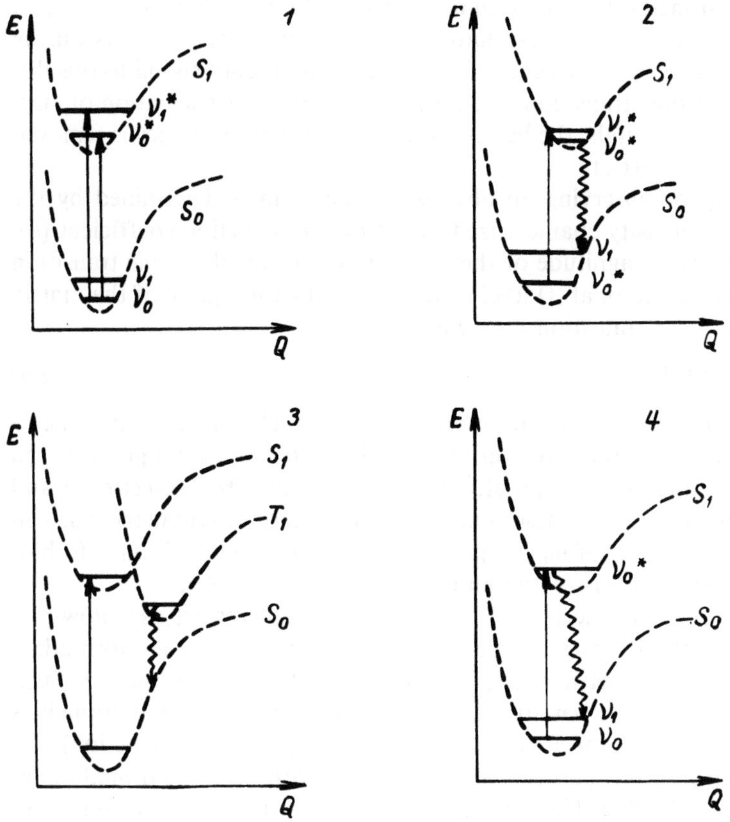

Fig. 2.4. The origin of spectra: *1* absorption; *2* fluorescence; *3* phosphorescence; *4* resonance Raman scattering

don principle, since the time of an electronic transition ($\tau_e \sim 10^{-16}$ s) is considerably shorter than the characteristic time of nuclear displacement ($\tau_n = 10^{-12}$ to 10^{-14} s), electronic transitions occur with the nuclei being practically fixed. This means that in the diagrams showing the adiabatic potential curves for the ground and excited states, the transitions can be represented by a vertical line between the potential curves of the upper and lower electronic states, i.e. with the nuclear coordinates Q being fixed (Fig. 2.4). The absorption line corresponds to a series of all possible vertical transitions. The Franck-Condon principle accounts, in particular, for the differences between the absorption and emission (fluorescence) spectra: in an excited state the nuclear configuration is somewhat different from the ground-state configuration (Fig. 2.4). Since electronic transitions are accompanied by a change in the vibrational state of the system's nuclei and there is a large set of vibrational sublevels in the system, the lines in the absorption spectra of complexes smear out into phonon bands.

In the majority of cases, the position of the absorption band maximum (ν_{max}) does not correspond to the difference between the energies at the minima of the adiabatic potential curves of the excited and ground states. However, since the approximation methods of quantum chemistry, the crystal field theory and the ICAO-MO method do not take into account nuclear displacements during oscillations, the values of ΔE calculated by these methods correspond to this difference. If the electronic transitions are associated with the participation of electronically degenerate orbitals, the bands are split when the degeneracy is removed due to the Jahn-Teller effect.

The possibility of observing an absorption spectrum is determined by the magnitude of line intensity characterized by the molar extinction coefficient (ε).

Quantitatively, the magnitude of the line intensity of an electronic transition from the initial (i) to the final (f) level is determined by the square of the matrix element of the dipolar transition moment (m_0):

$$\varepsilon \sim m_0^2 = |\langle \Psi_i | \vec{M} | \Psi_f \rangle|^2 , \quad (2.9)$$

where $\vec{M}$ is the electronic transition moment operator. This quantity provides a quantitative characterization of the variation of the dipole moment upon electron phototransfer; Ψ_i and Ψ_f incorporate, in a general case, both electronic and nuclear wave functions. Since these functions possess certain symmetry, one can establish whether the value of m_0 is equal to zero or not (the transition is forbidden or allowed) on the basis of the known symmetry roles [90].

The selection rules for electronic transitions in complexes are as follows:

1. Transitions with a change in the multiplicity of the system are forbidden. For instance, transitions between triplet and singlet states are forbidden. Such transitions are allowed if the wave functions of the states of different multiplicities are mixed (in the presence of a sufficiently effective spin-orbit interaction).

2. Electronic transitions between electronic states with the same orbital quantum number l are forbidden. This applies to all d-d transitions in metal complexes with a centre of inversion. Such transitions are allowed if the electronic transition is accompanied by the excitation of odd nuclear combinations. In such cases $\varepsilon \sim 10$ to $10^3\ M^{-1}\ cm^{-1}$. Completely allowed are transitions between electronic states of different parity and, in particular, transitions between bonding and antibonding levels. Allowed transitions are those between $\pi \rightarrow d$, $\sigma \rightarrow d$ and $d \rightarrow \pi^*$ levels localized, as a rule, at different atoms in complexes. Such transitions with a charge transfer are characterized, like other allowed transitions, by high band intensities ($\varepsilon = 10^3$ to $10^5\ M^{-1}\ cm^{-1}$).

A joint analysis of the experimental absorption spectra and the results of quantum-mechanical calculations enables one to solve a number of problems concerning the electronic structure of transition-metal complexes: to determine the order of arrangement of the terms and the magnitude of splitting between the terms; to establish the symmetry in the arrangement of the ligands; and to judge the vibronic properties of complexes, etc. The presence of intense, characteristic bands in the absorption spectra is of great assistance to experimentalists dealing with enzymes.

2.5.2 Optical Rotation and Circular Dichroism. Magnetic Optical Rotation and Magnetic Circular Dichroism

When molecules are irradiated with polarized light, i.e. electron waves in which the polarization vector is localized in a single plane, under certain conditions two interconnected phenomena arise [91–94]. One phenomenon (optical rotation, OR) pertains to the rotation of the plane of polarization of plane-polarized light of certain frequency ν which has passed through the sample at an angle φ_ν given by:

$$\varphi_\nu = a_{OR}(n_L - n_R) , \tag{2.10}$$

where n_L and n_R are the indices of refraction of the two components of plane-polarized radiation for a wave whose plane of polarization is rotated to the left (L) and to the right (R) respectively; a_{OR} is a coefficient. The other phenomenon is circular dichroism (CD); it is associated with differences in the extinction coefficient between the left- (ε_L) and right-hand (ε_R) circularly polarized light:

$$\theta_\nu = b_{CD}(\varepsilon_L - \varepsilon_R) . \tag{2.11}$$

According to the existing theory [91–93],

$$\varphi_\nu \sim m_0 \mu_0 f_{OR}(\nu) \tag{2.12}$$

and

$$\theta_\nu \sim m_0 \mu_0 f_{CD}(\nu) , \tag{2.13}$$

where μ_0 is the matrix element of the dipole magnetic moment of a given electronic transition; the functions of the line shapes f_{OR} and f_{CD} are interconnected by the Kroning-Kramers relations. As follows from formulas (2.12) and (2.13), optical rotation and circular dichroism are determined by the same selection rules as the absorption, but here the selection rules for the dipole magnetic moments are added. It can be strictly shown that the product $m_0\mu_0$ is not equal to zero only in those cases where the molecule has no centre of inversion or plane of symmetry, i.e. if it is totally asymmetric.

This property opens up interesting possibilities for the study of the electronic structure of asymmetric centres, which include the active sites of many metalloenzymes and metal carriers. The presence of characteristic bands in the OR and CD spectra, which may have either a positive or negative sign (depending on whether ε_L or ε_R is greater), provides a better resolution of the lines in these spectra than in the absorption spectra.

Even greater possibilities can be offered by methods based on the measurement of the rotation of the polarization plane in a magnetic field (MOR) and of circular dichroism in a magnetic field (MCD) [92–94]. Three effects are distinguished in the MOR and MCD spectra: the A, B and C effects.

The A-effect manifests itself in the case of electronic transitions with a degenerate excited state. Under the influence of a magnetic field (H_0) the degeneracy is removed. In the simplest case, where the ground state is non-degenerate and the excitation is two-fold degenerate (Fig. 2.5), the theory predicts a difference in the extinction coefficients ε_L and ε_R. This will give rise to a shift-

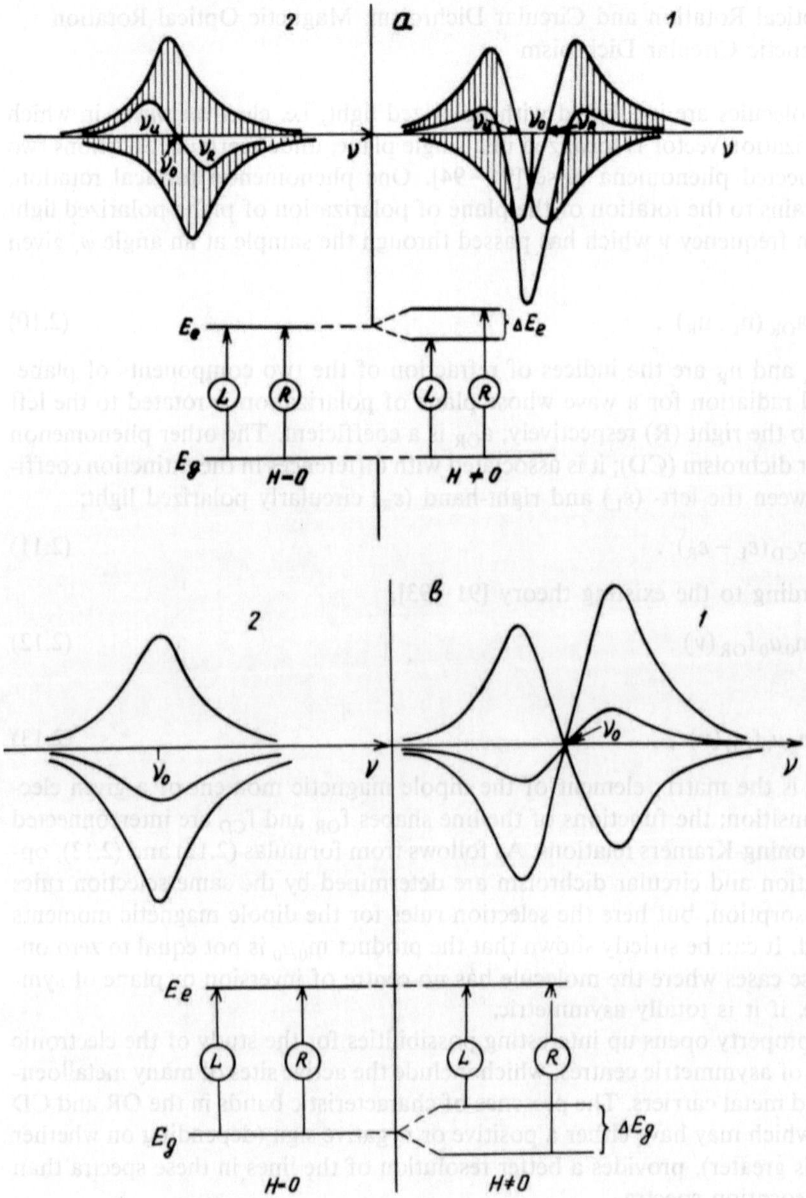

Fig. 2.5 a, b. The origin of the A-type (a) and C-type (b) effects of the magneto-optical rotation dispersion (*1*) and magnetic circular dichroism (*2*) [92–94]. Shown at the *centres* are the resulting spectra. *L* and *R* the left- and right-hand circularly polarized components of radiation respectively

ing of the absorption bands for the left- and right-hand components in an opposite sense as compared with the absorption band in the absence of a magnetic field. The resultant band has a characteristic shape (Fig. 2.5) and may have different signs, depending on the value of ε_L and ε_R. The shifts in MOR spectra occur in an analogous manner.

The quantum-mechanical theory predicts the following dependence of circular dichroism on the magnetic field strenght (H_0):

$$\theta_M \sim m_0^2(\mu_e - \mu_g)(\delta f_M(\nu)/\delta\nu) H_0 , \qquad (2.14)$$

where μ_e and μ_g are the matrix elements of the magnetic moment operator for the ground and excited states respectively.

Effects of the B-type occur due to the creation of asymmetry by the magnetic field in the molecule under study. In such a case:

$$\theta_M \sim m_0^2 \frac{\delta m_0^2}{\delta H_0^2} H_0 . \qquad (2.15)$$

Effects of the C-type are caused by the removal of the degeneracy of the ground states in an induced magnetic field (Fig. 2.5). In such cases it turns out that the extinction coefficients ε_L and ε_R will depend on the population levels in accordance with the Boltzmann law and, hence, on the temperature:

$$\theta_M \sim \frac{m_0^2 \mu_0}{k_B T} f_M(\nu) H_0 . \qquad (2.16)$$

As can be seen from Eqs. (2.14–2.16), which are valid at field strengths of up to about 10 T, all three effects depend linearly on H_0. It should also be noted that the Kroning-Kramers relations allow one to pass over easily from MCD to MOR spectra.

The aforementioned leads to the conclusion that the MCD and MOR methods provide us with unique possibilities for the study of the structure of transition-metal complexes: to evaluate the magnetic moments of excited states, to study degenerate states, to establish the spin state and to analyze the symmetry of the complexes, etc. The presence of positive and negative effects and narrow characteristic bands and the dependence of the intensity of some bands on temperature sharply increase the resolving power and the information content of the methods as compared with the OR and CD spectra.

2.6 Infrared and Raman Spectroscopy. Resonance Raman Scattering

In molecules, the nuclei occupy the mid-position (Q_0), which is determined by the minimum energy of interaction of all the nuclei and electrons of the system. Within the framework of the concepts of classical motion, the deviation of the nuclear positions from the mean value (ΔQ), e.g. under the influence of thermal fluctuations, causes and increase in the energy of the system and the nuclei strive to return to the original position. Thus, nuclear vibrational processes arise [90, 95]. If the restoring force (f_Q) is proportional to the displacement, then the vibration is termed harmonic and the vibrating system is termed the harmonic oscillator. The proportionality factor between f_Q and ΔQ is called the force constant (K_Q).

In a diatomic molecule, the frequency of the harmonic oscillation (ν_n) is related to the reduced mass of the molecule (M) and the quantity K_Q as follows:

$$\nu_n = (K_Q/M)^{1/2}/2\pi \ . \tag{2.17}$$

In the case of the quantum harmonic oscillator, the value of ν_n, like the energy of the system, may take only discrete values determined by the vibrational quantum number v = 0, 1, 2, . . .:

$$\nu_{n(v)} = h\nu_{n0}(v+1/2) \ . \tag{2.18}$$

Here, vibrations with v = 0 occur even at absolute zero temperature.

For the vibrations of the oscillator a selection role exists, $\Delta v = \mp 1$, which in real molecules, however, is partially violated because of the anharmonicity of the vibrations. Transitions at $\Delta v > 1$ are called overtones of the fundamental or normal modes. In polyatomic molecules, the complex vibrational process may be resolved into a combination of n-collective normal harmonic vibrations. For a non-linear molecule consisting of n nuclei, $n = 3m-6$. If the parallel vibrations differ substantially in frequency, they may be regarded as independent. If the frequencies of two normal modes with frequencies ν_n^0 are equal and a sufficiently strong dipolar interaction occurs between the modes of vibration, then, as a result of the resonance quantum-mechanical effect (the Fermi resonance), this degenerate vibration splits into two modes with frequencies less and greater than ν_n^0. The magnitude of splitting is the higher, the stronger the interaction.

In determining the number of vibrations the symmetry of vibrations is of substantial importance. At a high degree of symmetry part of the vibrations is degenerate.

Theory predicts that when the system passes over to a vibrational excited state, the absorption of electromagnetic radiation may be effective only if the vibration is accompanied by a change in the electric dipole moment (polarization) of the nuclei. This condition determines the selection rules for the absorption of radiation in the infrared region.

The vibrational processes in molecules are also reflected in the Raman spectra. When the substance is irradiated at a frequency far from the frequency of its absorption, additional (satellite) lines may appear in the scattered light. The origin of such lines may be accounted for qualitatively by the fact that during the interaction of electromagnetic radiation with the molecule some of the radiant energy is transferred to the excited vibrational levels and part of the energy is released from the excited levels. In a Raman spectrum, only those vibrations are active at which the polarizability of the system is changed. In a general case, for molecules possessing a centre of symmetry there are no active vibrations simultaneously in the infrared and Raman spectra.

The vibrational processes in metal complexes differ in some respects from those in organic and inorganic systems. The participation of the heavy nuclei of metal atoms ensures the appearance of vibrations with low frequencies of ν_n = $400-40\ cm^{-1}$. The presence of delocalized molecular orbitals common for metal atoms and ligands leads, as a rule, to the absence of characteristic group vibrations in vibrational spectra. Finally, a set of vibration frequencies in a complex as a whole is very sensitive even to a partial replacement of the ligands.

In metalloenzymes and metal carriers, the active sites incorporate only a small part of the atoms of the macromolecule and against the large number of bands in the vibrational spectrum, it is usually difficult to determine the frequencies corresponding to the vibrations of these atoms. Therefore, the considerably more selective method of resonance Raman scattering (RRS) has found the widest application. The selectivity of the method is due to the fact that the spectra do not display all the vibrations of the system but only those associated with the electronic excitation of the chromophore being studied.

The principles of the method are as follows. The sample is irradiated with monochromatic light of frequency ν_0 in the absorption band of the complex. In the scattered light relatively intense, narrow RRS bands are observed with frequencies shifted relative to ν_0. The origin of these bands is ascribed to the electronic transitions of the excited level (E_e) to the first vibrational level (v_1) of the ground state (E_g) (Fig. 2.4). The value of the intensity of resonance-scattered light is given by:

$$I_s \sim \frac{m_0' m_0''}{\nu - \nu_0 + \Delta_1^0} , \tag{2.19}$$

where m_0' is the electric dipole moment operator of the transition from the ground electronic and vibrational levels $E = E_g$, $v = v_0$; m_0'' is the corresponding operator for the transition from the excited level $E = E_e$, $v = v_i$ to the level with $E = E_g$, $v = v_1$; Δ_1^0 is a constant for a given system which characterizes the width of the vibrational levels. The high intensity of resonance scattering is provided by large values of m_0' and m_0'' for allowed transitions and by the smallness of the term $\nu - \nu_0$ in the vicinity of the absorption band.

2.7 Electron Paramagnetic Resonance

2.7.1 General

The paramagnetic properties of radicals and transition-metal complexes are due to the intrinsic momentum of an unpaired electron (to the spin S). In a constant magnetic field of strength H_0, in systems with $S = 1/2$ the electron spins are oriented in two positions: along the field and against it (the Zeeman effect). Because these orientations differ in the extent of interaction with the magnetic field, two Zeeman energy levels arise in the system which are differently populated [90, 96–100] (Fig. 2.6). The superposition of a perpendicular electromagnetic radiation of frequency ν leads to a reorientation of the spins and to the equilibration of the level populations if its energy ($h\nu$) is equal to the energy of spin reorientation ($g\beta_e H$):

$$h\nu = g\beta_e H , \tag{2.20}$$

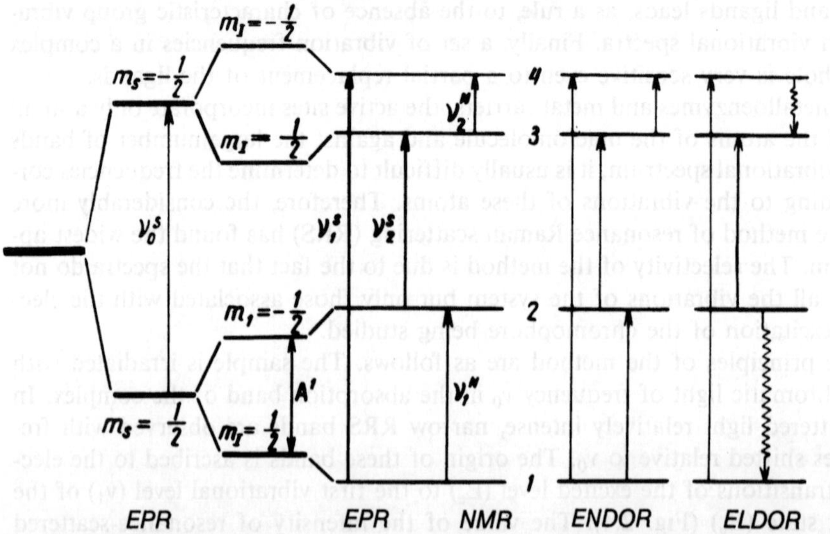

Fig. 2.6. The transitions between the energy levels used in EPR, NMR, ENDOR and ELDOR: ν_0^s, ν_1^s, ν_2^s are the frequencies of transitions between the Zeeman electron levels; ν_1^N and ν_2^N are the frequencies of transitions between the Zeeman nuclear levels; A^1 is the hyperfine interaction constant

where $\beta_e = 0.93 \times 10^{-16}$ J T^{-1} (T = tesla) is the Bohr magneton; g is a factor equal to $g_e = 2.0032$ for a free electron. In real paramagnetics, the effective f-factor g_{eff} deviates from this value because of the effect of orbital magnetism. The EPR spectrum is usually the dependence of the intensity of the first derivative of the absorption line on the magnetic field strength H at fixed frequency ν (Fig. 2.7).

The possibility of recording EPR spectra and the line shape of the spectrum is largely determined by the return (relaxation) of the spin system after the primary absorption act to a state of equilibrium. The most important characteristics of this process are the *spin-spin relaxation time* (T_2) and the *spin-lattice relaxation time* (T_1).

The spin-lattice relaxation is due to the interaction of electron spins with the magnetic and electric fields of the environment (the lattice) which are fluctuating both in magnitude and direction. This interaction results in a change in the population of the Zeeman levels and, hence, in the average energy of the spin system. The fluctuating local magnetic fields produced by the surrounding nuclear and electronic magnetic moments are factors responsible for the spin-spin relaxation. Without changing the average energy of the spin system at the Zeeman levels, this type of interaction causes a spread in the frequencies of the resonance transition between the levels and appears in the EPR spectra as a broadening of the absorption lines. In the simplest case of the Lorentz line shape, the quantity T_2 is connected with the value of line width between the points of maximum slope ($\Delta\nu_m$) in Hz by the following relation:

$$T_2^{-1} = \frac{\Delta\nu\sqrt{3}}{4\pi} . \tag{2.21}$$

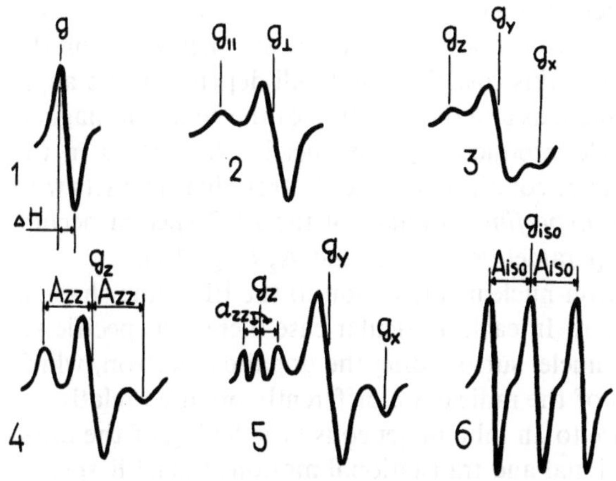

Fig. 2.7. Examples of the EPR spectra of paramagnetic centres differing by a set of parameters, the g-factor and the hyperfine structure (A): *1* isotropic centre; *2* centre with axial symmetry of the g-tensor; *3* centre with rhombic symmetry of the g-tensor; *4* spectrum of the nitroxide radical with a nearly axial symmetry of the A-tensor in a magnetically diluted rigid matrix taken on the EPR spectrometer at a microwave frequency of 3 cm; *5* spectrum of the nitroxide radical taken on the EPR spectrometer at a microwave frequency of 2 mm [114]; *6* spectrum of the nitroxide radical in a solution with averaged values of g_{iso} and A_{iso}

The values of T_2 and T_1 define another important property, which determines the possibility of observing the resonance, i.e. the ability of the EPR line to be saturated. The physical meaning of the saturation phenomenon means that in the case of slightly effective relaxation (low values of $1/T_1$ and $1/T_2$) the absorbed microwave radiation raises more and more spins to a state of increased energy, as a result of which the populations of the upper and lower levels are equilibrated and the rate of microwave absorption becomes equal to the rate of radiation. The dependence of the amplitude of the individual line on the absorption frequency is described by the equation:

$$A = a_2T_2(1+b_2T_2^2+c_2^2H_1^2T_1T_2)^{-1}H_1 \ , \qquad (2.22)$$

where a_2, b_2 and c_2 are coefficients; H_1 is the intensity of microwave radiation. The equation can be used to determine T_1 and T_2. For this purpose, use is also made of pulse techniques described in the literature [101–103].

An unpaired electron in a spherically symmetric s-state undergoes a contact or Fermi interaction with the nucelar spin. This interaction leads to the splitting of the EPR line into the components corresponding to the different orientations of the nuclear spin relative to the constant magnetic field. Such is the mechanism of formation of the so-called *isotropic hyperfine structure* (IHS). Because of the spherical symmetry of the electron density in the s-state, the result of this type of electron-nuclear interaction does not depend on the angle of rotation of the paramagnetic material relative to the magnetic field. Upon interaction with one nucleus with a quantum magnetic number I each of the lines must split into $2I+1$ lines. In such cases, only transitions with $m_I = 0$ are allowed (Fig. 2.6). In the case of equivalent nuclei with $I = 1/2$ the relative intensities obey a binomial distribution.

Another type of electron-nuclear interaction occurs when an unpaired electron resides on an orbital which does not possess spherical symmetry, e.g. on the p- or d-orbital. It is evident that in this case the result will depend on the angle between the direction of the orbital axis and the direction of the constant magnetic field. An additional local field produced by the nuclei will, just as in the preceding case, change the resonance condition for the electron that interacts with the nuclei. Thus, an *anisotropic hyperfine structure* of the EPR spectra occurs, which is characterized by three parameters: A_x, A_y and A_z (Fig. 2.7).

The contribution of an electron-nuclear interaction to the EPR line shape is accounted for in an analogous way. In each particular case there is a specific set of local-fields, generated by the nuclei surrounding the unpaired electron, which results when the molecular axes of the radical are differently oriented relative to the magnetic field. This gives rise to an inhomogeneous broadening of the lines.

Thus, in the absence of rotational and translational motions the EPR spectra of frozen, vitrified solutions are superpositions of the "microspectra" of an enormous number of particles with different degrees of orientations of the HFS and g-factor tensors relative to the external magnetic field. The motion of the particle leads to an averaging of the interactions and a simplified spectrum.

Analysis of electron-nuclear interactions is of great importance in the interpretation of EPR spectra. It permits one to reveal the electron delocalization of the electron over several nuclei and to elucidate the nature of ligands. In a number of cases it is useful to replace a natural isotope. For example, the replacement of H by D results in a sharp decrease of the HFS (hyperfine structure) constant due to the small value of the magnetic moment of the latter. The replacement of the natural isotope [^{56}Fe] by [^{57}Fe] leads to the splitting of the lines into two components.

The shape of the EPR spectrum is also substantially influenced by the anisotropy of the g-factor. The physical significance of this effect is that the unpaired electron in a radical or complex is acted on by the asymmetric electronic environment and the effect of orbital magnetism, which is responsible for the deviation of the g-factor from the value 2.0023, will depend on the orientation of the radical. The g-factor anisotropy is characterized in a general case by three values: g_x, g_y and g_z (Fig. 2.7).

The effect of the anisotropy of the g-factor and of the heterogeneity of the spectra can be effectively elucidated by using radiospectrometers which operate at various wavelengths: 2 mm, 8 mm, 3 cm and 36 cm. According to Eq. (2.20), the change of frequency must be accompanied by a change in the relative position of the lines, which differ in the g-factor, and not of the lines with different hyperfine structures. A spectacular notion of the advantages of radiospectrometers based on shortwaves is given in Fig. 2.7, which presents the spectra of the nitroxide radical recorded by 2-mm and 3-cm radiospectrometers [114].

2.7.2 Specificity of EPR Spectroscopy of Transition-Metal Complexes

The presence of a system of the d-shells of transition-metal atoms, which are partially filled with electrons, creates quite a new situation in complexes as compared

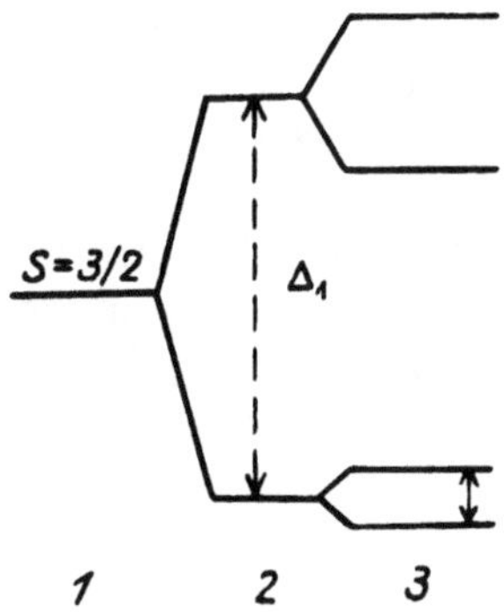

Fig. 2.8. Energy levels in a system consisting of three interacting spins [104]. *1* Spins do not interact; *2* splitting in the zero field with energy Δ_1; *3* Zeeman splitting in a magnetic field. The *solid arrow* shows the transitions detected by the EPR method

with radicals. This is associated with the possible presence of several unpaired electrons in high-spin complexes [24, 90, 96–100].

According to the Kramers theorem, if the number of unpaired electrons is odd, then in the absence of an external magnetic field the magnetic levels of the system are degenerate. The observation of electron paramagnetic resonance is possible only when this field is applied. In a system with an odd number of electrons the interaction between them can either completely or partially remove the degeneracy (the splitting in a zero field). In such systems the EPR spectra can, in principle, also be obtained in the absence of a magnetic field. However, in this case, too, the magnitude of the magnetic field may play a decisive role: an increase in the field strength H_0 increases the energy gaps between the initial split levels and splits the levels that have retained their degeneracy. The EPR spectra will display only those transitions, whose frequency corresponds to the microwave range. This is illustrated in Fig. 2.8, which also shows the behaviour of the Zeeman levels of a three-electron system with S = 3/2 [104].

As can be seen in Fig. 2.8, as a result of the splitting in the zero field (Δ_1) and in the magnetic field, two pairs of levels are generated for which the calculation predicts the following values of the g-factor: $g'_x = 0.34$, $g'_y = 0.32$ and $g'_z = 6.0$ for the upper pair and $g_x = 3.66$, $g_y = 4.32$ and $g_z = 1.98$ for the lower pair. Since the values of the g-factor for the upper pair shift the EPR lines to a practically unobservable region, only those lines can be recorded in practice which refer to the transitions between the lower levels. The line with $g_z = 6$ is observable but has a lower intensity. Another prediction of the theory is that at temperatures $k_BT > \Delta_1$, the integral line intensity will fall due to the transfer of some of the spins to the upper levels, the transitions between which are unobservable.

The second specific feature of transition metals in the important role of spin-orbit interactions.

In the presence of relatively low lying, unfilled shells this type of interaction has a substantial bearing on the magnetic properties of the complex and, in particular, on the magnitude of the g-factor.

For the Fe, Cu, Mn and Mo complexes considered in this monograph the effect of the crystal field is usually stronger than that of the spin-orbit interaction. The perturbations that act on the electron shell of the central atom have the following energy values (in order of magnitude): $10^4\,cm^{-1}$ for the crystal field; $10^2\,cm^{-1}$ for the spin-orbit interaction; and $1\,cm^{-1}$ for spin-spin interactions and the interaction with an external magnetic field. The last three types of interactions are therefore regarded as perturbations. Under these conditions, the total spin (S) and orbital (L) momnets are no longer coupled and the transition involves a change in the spin magentic quantum number by unity: $\Delta m_s = \mp 1$.

The third specific feature of transition-metal complexes is the substantial effect of the ligand environment. This effect manifests itself through the splitting (Δ) and the degree of covalence of the ligand-metal bond. In particular, for the Cu^{2+} ion in a tetragonal field the values of the g-factor are given by the equations:

$$g_{\parallel} = 2 - \frac{8\lambda\alpha^2\beta^2}{\Delta_{xy}}; \tag{2.23}$$

$$g_{\perp} = 2 - \frac{2\lambda\alpha^2\beta'^2}{\Delta_{xy,yz}}; \tag{2.24}$$

where Δ_{xy} and $\Delta_{xy,yz}$ are the splittings between the orbitals respectively; λ is the spin-orbit coupling constant which is equal in this case to 270 cm^{-1}; α, β and β' are fractions of the unpaired electrons on the $d_{x^2y^2}$, d_{xy} and $d_{xz,yz}$ orbitals respectively.

The presence of a spin-orbit interaction in the complex is a necessary condition for the effect of the external electric field on the g-factors of the EPR spectra, the effect of the *linear electric field effect* (LEFE). Another necessary condition is the absence of a centre of inversion in the complex. This makes it possible to use LEFE for the study of the symmetry of the arrangement of ligands around the central atom [105, 106].

Other specific features of transition-metal complexes are associated with the delocalization of the spin, densitiy over the nuclei of the central atom and over the ligands, which gives rise to a large variety of HFS effects. Finally, the complexes exhibit a wide range of spin-lattice relaxation times T_1 and their temperature dependencies. As an example, we give the following values of T_1 (in seconds) for aqua transitional-metal complexes at T = 77 K: $\approx 10^{-8}$ for Cu^{2+}; $\approx 10^{-8}$ for Mn^{2+} and Fe^{3+} and 1.5×10^{-12} for Fe^{2+}.

The low values of T_1 (10^{-11} to 10^{-13} s) for biological iron complexes at physiological temperatures do not allow the EPR spectra to be observed under these conditions. For these spectra to recorded low temperatures are required (below 77 K).

In concluding this section, let us sum up the interactions in complexes with an axial symmetry using the apparent Hamiltonian ($\hat{H}_{app}$):

$$\hat{H}_{app} = D[S_z'^2 - \tfrac{1}{3}S'(S'+1)] + g_{\parallel} H_z S_z' + g_{\perp}(H_x S_x' + H_y' S_y') \\ + g_{\parallel} A_{\parallel} S_z' + A_{\perp}(S_x' I_x + S_y' I_y) - \gamma_N \beta_N H_0 I\,. \tag{2.25}$$

Here, the first term characterizes the splitting in the zero field, the second and third terms characterize the Zeeman splitting, the fourth and fifth the hyperfine structure and, finally, the last term defines the direct interaction of the magnetic field with the nuclear magnetic moment. It should be emphasized that, in view

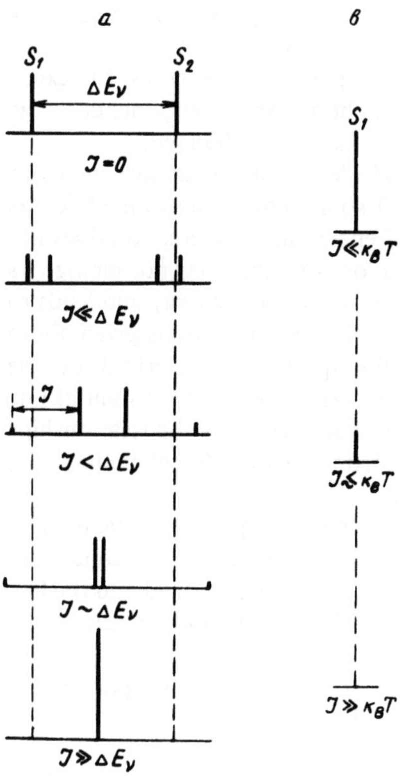

Fig. 2.9 a, b. The position and intensity of the EPR lines of two spins, S_1 and S_2, upon their approach, as functions of the magnitude of the spin-spin interaction (J), energy splitting between the lines, $\Delta E_\nu = h(\nu_1 - \nu_2)$ and thermal energy k_BT: **a** $\Delta E_\nu \neq 0$; **b** $\Delta E_\nu = 0$

of what has been said above, the value of the apparent spin S′ in the apparent Hamiltonian differs in a general case from the value of the true spin S.

2.7.3 Effect of Spin-Spin Interactions of the Parameters of EPR Spectra

The principal types of interaction between the spins localized at different atoms of transition metals or radicals are the exchange and dipolar interactions considered above [90, 96–100].

Let us first consider the case of interactions with $S = 1/2$, which have the same parameters (g, A) and are characterized by an isotropic exchange with an exchange integral J. The position and total line intensity when the spins are brought closer together or T is changed will not change to the temperature at which J becomes comparable with k_BT (Fig. 2.9). As the temperature continues to drop, the intensity of the spectrum will begin to fall if the spins in the paired (singlet) state have a lower energy than in the unpaired (triplet) state. In the presence of an anisotropic exchange or dipolar interaction with the quantity J′, the line splits into two, each of the lines being shifted by an amount 3/4 J′ (Fig. 2.9).

For different ions with different resonance energies (ΔE_ν), as the spins approach each other and the interaction (J) is enhanced, the following sequence of changes in the spectrum must be observed (Fig. 2.9): (1) the splitting of each of the levels by an amount approximately equal to J (at $\Delta E_\nu \gg J$); (2) the intensity of the extreme external components fall and they move apart from each other, the internal components increase and come closer together ($\Delta E_\nu \sim J$); (3) the external pair of components practically disappears, the internal components merge and in the spectrum a single line is observed ($\Delta E_\nu \ll J$) is observed.

Such a difference in the behaviour of equivalent and non-equivalent (with respect to the position in the EPR spectrum) spins may be accounted for as follows [100]. In the first case, the exchange of energy upon interaction does not alter the average energy of the spins and therefore the line does not change its position. Secondly, the effective interaction averages the spin energy and leads to the appearance of one "average" resonance line. The considerations given above may also prove useful in the analysis of the spectra characterized by the anisotropy of the g-factor and the hyperfine structure and by a substantial line width. For portions of the spectrum having the same frequency the relationships observed for identical ions are valid, and for portions with different frequencies, relationships observed for different ions hold.

Spin-spin interactions affect the line shape of the EPR spectrum. The interaction between a given spin and the other spins depends on a set of distances between them, on the angles between the direction of the magnetic field and the line that connects the given pair, etc. If the local fields assume random values, the EPR spectrum line has a Gaussian form.

Upon intensive exchange at frequency $\nu_{ex} = J/h$ the central spin part of the line assumes a Lorentz shape with the half-width given by:

$$\Delta\nu_{1/2} = \langle\Delta\nu_0^2\rangle/\nu_{ex} \ , \tag{2.26}$$

where $\langle\Delta\nu_0^2\rangle$ is the value of the second moment of the parent Gaussian line. Of paramount importance for the practical problems of determination of the distance between the paramagnetic centres is the determination of the contribution of the dipole-dipole interaction to the quantities T_1 and T_2 for the spin observed. The energy of such an interaction depends on the distance between the spins, their magnetic moments and on the values of the spin-spin relaxation time (τ_2) and the spin-lattice relaxation time (τ_1) of the second paramagnetic. A further important factor is the product of the relaxation time $\tau_{1,2}$ to the interaction energy expressed in frequency units ($\Delta\omega$).

At $\Delta\omega^2\tau_1^2 \gg 1$ the effect of the second spin may be regarded approximately as an interaction with a permanent dipole moment; at $\Delta\omega^2\tau_1 \ll 1$, because of the weak interaction, it may be considered a weak perturbation.

As shown by Kulikov [107, 108], of special interest for the solution of structural problems is the case where one paramagnetic centre (a nitroxide radical or an ion of a transition metal) has times T_1 and T_2 long enough for the line to be observed and the saturation curves to be obtained, and the second centre is characterized by short relaxation times, i.e.

$$\tau_1\mu\gamma/r^3 \ll 1 \ . \tag{2.27}$$

Under these conditions formulas have been derived for the dipolar contribution to the relaxation time of the spin being observed [see Egs. (3.11 and 3.12)].

According to these formulas, from the experimentally measured values of $\Delta(1/T_1)$ and $\Delta(1/T_2)$ and the known values of μ, τ_1 and $\Delta\omega$, it is possible to calculate the distance r. Estimates shows that under the most favourable conditions, when $\tau \sim \Delta\omega_i^{-1}$, the maximum possible (for determination) distances to the second paramagnetic of spin S_2 are as follows: $r_{max} = 10$ nm for $S_2 = 1/2$ and $r_{max} = 15$ nm for $S = 5/2$.

For cases where r is known, e.g. from X-ray data, Eqs. (3.11) and (3.12) can be used to determine the magnetic parameters of the EPR spectrum of the unobservable centre (μ or $\tau_{1,2}$), In principle, we may also employ the approach in which the value of r is determined from the maximum of the temperature dependence at $\tau \sim \Delta\omega_i^{-1}$; at other temperatures, formula (3.11) is used for the determination of τ_1.

Various applications of the approach under consideration will be considered in more detail in Chapters 3, 7 and 9.

One of the effective channels of spin-lattice relaxation in transition-metal complexes is the decoupling of spins with a transition of one or several of them to higher levels by a two-phonon mechanism (the Orbach effect) [109, 110]. Since such a transition is associated with the overcoming of the energy of the exchange interaction between the spins, a quantitative study of the effect of temperature on T_1 allows one to determine the value of the exchange integral (J). In particular, as has been shown in the literature [110], in anti-ferromagnetic complexes, in which the change of the spin state $S = 1/2 \rightarrow S = 3/2$ is possible, the following relation is fulfilled:

$$\log T_1 = 1.3|J|/T + C \; , \tag{2.28}$$

where C is a constant which is independent of J.

2.8 Electron-Nuclear and Electron-Electron Double Resonance. Magnetic Nuclear Modulation

In double-resonance methods, the sample place in a magnetic field is subjected to the simultaneous action of two different radio frequencies. In the electron-nuclear double-resonance (ENDOR) technique, the sample is subjected to microwave radiation (with a frequency of about 10^{10} Hz) and radio frequency radiation (10^6 to 10^8 Hz). In the electron-electron double-resonance (ELDOR) technique, the microwave radiation is applied at two frequencies. The principles of the ENDOR and ELDOR techniques can be illustrated by an example of the hyperfine interaction of spin $S = 1/2$ with a nuclear spin of $I = 1/2$ [111–113].

Figure 2.6 shows a diagram of energy levels, whose position is specified by a Hamiltonian:

$$\hat{H} = g\beta_e H m_2 + A' m_s m_I - g_N \beta_N H m_I \; , \tag{2.29}$$

where g and g_N are the electronic and nuclear g-factors respectively; β_e and β_N are the electronic and nuclear Bohr magnetons ($\beta_e = 1836\,\beta_N$); A is the hyperfine interaction constant; H is the magnetic field; m_s and m_I are the projections of the electronic and nuclear spins on the direction of the magnetic field.

The arrows represent the transitions observed in EPR and NMR. The transitions observed in EPR spectroscopy are: $\Delta m_s = \pm 1/2$ and $\Delta m_I = 0$ at frequencies ν_1^s and ν_2^s. These transitions differ from each other by an energy A. The transitions observable in NMR are: $\Delta m_s = 0$ and $\Delta m_I = 1$ at frequencies $\nu_1^N = 0.5\,A' + g_N \beta_N H$ and $\nu_2^N = 0.5\,A' - g_N \beta_N H$.

The ENDOR spectrometers are based on EPR and equipped with a special resonant cavity, which enables the sample to be exposed to radio frequency radiation.

In the ENDOR method one EPR line, for example, with a frequency ν_2^s, is saturated with strong microwave radiation, so that the population of levels 2 and 3 are equilibrated and the EPR signal is weakened due to the saturation. The sample is simultaneously subjected to a strong radiofrequency field. If the frequency of this field is ν_1^N or ν_2^N, the populations of levels 2 and 3 return to equilibrium and the EPR line of frequency ν_2^N increases in intensity. The ENDOR spectrum is the dependence of the amplitude of the saturated EPR line on the frequency of radio frequency radiation. In the case under consideration the ENDOR spectrum consists of two lines of frequencies, ν_1^N and ν_2^N. Knowing these frequencies, one can calculate the quantities A' and g_N. The ENDOR technique base an advantage in the analysis of complex, poorly resolved EPR spectra, which are a single, structureless line. Even in such a case, the method enables one to determine the hyperfine interaction constant and thus increases the amount of information provided by the EPR method and its resolving power. The ENDOR spectra are always less complicated than the EPR spectra and their interpretation is simpler.

In the ELDOR method, one EPR line (e.g. of frequency ν_2^s) is observed in a weak microwave field, and the other transition (e.g. of frequency ν_1^s) is saturated with a strong microwave field. If the transitions marked by wavy arrows are possible in the system, the saturation of the line of frequency ν_2^s will cause the saturation of the ν_2^s line and a decrease in its intensity.

The ELDOR spectrum is usually the dependence of the amplitude of the EPR line at fixed frequency of the saturating, powerful microwave radiation and at the frequency of the radiation on the intensity of the constant magnetic field strength (H). In the case under consideration, the ELDOR spectrum is composed of two lines which differ in energy by an amount 2A'. Like ENDOR, the ELDOR method enhances the resolving power of the EPR technique in the case of complex spectra. Moreover, the method makes it possible to study the mechanism of nuclear and electronic relaxation (some of the paths are shown in Fig. 2.6 by wavy lines).

In the absence of these mechanisms the ELDOR signal is equal to zero. In the case of nitroxide radicals, for example, three mechanisms exist: (1) nuclear relaxation; (2) relaxation due to the rotation of the radical, which modulates the anisotropic components of the hyperfine interaction; (3) relaxation due to spin exchange.

The interaction between an electron spin and the nuclear moments of the surrounding atoms manifests itself also in the nuclear magnetic modulation effect (NME) [102, 115]. As a result of this effect, a non-monotonic decrease in the spin echo signal is observed. The fall-off modulation function is described by the equation:

$$V_{mod} = (1-\tfrac{1}{2}K)+\tfrac{1}{2}K \times [\cos\omega_\alpha t+\cos\omega_\beta t-\tfrac{1}{2}\cos(\omega_\alpha+\omega_b)t-\tfrac{1}{2}\cos(\omega_\alpha-\omega_\beta)t] \ , \quad (2.30)$$

where K is a parameter which characterizes the extent of modulation; ω_α and ω_β are the magnitudes of the hyperfine splitting expressed in frequency units:

$$\omega_{\alpha\beta} = \{[\omega_I \pm A' \pm \tfrac{1}{2}F(1-3\cos^2\theta)]^2+[\tfrac{3}{2}F\sin\theta\cos\theta]^2\}^{1/2} \ , \quad (2.31)$$

where $\omega_I = g_N\beta_N H/h$; θ is the angle between the direction of the field H_0 and the line connecting the spins:

$$F = \mu_N g_N \mu g/\hbar r^3 \ . \quad (2.32)$$

The most valuable information on the structure of the complex is contained in the parameters K and F. As can be seen from Eq. (2.32), the letter parameter (F) characterizes the effectiveness of the electron-nuclear dipolar interaction, which depends on the separation r. A quantitative analysis of F therefore allows this distance to be determined.

2.9 Nuclear Magnetic Resonance

The nuclear magnetic resonance (NMR) method will be considered here only in connection with the problem of the interaction of nuclear spins with electron spins of paramagnetic ions and radicals. The analysis of this interaction may be useful in the investigation of metalloenzymes and metal carriers. The dipolar and contact (Fermi) interactions are reflected both in the position of NMR lines and in the relaxation parameters T_1 and T_2 of the NMR spectra [27, 99, 116–123].

The factor responsible for a paramagnetic shift in the case of radicals and ions of transition metal of the first and second rows of the Mendeleev Table is usually the isotropic, contanct of Fermi interaction:

$$\delta_{NC} = \frac{\Delta H}{H_0} = -\frac{a_h g_e^2 \beta_e^2 S(S+1)}{3 g_N \beta_N k_B T} \ , \quad (2.33)$$

where a_h is the hyperfine interaction constant which is proportional to the spin electron density at a given nucleus.

In those cases where the population of the levels is changed with a change in temperature, e.g. upon transition of the spin system from the anti-ferromagnetic to the paramagnetic state, deviations must be observed from the law $\delta_{NC} \sim T^{-1}$ predicted by Eq. (2.3). The quantity a_h usually varies within the range of 10^8 to $10^9\ s^{-1}$, which leads to shifts exceeding the shifts from diamagnetic molecules. This makes it possible to study the nuclear environment of the ligands of the active sites of metalloenzymes.

Another mechanism of paramagnetic shift, called the pseudocontact shift, is due to the dipole-dipole interaction of the nuclear spin with the electron spin of the complex and depends on the magnetic anisotropy of the ligand field and the angle θ' between the radius connecting the spins and the principal axis of the complex. For molecules with axial symmetry:

$$\delta_{Npc} = \frac{\Delta H_0}{H_0} = -\frac{\beta_e^2 S(S+1)}{k_B T} (g_\parallel^2 - g_\perp^2) \frac{(3\cos^2\theta - 1)}{r^3} . \tag{2.34}$$

This expression may, in principle, be used to determine r (up to 1.5 nm) for systems with a strong anisotropy of the g-factor, e.g. for some lanthanides used as spin labels.

The nuclear-relaxation parameters are substantially influenced in the presence of unpaired electrons by the following factors: the distance between the spins (r), the magnitude of the spin quantum numbers of electrons (S) and nuclei (I), the lifetimes of the complex (τ_M), the correlation time of the dipolar electron-nuclear interaction (τ_c'), the unpaired electron density at the nucleus and the intensity of the chemical exchange between the nuclei. Here, we shall limit our consideration to the spin dipole-dipole interactions, which are most widely used in the study of the behaviour of substrates at the active sites of metalloenzymes.

In a rapid chemical exchange of ligands at paramagnetic active sites ($\tau_M \leqslant 10^{-4}$s) the experimentally measured rates of relaxation depend on the proton relaxation time of the electron spin (τ_c'), the ratio of the proton concentration in the paramagnetic complex to their total concentration (f_M), the contribution of diamagnetic interactions, the effects that arise upon direct contact with paramagnetic particles without formation of a complex with a metal and also on the ratio of the resonance frequencies of protons before and after the complex formation. The complete mathematical expression that takes all these factors into account has been analyzed in the literature [121, 122].

If the complex formation is not accompanied by a strong chemical shift, the quantity T_{2M} is directly related to the structure and molecular motion of the paramagnetic complex and is calculated from experimental data using the following equation:

$$\Delta\Delta H_{1/2} = \frac{f_M}{\pi} \frac{1}{T_{2M}} , \tag{2.35}$$

where $\Delta\Delta H_{1/2}$ is the broadening of the resonance line of protons caused by the formation of a paramagnetic complex. The quantity $\Delta\Delta H_{1/2}$ is defined as the width of a Lorentz line at half-height expressed in Hz. The contribution of the paramagnetic complex to the increase of the rate is determined from the rate difference in the presence and absence of the complex with account taken of the fraction complexed.

According to the theory developed by Solomon and Blombergen [124]:

$$1/T_{2M} = \frac{1}{15} \frac{S(S+1)\gamma_I^2 g^2 \beta_e^2}{r^6} \left[4\tau_c' + \frac{3\tau_c'}{1+\omega_I^2\tau_c^2} \right] ; \tag{2.36}$$

$$1/T_{1M} = \frac{2}{15} \frac{S(S+1)\gamma_I^2 g \beta_e^2}{r^6} \left[\frac{3\tau_c'}{1+\omega_I^2\tau_c'^2} \right] , \tag{2.37}$$

where τ_c' is the correlation time of the dipole-dipole interaction (the above formulas hold for $\tau_c' > 2\times10^{-10}$ s, which corresponds, for example, to Mn^{2+} and Cu^{2+} ions and nitroxide radicals); ω_I is the frequency of the Larmor precession of the nuclear spin; γ_I is the gyromagnetic ratio; g ist the g-factor. The functions in parentheses will be denoted as f_1 and f_2.

The effectiveness of the dipole-dipole interaction can, in a general case, be determined by three parallel processes: (1) molecular rotation (with the correlation time τ_r), which averages the magnetic fields created by electron spins at the point where the proton resides; (2) the relaxation of the electron spin which averages the interaction with the effective time τ_s; (3) the movement of the spins away from one another as a result of dissociation (with the time τ_M). In accordance with the equation for parallel processes:

$$\tau_c - 1 = \tau_r^{-1} + \tau_2^{-1} + \tau_M^{-1} \tag{2.38}$$

the effective correlation time will be determined by the shortes time of one of the processes mentioned above. Equations (2.36–2.38) open up unique possibilities for the study of the fine structure of paramagnetic complexes through the measurement of the times T_1 and T_2. Indeed, as follows from Eqs. (2.36) through (2.38), the quantities T_{1M} and T_{2M} depend on the three most important parameters: the electron spin S, the separation between the electrons and the nucleus of interest (r) and the correlation time (τ_c').

Most often it is necessary to determine the separation between the various protons of the ligand and the paramagnetics at an enzyme. Since the quantity S is known, as a rule, the task of calculating r is reduced to finding out which of the three processes (electron spin relaxation, rotation of the macromolecule, dissociation of the complex) leading to a disturbance of the dipole-dipole spin-nuclear interactions is the most rapid [see Eq. (2.38)]. For metalloenzymes the value of τ_s lies within the range 10^{-6} to 10^{-11} s, $\tau_r = 10^{-8}$ to 10^{-6} s and $\tau_M = 10^{-5}$ to 10 s (for enzyme-substrate and enzyme-inhibitor complexes); therefore, the term containing τ_M is usually disregarded.

To calculate the separation r from the extent of paramagnetic broadening of the EPR line, the following formula can be used:

$$r = 46 \left[S(S+1) f_2(\tau_c') \frac{f_M}{\Delta\Delta H_{1/2}} \right]^{1/6} \text{nm} , \tag{2.39}$$

which has been derived from Eqs. (2.36) and (2.37); $\Delta\Delta H_{1/2}$ is expressed in Hz; τ_c' is expressed in seconds. For instance, according to the data reported in the literature [122], for a spin label at a relatively rapidly rotating lysozyme molecule ($\tau_c' = 10^{-8}$ s) in aqueous solutions at room temperature ($a_3 = 2.3$):

$$r = a_3 [f_M/\Delta\Delta H_{1/2}]^{1/6} \text{nm} . \tag{2.40}$$

For the spin labels of macromolecules with a molecular mass equal to or exceeding 500000 or for very viscous media, the value of τ_c' is equal to 2×10^{-7} s, and an analogous calculation gives $a_3 = 4.2$. For a macromolecule with a value of τ_r close to the value of τ_s of the spin label, $a_3 = 3.9$. For complexes of macromolecules with Mn^{2+} one can use the value of $a_3 = 2.3$, in calculations in the derivation of which it was assumed that $\tau_c' = \tau_s \approx 4\times10^{-10}$ s.

For rapidly relaxing ions that are not observably by EPR (for example, for ions with $\tau_s = 10^{-11}$ s and $a_3 = 1.1$) paramagnetic broadening is difficult to observe even in those cases where the protons are near the paramagnetic centre.

Thus, the NMR method, especially when used in conjunction with EPR spectroscopy and the spin-label technique, is a very effective tool in the investigation of the structure of enzyme-substrate complexes.

In conclusion, we shall enumerate the basic possibilities of the spectral methods for the study of the active sites of metalloenzymes. These methods enable one (1) to establish the valence and nature of ligands and the symmetry of a metalloenzyme complex; (2) to study quantitatively the effectiveness of the interaction of transition-metal atoms between one another and with other groups and to calculate the distance between them; (3) to determine the intramolecular dynamic characteristics of the enzyme and the parameters of the chemical exchange of ligands.

Unfortunately, in order for these possibilities to be realized, certain conditions must be satisfied, conditions that are not always observed in the study of particular biological systems. Therefore, new approaches are being sought for and attempts are being made to increase the number of sites of the molecule having convenient spectral properties. This can be achieved, in particular, by replacing diamagnetic metal atoms by copper, iron, cobalt, manganese, lanthanides and other paramagnetic atoms, through the introduction of Mössbauer, luminescent, electron-dense (electron-scattering) and spin labels. These approaches will be discussed in the next chapter.

CHAPTER 3

Physical Label Techniques

The structural and dynamic problems involved in the study of metalloenzymes can be solved by using new methods based on the introduction of various labels into these structures. The principle of this approach consists in the chemical modification of the chosen functional groups by means of special compounds (labels), whose properties make it possible to trace the state of the biological matrix by means of physical methods. Four types of compounds are used as such labels:

1. Paramagnetic centres (stable nitroxide radicals and ions with unpaired electrons) [24, 99, 125–130].
2. Luminescent labels, molecules which quench luminescence, and other chromophores [24, 99, 125, 131–134].
3. Electron-scattering atomic groupings, e.g. polymercurials, which effectively scatter electrons and X-rays [24, 99, 125, 133–142].
4. Mössbauer atoms (^{57}Fe) which give NMR spectra [24, 99, 125, 143–146].

Three types of problems may be distinguished which can be solved by means of labelling techniques:

1. The determination of the distance between functional protein and other groups, including those present in active sites of enzymes. The estimation of the depth of penetration of paramagnetic centres in biological matrix.
2. The study of the microrelief and molecular mobility of various layers of water-protein, membrane and other matrices in the vicinity of the labels introduced, including the dynamic state of the active sites of enzymes.
3. Recording of conformational changes in proteins, enzymes and other structures, including the effects that spread over considerable distances from the active sites (allosteric and transglobular effects).

The methods that rely on the use of labels, though being inferior in accuracy to X-ray structural analysis, have nevertheless a number of merits:

1. It is possible to work with whole, native cellular and subcellular structures with aqueous, diluted solutions of proteins, enzymes and other objects.
2. Information can be obtained on the molecular, dynamic characteristics of various regions of the biological matrix under investigation.

3. It is possible to probe the active-site structure of relatively high-molecular enzymes, including those present in complex biological structures, e.g. membrane structures.
4. The measurements are fast and widely available.

Naturally, new methods cannot substitute the well-known techniques that have become classical. Their application, however, allows one, in some cases, to obtain unique information on the structure of the objects under study and, in other cases, is an independent means of testing the results obtained by conventional methods.

3.1 The Spin-Labelling Technique

3.1.1 General

The spin-labelling technique was first applied by McConnell [147]. The spin labels most often used are stable nitroxide radicals (NR); complexes of paramagnetic metals are employed less frequently. The basic source of information is the paramagnetic nitroxide group of radicals which is chemically stabilized by substituents (R) at alpha-carbon bonds:

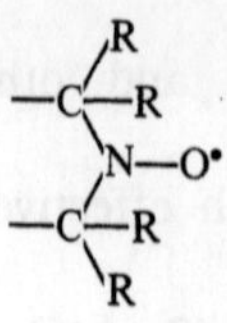

At present, extensive use is made of the following derivatives of nitroxide radicals [24, 99, 147–153]:

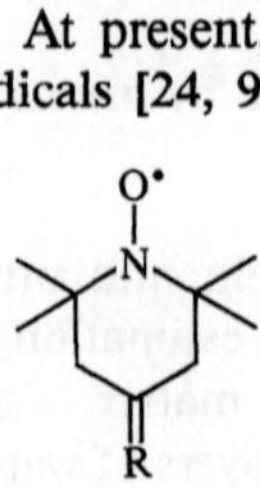

I

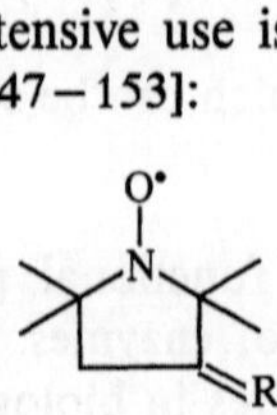

II

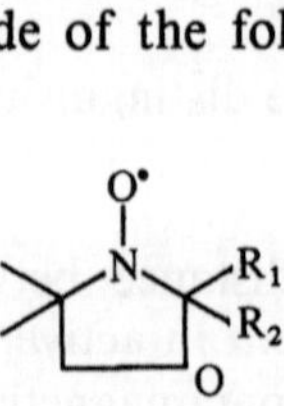

III

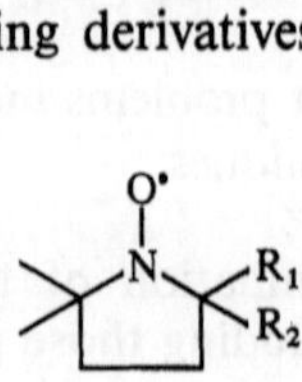

IV

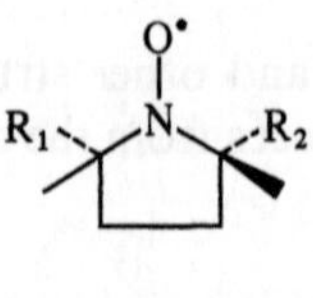

V

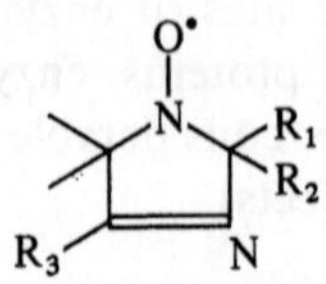

VI

where I = piperidines; II = pyrrolidines; III = piperidine derivatives of nitroxide radicals; II = pyrrolidine derivatives; III = doxyl derivatives; IV = proxyl derivatives; V = azethoxyl derivatives; VI = imidazoline derivatives.

The actual efficiency of the spin-label technique depends to a large extent on the possibility of specific addition of NR by a reaction with the free valence remaining intact [154]. Therefore, wide use is made of labels containing, as an active group, fragments of known reagents, whose selective action has been extensively tested by experimentation.

Today, there is a wide range of spin labels capable of reacting more or less specifically with the functional protein groups. In each of these labels three main components may be distinguished: (1) a paramagnetic fragment which gives an intense EPR signal; (2) a chemically active residue, which participates in a covalent or coordination addition to a certain protein group; (3) a linking atomic group (a "leg"), whose structure (length, flexibility, hydrophobicity and polarity) determine, to a large extent, the behaviour of the label after addition to the protein macromolecule.

Below the formulas of some spin labels and spin probes are given. According to existing terminology, a spin label is defined as a compound which becomes covalently linked to the object under investigation; spin probes either add on non-covalently or diffuse freely in the medium (R = piperidine or pyrrolidine nitroxide residues):

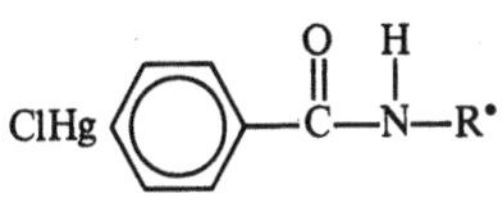

VII

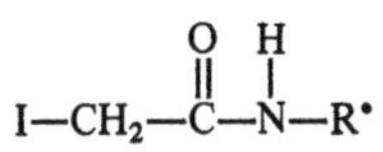

VIII

IX

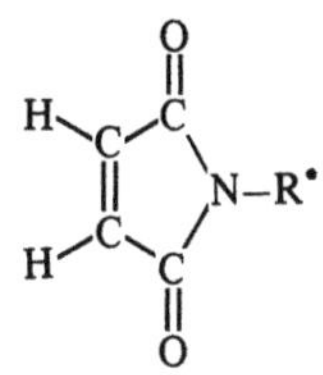

X

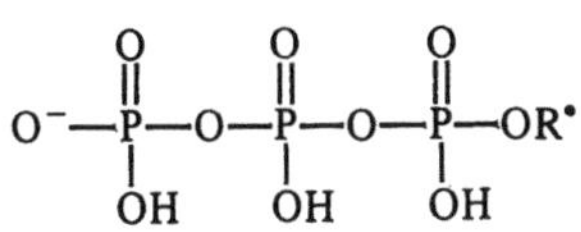

XI

$N\equiv C-R^{\bullet}$

XII

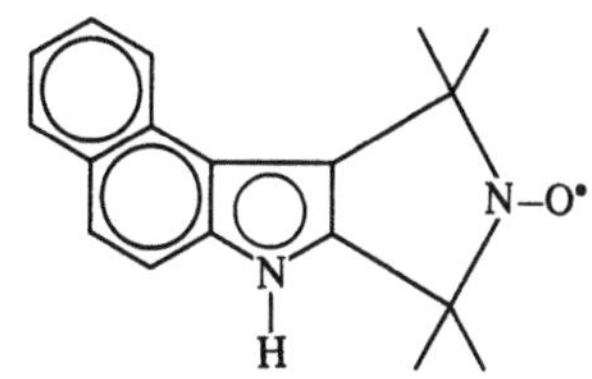

XIII

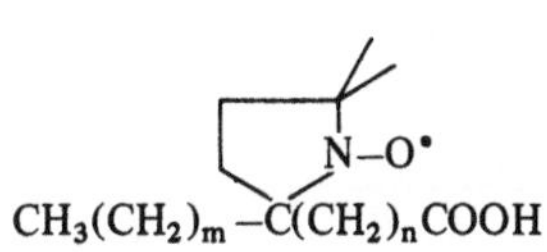

XIV

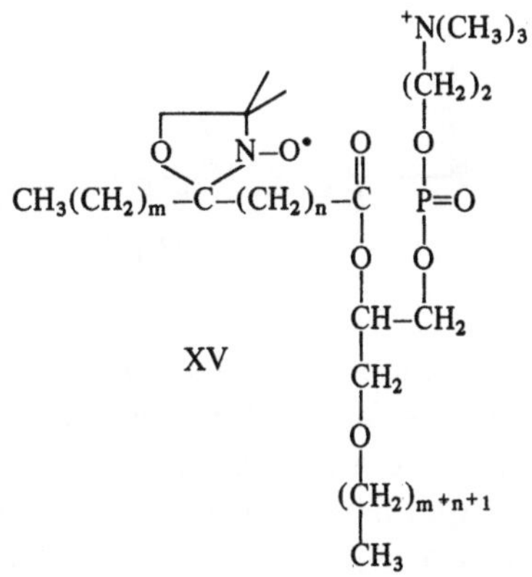

XV

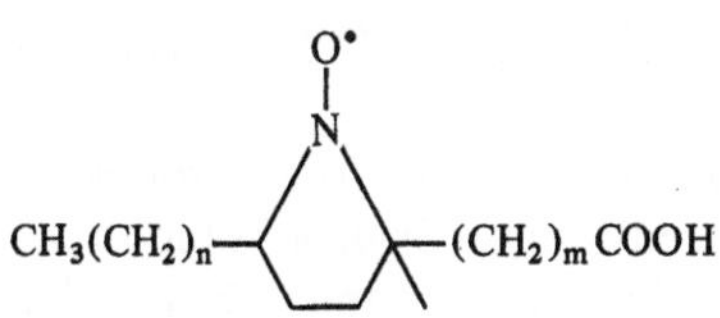

XVI

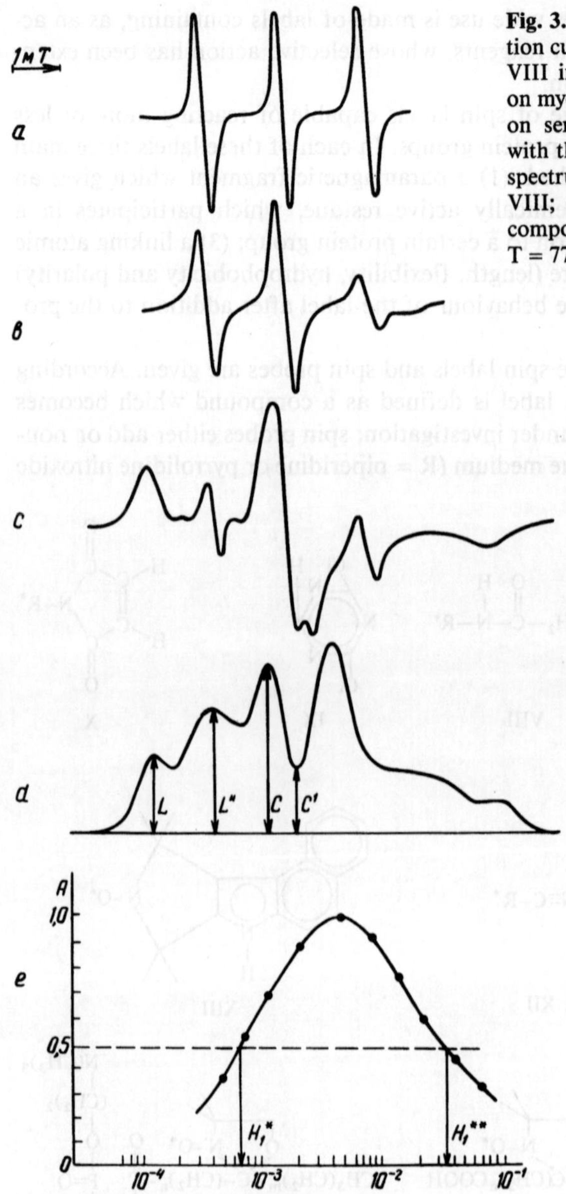

Fig. 3.1 a–e. The EPR spectra and saturation curves for nitroxide radicals: **a** radical VIII in aqueous solution; **b** radical VIII on myoglobulin in solution; **c** radical XIII on serum albumin in solution admixed with the unattached radical; **d** absorption spectrum with a phase shift in radical VIII; **e** saturation curves for the central component of radical VIII in alcohol at T = 77 K

Compounds VII–XII contain fragments of conventional (in biochemistry) reagents for the functional groups of proteins; probes XIII–XVI are used to study the membranes.

Spin labels and probes can be selectively added to the functional groups of proteins, lipids, nucleic acids, sugars, etc., introduced into the hydrophobic zones of membranes or localized on the electrostatically charged portions of molecular systems.

The typical EPR spectra of nitroxide radicals are given in Fig. 3.1. Analysis of these spectra permits one to obtain the following information about the local properties of spin-labelled systems.

1. Calculation of the parameters of rotational diffusion of NR permits the determination of microviscosity and the study of the microrelief of the environment. The slightest changes of these properties, e.g. those brought about by conformational or phase transitions, are distinctly recorded from the corresponding changes in the EPR spectra.
2. Determination of the distances between the spin-labelled portions from the magnitude of the spin-spin interaction between the radicals and between the radicals and the paramagnetic atoms of the metal (PM).
3. Estimation of the micropolarity of the environment.
4. Quantitative study of the redox properties of the medium on the basis of the reduction rate of the radicals.
5. Elucidation of the distribution of spin-labelled, physiologically active compounds in biological structures.

These possibilities of the spin-label method are an invaluable aid to experimentalists working with complex biological systems.

3.1.2 Effect of Rotational Diffusion of the EPR Spectra of Nitroxide Radicals

The theory of the EPR spectra of stable nitroxide radicals in solution is based on the McConnell model [155]. In constructing the model it was assumed that the dependence of the shape of the EPR spectra of NR on the correlation time of rotational diffusion (τ_c) is due to the averaging effect of rotation on the hyperfine structure and the g-factor.

The following types of motion of nitroxide radicals are distinguished in a theoretical treatment of the McConnell model: (1) rapid motion (10^{-10} to $2\times10^{-9}\,s^{-1}$); (2) slow motion (2×10^{-9} to $10^{-7}\,s^{-1}$); and (3) very slow motion (10^{-6} to $10^{-3}\,s^{-1}$). The theory of the line width and line shape of EPR spectra in the rapid-motion region has been developed by Kivelson [156] and Freed and Frenkel [157]. There are a large number of works devoted to the analysis of EPR spectra in the region of slow motion [157–160]. The results of investigations of very slow motion are also available [160].

A further important parameter in the theory of EPR spectra is the type of particle motion. The simplest motion is isotropic rotation, where the frequency of reorientation is the same in all directions. Most paramagnetic probes, however, are asymmetric particles; a theory in which the anisotropy of rotation as applied to nitroxide radicals is taken into account has been developed by Vasserman et al. [161].

In the region of slow motion, the shape of EPR spectra is significantly influenced by the mechanism of radical reorientation.

Three mechanisms of motion are commonly dealt with: (1) the model of constant Brownian motion; (2) the random-jump model which assumes instantaneous reorientations of the particles alternating with a relatively prolonged state of rest; and (3) the intermediate model of free diffusion [157].

The validity of the theory is determined to a large extent by how closely the models on which it is based correspond to the actual motion in the systems under study.

For the region of rapid motion the following formula may be used:

$$\tau_c^{-1} = 1.2\times10^9\,[(\sqrt{h_{+1}/h_{-1}}-1)\Delta H_{+1}]^{-1}\,s^{-1}\,, \tag{3.1}$$

where ΔH_{+1} is the width of the m_{+1} component in mT; h_+ and h_- are the intensities of the spectrum components with $m = +1$ and -1.

Methods for taking the anisotropy of rotation into account have been described in the literature [157, 161, 164].

In order to calculate τ_c in the region of slow rotations, one has to compare the experimental spectra with the spectra calculated with the aid of electronic computers. The results of such calculations depend on the choice of the motion model chosen [157]. For instance, for isotropic free diffusion we have:

$$\tau_c = 8\times10^{-10}(1-A_{zz}/A_{zz}^0)^{-1.16}\,s\,, \tag{3.2}$$

where A_{zz} and A_{zz}^0 are the parameters of the hyperfine structure of the spectrum under given conditions and in the absence of any motion respectively.

A very promosing method of studying the anisotropy of motion in this region is submillimetre spectroscopy, noted for its high resolving power [114] (see Figs. 2.7 and 2.8).

The specific feature of EPR spectroscopy in the 2-mm range is represented by the spectacular demonstration of the complicated molecular dynamics of the nitroxide label or probe in the form of an EPR signal (Fig. 3.2). Since in the EPR spectra of nitroxide radicals recorded in this range a complete resolution of all the canonical components [162, 163] is achieved, the specific features of slow molecular motions, e.g. the anisotropy of rotations, are manifested in non-identical relaxational changes (shifts and broadenings) of the various components of the spectrum.

Calculations of the EPR spectra in the 2-mm range based on a rigorous theoretical approach (the method of the stochastic Liouville equation) have shown that not only the frequencies and orientation of the axes of anisotropic rotations (relative to the principal magnetic axes) have an effect on the shape of the spectrum (Fig. 3.2a). It is essential that the character of random rotations of the radical also manifests itself in the form of a signal. Upon Brownian diffusive rotation, for example, the shifts and broadenings of the z-component are approximately the same in magnitude; the jumps to large angles broaden the component, which practically remains unshifted. This allows one to determine in an analogous experimental situation the model of molecular rotations with the aid of a simple procedure of comparing the experimental dependence of the broadening on the shift of the z-component with theory.

The theoretical dependencies of the spectral parameters (shift and broadening of the high-field z-component) on the correlation time τ_{iso} [163] allow us to estimate, with good accuracy, the frequency of isotropic roations (Fig. 3.2b). The

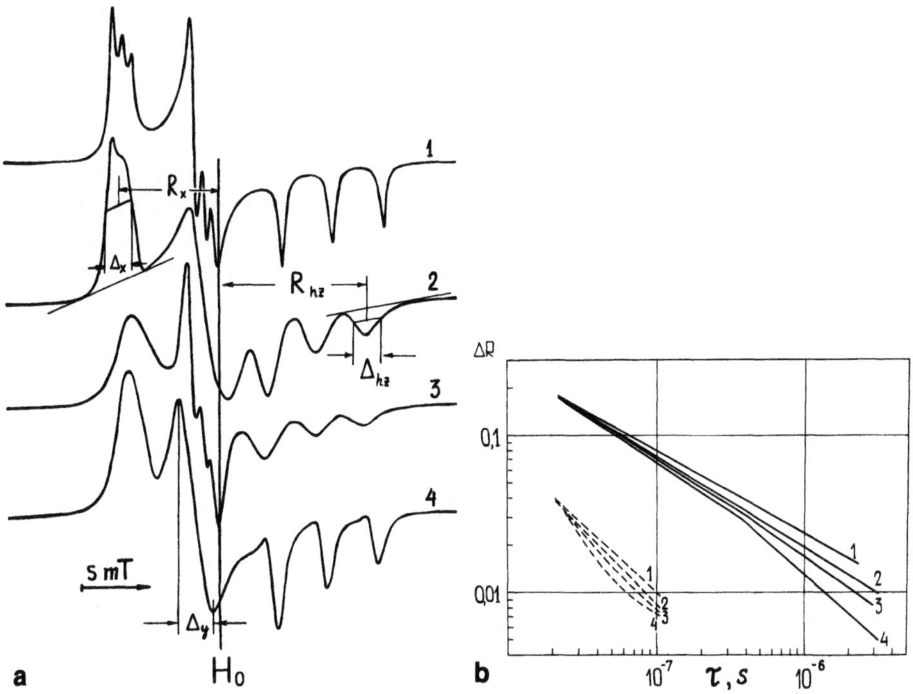

Fig. 3.2a, b. Theoretical 2-mm EPR spectra of the nitroxide radical (**a**) and theoretical dependence of the relative shift $\Delta R = (R_{hz} - R_h^*)/(R_h^* - H_0)$ on the isotropic correlation time τ_{iso} (**b**) [163].
a *1* absence of rotation; *2, 3, 4* axisymmetric rotations with the correlation times $\tau_\perp = 2 \times 10^{-7}$ s, $\tau_\| = 2 \times 10^{-8}$ s and orientations of the symmetry axis of fast rotations about the x, y, z-axis of the tensor respectively. In calculations use was made of the following parameters: $g_{xx} = 2.0084$; $g_{yy} = 2.0060$; $g_{zz} = 2.0022$; $A_{xx} = A_{yy} = 6.73$ Hz, $A_{zz} = 3.725$ mT; the width of the spin packet is 0.3 mT, $H_0 = 4.9500$ T.
b *1–4* values of $\Delta^0_{hz} = 0.17$, 0.5, 0.8 and 1.1 mT for the Lorentz line shape of the spectrum; the *solid lines* correspond to the model of Brownian rotation; the *dashed lines* represent the large-angle jump model

simplified evaluation of the mean correlation time $T = \sqrt{T_{||} T_\perp}$ proposed by Antsiferova and Lyuboshevskaya makes it possible to avoid using the laborious procedure of fitting a theoretical spectrum to each experimental spectrum and provides a method of obtaining the frequency characteristics of the rotations of the spin label or probe in biological systems.

With correlation times ranging from 10^{-6} to 10^{-3} s the methods mentioned above are unsuitable because of the insensitivity of the shape of the EPR spectrum to the frequency of radical reorientation. In this region, the mobility is studied by making use of the phenomenon of saturation transfer. The essence of this phenomenon is that the portion of the EPR spectrum corresponding to a certain radical reorientation is saturated with a microwave field. If the τ_c of radicals is comparable with or shorter than the spin-lattice relaxation time, then, as a result of the reorientation, the saturation is transferred to another portion of the spectrum. Several methods have been developed which permit the estimation of

τ_c from the intensity of saturation transfer through the analysis of the saturability of various portions of the spectrum, i.e. ELDOR and ENDOR techniques, absorption with a phase shift and dispersion [160].

The sensitivity of the saturation curve parameters of nitroxide radicals to very slow motions of the environment in the region of $\tau_c = 10^{-2}$ to 10 s has recently been demonstrated [166]. In particular, the magnitude of the effective spin packet ΔH_1, as determined by the method described in the literature in vitrifying liquids and in proteins, increases sharply with increasing mobility of radical pairs [165, 166] and electric dipoles in the region of excited phosphorescing chromophores.

3.1.3 The Double-Label Technique

In solving a number of important problems involved in enzyme catalysis, it is necessary to know the distance between certain functional groups of enzymes. To this end, one must first of all decipher the three-dimensional arrangement of the groups present in the active site. The study of allosteric transitions and other long-range effects also requires knowledge of the distances between the active site and the site of allosteric action. Experimental investigations of the structure of multicentre metalloenzymes containing several metal atoms per macromolecule are beset with serious difficulties.

A technique making use of double paramagnetic labels has been proposed for deciphering the structure of biological materials, primarily enzymes [168]. The principle of this method is that the various groups in a protein macromolecule can be modified by spin labels of identical or different structure. The EPR spectrum of spin-labelled proteins allows one, under certain conditions, to determine the distance between the modified groups of the protein. The method has been further developed by other investigators [108, 125, 170–173].

As has already been noted (see Sect. 2.7.3), the EPR spectra of paramagnetic centres exhibit under certain circumstances a subtle response to the approach of other paramagnetic groups. Two types of interaction between paramagnetic centres are distinguished: (1) dipole-dipole interactions associated with the fact that the magnetic dipole of one paramagnetic group induces al local magnetic field at the site of another paramagnetic group; and (2) exchange interactions caused by the overlap of the orbitals of unpaired electrons upon approach of paramagnetic particles to one another.

In model and biological system, the interacting radicals are, as a rule, oriented randomly and there is a set of distances between the dipoles. The g-factor anisotropy and the hyperfine structure anisotropy exert an influence on the EPR spectra and there is fast motion of the paramagnetic centres which averages the dipole-dipole interaction.

The analysis of EPR spectra is most conveniently performed under conditions in which the molecular motion of the spins is either practically stopped (frozen solutions) or slowed down to such an extent that it no longer affects the EPR spectra (when the centres are rigidly bound to very large macromolecules or in highly viscous media).

The following methods used to calculate the distances between paramagnetic centres from the magnitude of the dipole-dipole interaction should be noted: (1) empirical calibration; (2) the method of second moments; (3) measurements of the dipole-dipole splitting; (4) analysis of the broadening of single spectral components; (5) comparison of experimental and calculated spectra.

The empirical calibration method is based on a comparison of the shape and parameters of the EPR spectrum of the object being studied with the spectrum of a sample with a known arrangement of spins. The data on solid solutions of radicals and complexes or chemical compounds of known structure, such as polyradicals, may be used as standard data.

In the method of second moments, use is made of the well-known van Vleck equation [100] which relates the dipole-dipole component of the second moment in the line (ΔM_2) to the parameters of the magnetic moments of several spins (N) and the distance (r_{ik}) between them:

$$M_2 = a_2 \gamma_e h^2 S(S+1) N^{-1} \sum r_{ik}^{-1/6} , \tag{3.3}$$

where γ_e is the gyromagnetic ratio; S is the spin of the interacting particles; a_2 is a coefficient.

For a pair of radicals the following formula can be used:

$$r = 1.08 \Delta M_2^{-1/6} . \tag{3.4}$$

Here and afterwards all the line width parameters are expressed in mT and r in nm.

In magnetically diluted polycrystalline samples (biradicals on macromolecules, the active sites of enzymes, radical pairs obtained as a result of radiolysis, etc.) the dipole-dipole interaction may lead to the splitting of the line by an amount D_d. In the absence of exchange, the distance between the pairs can be calculated from the formula:

$$r = 1.41\, D_d^{-1/3} . \tag{3.5}$$

If the spectrum shows a component, whose width varies as the spins are brought closer together, the distance can be estimated from the magnitude of the broadening ΔH_d by using the following formula:

$$r = 1.63\, \Delta H_d^{-1/3} . \tag{3.6}$$

This formula holds, strictly speaking, only for the cubic lattice. The numerical coefficient, however, depends only slightly on the manner in which the spins are arranged.

In a number of cases, the EPR spectra can be calculated theoretically. The effect of the dipole-dipole interaction on the line shape under the influence of a paramagnetic group has been analyzed by Taylor et al. [172]. The contribution of the interaction to the line width was taken into account using the formula:

$$\Delta H_d = g \beta \mu^2 \tau_s / r^6 , \tag{3.7}$$

where τ_s is the correlation time of the dipolar interaction, e.g. the electron-spin relaxation time of the centre causing the "broadening" [at a distance of r = 1 nm, Eq. (3.7) is valid at $\tau_s < 10^{-9}$ s]. The theoretical curves obtained by Taylor et al. permit the estimation of r if the other parameters in Eq. (3.7) are known.

The distance between the nitroxide radicals can be determined from the relative intensity of the allowed (I_1, $\Delta m_2 = 1$) and forbidden (I_2, $\Delta m_s = 2$) transitions [170]:

$$r = 0.18\,(I_1/I_2)^{1/6}\ . \tag{3.8}$$

At distances ranging from 1.3 to 3.0 nm it is convenient to use the Kokorin-Zamaraev parameter d_1/d (Fig. 3.5), which is connected with r by means of the following empirical relation:

$$r = 0.93 + 0.077/(d_1/d + 0.036\,A^0_{ZZ} - 1.76)\ . \tag{3.9}$$

The values of A^0_{ZZ} can be determined from the same EPR spectrum (Fig. 3.1). Because of the strong dependence of the line shape on the parameters measured, the accuracy of its determination is rather high (0.1 – 0.2 nm), However, the uncertainty in the orientation of nitroxide fragments and the possibility of the existence of several conformations of the labels appreciably lower the efficiency of the method. The effectiveness of the method can be raised by the combined analysis of experimental and calculated spectra with account taken of the possible variants of interaction.

As has been shown by Kulikov [107, 108, 173], interesting possibilities are offered by the investigation of the paramagnetic contribution to the value of the spin-lattice relaxation rate $1/T_1$ (Chap. 2).

These possibilities are especially strongly manifested in the case of nitroxide radicals, for which the values of $1/T_1^0$ and $1/T_2^0$ differ significantly in the absence of foreign paramagnetic groups, the value of $1/T_1^0$ being much lower than that of $1/T_2^0$. For example, at a temperature of 77 K in a water-glycerol mixture (1 : 1) the values of $1/T_1^0$ and $1/T_2^0$ are equal to 2.7×10^3 s and $\approx 10^{-7}$ s respectively. Since at small τ_s the dipole-dipole interactions make an approximately equal contribution to $1/T_1^0$ and $1/T_2^0$ (Eqs. 3.10 and 3.11), it follows that the value of $1/T_1^0$ is considerably more sensitive to the interaction than the quantity $1/T_2^0$, which is associated with the line shape.

The quantity $1/T_1$ can be determined experimentally by using pulse techniques, i.e. the pulse-saturation technique and the spin-echo technique [100 – 103, 160] and also the methods based on the analysis of the saturation curves [107, 108, 160, 167, 173]. The pulse-saturation and spin-echo methods have the advantage of a more direct determination of $1/T_1$, but they entail certain experimental difficulties, especially in the case of the relaxation times (for more details, the reader is referred to the literature [100 – 103]).

In order to determine T_1 and T_2 from the saturation curve obtained under conditions of rapid passage, one can employ the method proposed by Safronov et al. [167]. The determination of r is carried out with the aid of the following formulas [107, 108]:

$$1/T_1 = \frac{\mu^2\gamma_e^2}{6r^6\tau_s}\left[\frac{4}{5\,\omega_1 - \omega_2)^2} + \frac{24}{5\,(\omega_1+\omega_2)^2} + \frac{12}{5\,\omega_1^2}\right] \tag{3.10}$$

$$\Delta\,1/T_2 = \frac{4}{15}\,\frac{\mu^2\gamma_e^2}{r^6}\,\tau_s\ , \tag{3.11}$$

where ω_1 and ω_2 are the resonance frequencies of the interacting spins; τ_s is the relaxation time of the second paramagnetic group; μ_s is the magnetic moment of the paramagnetic group; γ_e is the gyromagnetic ratio for the electron.

The formulas were derived on the condition that $(\omega_1-\omega_2)^2\tau_s$ is much smaller than unity; averaging is carried out over the angle between the centres and the direction of the magnetic field. The application of this method requires knowledge of the value of the electron-spin relaxation time τ_s of the second paramagnetic group. Approaches have, however, been suggested which allow one to circumvent this condition. Thus, for example, the quantity, τ_s is dropped from the product of $1/T_1$ and $1/T_2$, a circumstance that can be used for calculations. An ingenious method of finding τ_s relies on the experimental dependence of $1/T_1$ on τ_2 [107]. Since for ions in solids the quantity τ_s decreases monotonically with increasing temperature, one can find experimentally the temperature at which $\tau \sim \Delta\omega^{-1}$, where $\Delta\omega$ is the difference of the resonance frequencies of the spectrum. It can easily be shown that under these conditions the maximal dipolar contribution is equal to:

$$(\Delta\, 1/T_1)_{max} \approx \mu^2\gamma_e^2/15\, r^6\, \Delta\omega \ . \tag{3.12}$$

These methods have been tested on a system with known location of the label, i.e. hemoglobin samples with the β-93 SH group modified by various radicals [107]. The values of r obtained agree well with the X-ray diffraction data. Analysis of the exchange interactions between the nitroxide fragments which occur when they are brought closer together, may be useful for the solution of a number of structural problems [24, 170]:

1. Biradicals and polyradicals may serve as labels and probes, when their EPR spectra are extremely sensitive to the state of the microenvironment.
2. The mutual arrangement of two or more labels on the macromolecule, which permits direct contact between the nitroxide fragments, can be revealed by analysis of the spectrum parameters and their temperature dependence.
3. Polyradical systems may be regarded as model systems of the exchange interaction between the spins complicated by the g-factor anisotropy. The theoretical and experimental spectra of such systems can be used for a comparison with the spectra of polyatomic paramagnetic centres characterized by the EPR parameters close to those of the radicals.

3.1.4 The Spin Label-Spin Probe Method

In this section we shall analyze new possibilities offered by the use of spin probes, which are chemically inert paramagnetic species capable of diffusing freely in solution. The spin-probe technique has been successfully employed for studying the exchange relaxation, the magnetic and other properties of paramagnetic complexes in solution [174]. It has been suggested by Likhtenstein and Grebenshchikov that this method be used for probing the paramagnetic active centres of enzymes and also for investigating the microstructure of proteins in the vicinity of the added spin labels (Fig. 3.3) [173].

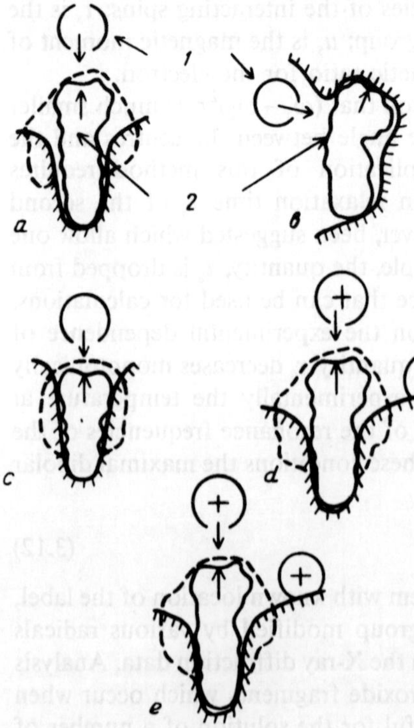

Fig. 3.3 a–e. The spin probe-spin label approach [24, 99]: *1* spin sprobe; *2* spin label. The effect of the microrelief (**a–c**) and charge distribution (**d, e**) on the nature of the label-probe interaction

The magnitude of braodening ΔH_0 of the EPR line of a stable nitroxide radical in solution is related, within a certain concentration range (10^{-3} to 10^{-1} M), to the rate constant of exchange relaxation (k_{ex}) upon encounter between the radical and a paramagnetic centre by the following, simple equation [174]:

$$\Delta H_0 = 6.5 \times 10^{-8} k_{ex} C , \tag{3.13}$$

where C is the concentration of the second paramagnetic species expressed in M; the broadening ΔH_0 is expressed in oersteds and k_{ex} in $M^{-1} s^{-1}$, which allows one to determine the value of k_{ex} experimentally within a certain range from the change in the line width of the EPR spectrum of the radical. A method has recently been worked out for the determination of k_{ex} from the extent of the change in the rate of the spin-lattice relaxation of the radical in the presence of a paramagnetic species ($\Delta 1/T_1$) [175]. Since the parameter $1/T_1$ is more sensitive than I/T_2, especially in the region of slow rotation, the method allows one to widen the range of k_{ex} values being determined.

For a particular pair of paramagnetic species the value of k_{ex} depends on their electronic structure, microviscosity, steric hindrances and the distribution of electrostatic charges in the region of encounters (Fig. 3.2). Hence, after a preliminary, empirical calibration the method can be used for an experimental study of these factors in biological and other objects.

In principle, both types of spin-spin interactions can occur in solutions: the dipole-dipole interaction and the exchange interaction. The authors, who have carried out a detailed study of the behaviour of paramagnetic particles [173, 174], have come to the conclusion that in relatively low viscosity media the major contribution to the EPR line width is introduced by exchange interactions upon encounters. The probability of a relaxation during the encounter of paramagnetic particles is determined, in particular, by the magnitude of the exchange integral J. According to the available data [170, 173], the exchange integral J depends on the unpaired electron density at sites of direct contact between he colliding particles. The value of J is of the order of $10^{11}\ s^{-1}$, i.e. it is close to the inverse of the lifetime of the colliding complexes in non-viscous solutions (τ_{col}). The probability of relaxation (P_{ex}) upon biomolecular collisions is given by the formula:

$$P_{ex} = f_{ns} J^2 \tau_{col}^2/(1 + J^2 \tau_{col}^2) \ , \tag{3.14}$$

where f_{ns} is the nuclear-statistical factor equal to 2/3 for nitroxide radicals; P_{ex} is defined by the following equation:

$$k_{ex} = P_{ex} K_d \ , \tag{3.15}$$

where K_d is the number of encounters in solution and can be estimated by using, in particular, the well-known formulas for diffusion processes. By this definition, the quantitiy P_{ex} incorporates a factor (not known in the general case) which characterizes the steric factor).

A more general theory for exchange relaxation during the encounter between paramagnetic particles in solution has been developed [174]. This theory takes into account the influence of such factors as the paramagnetic relaxation time of electron spins (T_1), the value of the magnetic spin numbers of paramagnetic particles, S_1 and S_2, and also the difference between the resonance frequencies of the spins (δ).

According to the theory, the relaxation probability (P_{ex}) during an encounter is mainly determined by the relationship between the time T_1, τ_{col} and the exchange integral J. Experimental criteria for a strong ($P_{ex} \sim 1$) and a weak ($P_{ex} \ll 1$) exchange have been found. For instance, in the case of strong exchange, the temperature dependence of k_{ex} must be determined by the temperature dependence of the number of encounters, i.e. by the dependence of k_d on the viscosity of the medium. In the case of weak exchange, because of the opposing dependencies of k_d and τ_{col} on the viscosity, the quantity k_{ex} must be weakly dependent on the temperature.

The diffusion coefficient (D) is inversely proportional to the reduced radius (R) of colliding particles. The comparison of the accessibility of the spin label to paramagnetic probes takes into account, in a first approximation, the difference in the value of R between the macromolecules and the radical, which at the limit, approaches 2. Therefore, it is expedient to use as the accessibility parameter, the quantitiy k_w/k_{pr} ("microviscosity"), which characterizes the increase of the viscosity and steric hindrances in the region of the encounter between the probe and the radical on a protein (k_w and k_{pr} are the k_{ex} constants for the radical in water and the spin label on protein respectively).

Mention should be made of the following trends in the treatment of the experimental data obtained in investigations of globular proteins [99, 173]. For radicals with a relatively large distance from the polypeptide backbone chain to the piperidine ring, the value of $k_w/2k_{pr}$ is close to unity, which implies that the microviscosity in the region of the nitroxide group is close to the viscosity of pure water. For shorter labels the value of $k_w/2k_{pr}$ ranges from 1.5 to 2.5. Thus, the nitroxide fragments of the labels are relatively easily accessible to paramagnetic probes. Whereas the value of $k_w/2k_{pr}$ for different radicals on proteins and in solution varies insignificantly (by no more than two or three times), the rotational diffusion parameter τ_c varies at the limit by nearly 5×10^3 times.

Thus, the intramolecular rotations of the labels on proteins experience additional steric hindrances, which are practically absent in the case of a spin probe. This result is qualitatively described with the aid of the anisotropic rotation model of the nitroxide fragments of the labels.

The intensive development of the chemistry of stable nitroxide radicals [148–153] has made possible the use of electrostatically charged derivatives of NR as convenient objects for the study of the effect of the electrostatic factors on the spin-relaxation process.

Experimentation has shown that the Debye electrostatic equation [176] with $\varepsilon = 80$ describes well the experimental data. This equation can be used for an approximate estimation of the distance between the nitroxide group of the spin label on the protein and the nearest charged group, provided that this distance does not exceed 1.0–1.2 nm.

The dependence of the experimentally observed relaxation parameters on the nature of complex-forming metals, and on the number and nature of substituents in the porphyrin ring of the complexes has been examined by Grebenshchikov et al. [177]. The porphyrin comlexes of various metals enter into the composition of the active sites of a large number of proteins and enzymes, which has determined the choice of the objects to be studied.

As has been shown by Grebenshchikov et al. for iron- and copper-containing complexes with conjugated ligands, including hemin, high values of exchange rate constants upon encounter with the nitroxide radical ($k_{ex} = 1-3\times10^9\,M^{-1}s^{-1}$) have been observed, which provide an effective broadening of the EPR line of the radical. However, under identical conditions, the changes in the EPR spectra of spin probe solutions with ferrihemoglobin and oxidized cytochrome *c* are much smaller.

New possibilities in the study of the effect of the exchange on the rate of spin-lattice relaxation of spin labels have been reported [125, 175, 178]. In particular, the rate constants of the spin exchange of spin-labelled proteins and heme-containing proteins have been measured, such as hemoglobin and cytochrome ($k_{ex} \approx 3\times10^8\,M^{-1}\,s^{-1}$) and also the values of k_{ex} (O_2) for the process of interaction of spin probes in biological model membranes.

Such measurements have been impossible to carry out by methods using the EPR line shape parameters.

The phenomenon of spin relaxation of nitroxide radicals and other centres, under the influence of other paramagnetic species, can be used to establish the depth of immersion of the centres into biological matrices (e.g. proteins and membranes). For this purpose, it is necessary to study quantitatively the effect of the dipole-dipole interaction between the centres being studied and the paramagnetic ions randomly distributed in the bulk of the vitrified sample on the relaxation parameters of the centres, e.g. on the value of T_1, which is determined by the continuous wave saturation method.

A method for the determination of the distance of nearest approach (r_{min}) between a stable radical ($R^{\bullet}$) and a relaxing ion (RI) [171] has been developed. It is obvious that in the case of the penetration of $R^{\bullet}$ into the impermeable matrix, the quantity r_{min} is equal to the depth of immersion r_{im}. If the centre resides at a sufficient depth, then $r_{min} > \bar{r}$, where $\bar{r}$ is the average distance between the stable radical $R^{\bullet}$ and IR. As has been shown [171], in this case the contribution of the dipolar interaction to the rate of relaxation of $R^{\bullet}$ is given by the following formula:

$$(\Delta 1/T_1)_d = A_d \mu^2 \gamma^2 \tau_s C / r_{min}^3 , \qquad (3.16)$$

where C is the ion concentration in the solid solution; A_d is a factor dependent on the geometry of the surface, which impedes the approach of the radical and the ion, for example, if the surface is plane $A_d = 0.7$. If $R^{\bullet}$ is at a distance r_{im} from the surface of a sphere of radius R, then:

$$A_d = 16\pi/9(2 - r_{min}/R)^3 . \qquad (3.17)$$

Figure 3.4 shows the results obtained by testing the method in the case of $NR^{\bullet}$ immersed at different depths in the membranes of chromophores. As seen from

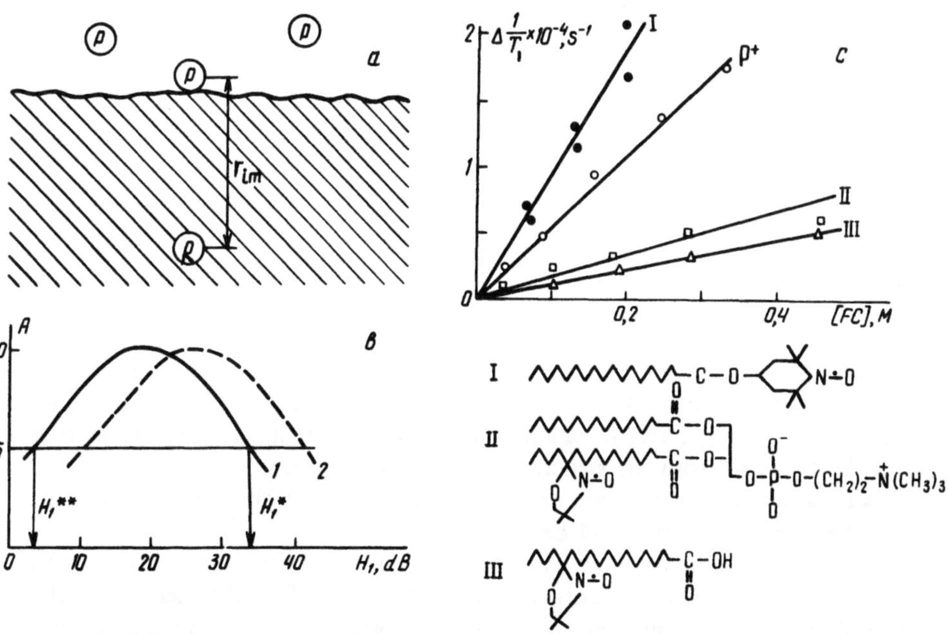

Fig. 3.4 a–c. Estimation of the depth of immersion (r_{im}) of the radical centre (R) in a biological matrix from the efficiency of the dipolar interaction with the paramagnetic centres (P) in a solid solution [107, 207, 171]; **a** diagram illustrating the principle of the method; **b** shape of the curves without interaction (*1*) and with interaction (*2*); **c** dependence of the dipole contribution $\Delta(1/T_1)$ on the concentration of ferricyanate (*FC*) for radicals in chromatophores of *R. rubrum*. *I–III* Spin probes; P^+ cation of chlorophyll

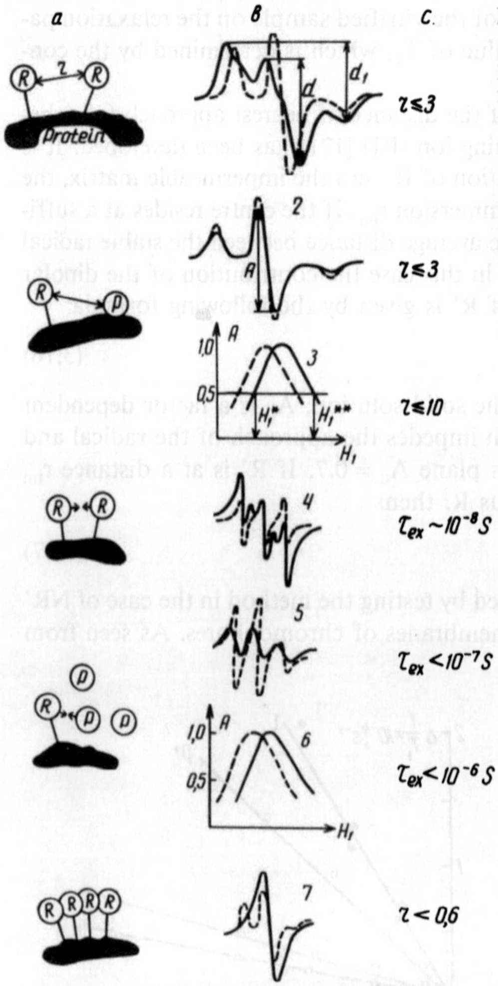

Fig. 3.5a–c. The double-label technique [24, 99]. Types of interactions between paramagnetic centres (**a**), the corresponding changes in the EPR spectra and saturation curves (**b**) and the range of measured distances (r, in nm) between the centres and of the effective times of spin exchange (τ_{ex}). *R* Nitroxide radical; *P* paramagnetic centre. The *broken lines* represent the non-interacting centres and the *solid lines*, the interacting centres. *1–3* Dipole-dipole interactions; *4–6* dynamic exchange interaction; *7* static exchange interaction

Fig. 3.4, the linear dependencies of $(\Delta 1/T_1)_d$ on C correspond to Eq. (3.16). The values of r_{im} for probes XIV (m = 3, n = 12) and XV (m = 3, n = 12) are equal to 2.2 and 2 nm respectively, which is in good agreement with the values that might be expected from the structure of these compounds.

Examples of the application of the method described above to the study of the active sites of enzymes will be given in Chapter 9.

Figure 3.5 presents all types of interactions used for structure investigations in paramagnetic systems, the corresponding changes in the EPR spectra and the range of measured distances between the paramagnetic centres. The double spin-label and spin label-paramagnetic probe techniques permit the determination of the distance between the labels from 0.3 to 6.0–8.0 nm, which practically covers the sizes of individual enzymes. As will be shown in the next section, the use of electron-scattering labels enables one to extend the range of distances to be measured to 10.0 nm upwards.

3.2 Luminescent Labels

3.2.1 Physical Principles

Luminescent labels (LL) have found wide application to the solution of a number of structural and dynamic problems in enzyme and membrane systems [24, 99, 131–134, 179–183]. This technique is based on luminescence phenomena: fluorescence and phosphorescence.

Upon absorption of light a molecular system (chromophore) undergoes a transition from the ground singlet state (S_0) to an electronically excited singlet state (S_1). The reverse transition may be accompanied by a secondary emission (fluorescence) or it takes place by a radiationless mechanism (Fig. 3.6). Having given part of its energy to the environment, the system may pass over from the state S_1 to an excited triplet state (T_1). The latter is inactivated in two ways: by emission (phosphorescence) and by a radiationless mechanism.

A significant difference between fluorescence and phosphorescence is that the first process is a $S_1 \rightarrow S_0$ transition allowed by quantum mechanics, since it does not change the multiplicity of the system. The second transition, $T_1 \rightarrow S_0$, involves a change in multiplicity and, hence, is forbidden. In accordance with this, the lifetimes of the excited states during fluorescence ($\tau_f^* = 10^{-8}$ to 10^{-9} s) are much shorter than the phosphorescence lifetimes ($\tau_{ph}^* = 10$ to 10^{-6} s).

The main parameters of both types of luminescence are: the luminescence intensity, I_0; the frequency of the maximum of the fluorescence and phosphorescence spectra, ν_{max}; the line shape of the spectrum, the quantum yield, φ_0; and the lifetime of the excited state, τ^*. Of the many mechanisms of the effect of the state of the medium that surrounds the chromophore on the luminescence parameters, the following two mechanisms are widely used: the effect of the polarity and molecular dynamics of the solvent on ν_{max} and I_0 and the radiationless energy transfer from the excited state to other chromophores. The latter process is recorded by the change of τ^*, φ_0 and I_0.

The excitation of a chromophore group is accompanied by a change in the electric dipole moment of the molecule. This involves a change in the energy of interaction with the surrounding molecules. If the characteristic time of the reorganization of the medium's dipoles (τ_r) is much longer than the lifetime of the excited state (τ^*), then the dipoles have no chance to follow the change of the light-induced electric field and the transition to the ground state is effected from the "unsolvated" level. In another limiting case, $\tau_r \ll \tau^*$, the interaction between the dipoles and the excited molecule lowers the energy of the system and luminescence quenching is effected from the "solvated" level. The decrease of the transition energy with decreasing τ_r manifests itself experimentally, in particular, in a change in ν_{max} (a relaxation shift).

In the simplest case of a homogeneous system the quantities ν_{max}, τ_r and τ^* are related to the limiting values of ν_{max}^0 and ν_{max}^∞ as follows (at $\tau_r \rightarrow 0$ and $\tau_r \rightarrow \infty$ respectively [184]:

$$\nu_{max} = \nu_\infty + (\nu_{max}^0 - \nu_{max}^\infty)\tau^*/(\tau^* + \tau_r) \ . \tag{3.18}$$

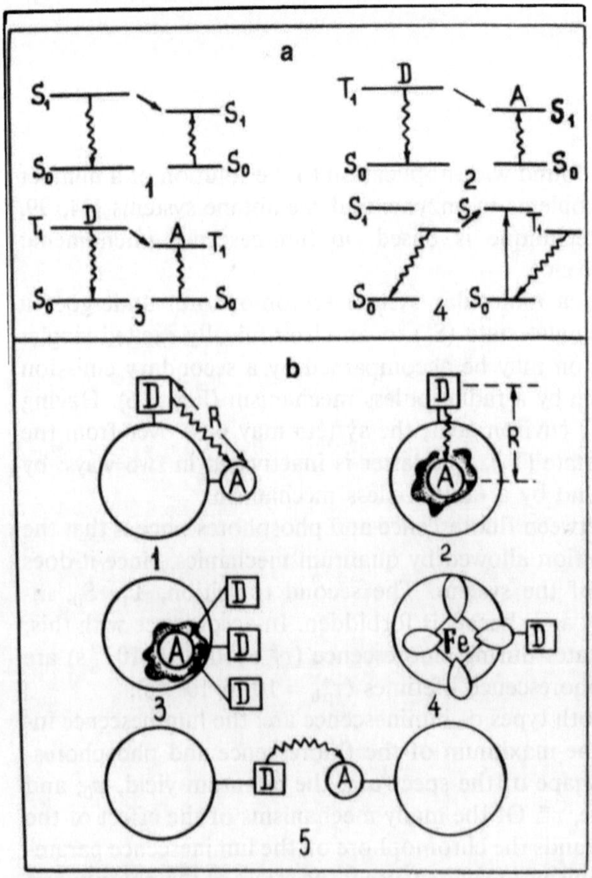

Fig. 3.6 a, b. The luminescent label technique. **a** The processes of energy migration and exchange used in the method. The *wavy arrows* indicate the path of energy migration from donor to acceptor. The *vertical arrows* indicate the direction of energy transfers inside the chromophores. *1* S→≶S transition; *2* T→S transition; *3* T→T transition; *4* T→T annihilation. **b** Types of structural and dynamic problems solvable by the method (*D* donor group; *A* acceptor group): *1* distance between labelled portions of the biological matrix and its variation; *2* distance between the active centre and the group at the surface; *3* depth of the centre; *4* delocalization of electron density over the protein globule; *5* the collision between molecules including macromolecules

In real systems, there is, as a rule, a set of values of τ_r and other parameters. Analysis of relaxation shifts in such systems requires special approaches. An approach that takes into account the distribution over the times τ_r will be described in Chapter 10.

The radiationless transfer of energy from an electronically excited molecule (the donor) to an unexcited molecule (the acceptor) may occur by two mechanisms: the dipole-dipole (inductive resonance) mechanism and the exchange mechanism. The necessary condition for energy transfer is that the molecule that receives the energy must have its energy levels overlapped with the levels of the donor molecule (Fig. 3.6). It will be appropriate to discuss these mechanisms separately.

3.2.2 Inductive-Resonance Energy Transfer

The singlet-singlet transfer of energy from the donor to the acceptor can be established experimentally from: the kinetics of quenching of the donor fluorescence in the presence of an acceptor; the decrease of the fluorescence quantum yield; the enhancement of the acceptor fluorescence upon excitation of the donor; and from depolarization of fluorescence (when the donor molecule and the acceptor molecule are identical). The theory of this phenomenon has been developed [185–189].

A quantum-mechanical description of the energy transfer process for the electric dipole-dipole interaction has been given by Forster [187]. According to this theory, the rate constant of energy transfer from the donor to the acceptor is given by:

$$k_{DA}^{s} = \frac{9000 \ln 10\, k_{or}^{2}}{128\pi^{5} n^{4} N \tau_{D}^{*} r^{6}} \int I_{D}(\nu)\, \varepsilon_{A}(\nu)\, \nu^{-4} d\nu \ , \tag{3.19}$$

where τ_D is the fluorescence lifetime of the excited state of the donor with respect to emission; N_A is Avogadro's number; n is the refractive index of the medium; r is the distance between the donor and the acceptor; $\int$ is the normalized overlap integral of the luminescence spectrum of the donor, $I_D(\nu)$, and the absorption spectrum of the acceptor, $\varepsilon_A(\nu)$; the quantity K_{or} is the orientation factor which characterizes the relative orientation of the vector directions of the transition dipole moments in the donor and the acceptor; this quantity is defined by the following equation:

$$K_{or} = |3 \cos \theta_1 \cdot \cos \theta_2 - \cos \theta_3| \ , \tag{3.20}$$

where θ_1 and θ_2 are the angles between the direction connecting the donor and the acceptor and the vectors of the transition dipole moments in the donor and the acceptor respectively; θ_3 is the angle between these vectors.

Knowledge of the factor K_{or}^2 is very important, but its calculation requires exact knowledge of the geometric factors and is commonly the most difficult task for the experimentalist. For a random and directionally equally probable orientation of the transition moment vectors, the donor and the acceptor, $K_{or}^2 = 2/3$. Methods for the calculation of K_{or} for various cases have been described in the literature [188, 189].

Let us introduce the concept of the critical distance for transfer of energy, r_{cr}, at which the probability of energy transfer is equal to the probability of the deactivation of the excited state of the donor. It is easy to show that substitution of the numerical coefficients into Eq. (3.20) yields the formula:

$$r_{cr}^{6} = 8.8 \times 10^{-31} \frac{K_{or}^{2} \varphi_{0}^{D}}{n^{4}} \int I_{D}(\nu) \varepsilon_{A}(\nu) \nu^{-4} \delta\nu \ , \tag{3.21}$$

where φ_0^D is the fluorescence quantum yield of the donor. Thus, for any donor-acceptor pair, one can calculate the critical distance for energy transfer if one knows the luminescence spectrum of the donor, the absorption spectrum of the acceptor and the value of K_{or}^2.

In order to calculate the distances r between the donor and the acceptor with the critical distance r_{cr} being known for a case where one donor interacts with one acceptor, the following formula holds:

$$r = r_{cr}(\varphi_0/\varphi - 1)^{-1/6} . \tag{3.22}$$

Calculations show that the range of r values measured is from 0.8 to 7.0 nm, depending on the value of the critical radius. Apart form pair interactions, other, more complicated cases may also be encountered. Expressions for the degree of quenching for various numbers of donor-acceptor pairs have been given in the literature [131].

If the donor is in a triplet state, then with a high probability of the triplet state of the donor and with a large overlap of the emission spectrum of the donor and the absorption spectrum of the acceptor, an effective T→S energy transfer is possible. The energy transfer occurs by an inductive resonance mechanism. With the transfer of energy taking place by this mechanism, the slow rate of transfer is compensated for by a long emission lifetime of the donor because a transition involving a change of multiplicity is quantum mechanically forbidden. Since the experimentally measured parameters are highly sensitive to the distance between the chromophore groups, the methods under consideration may be used for the study of the structure of enzymes. The method is especially useful in comparative investigations, e.g. in the study of conformational changes. The validity of such an approach has been demonstrated for a large number of cases [131–134]. For example, it has been shown [134] that the distance between the β-93 SH group of hemoglobin and the heme group as determined from formulas (3.21) and (3.22) is in good accord with the X-ray diffraction model of the protein. The distance between the chromophores of the protein and the fluorescing and quenching labels has been shown to be changed when pH is changed, urea is added and the ligands are bound by serum albumins [133, 134].

3.2.3 Triplet-Triplet Energy Transfer. Triplet Labels

The triplet-triplet transfer of energy was first detected by Ermolaev and Terenin [190, 191]. These authors and, later, other investigators found that this energy transfer can be measured quantitatively by the following methods [191]:

1. From the amount of quenching of the excited triplet states of the donor in the presence of the acceptor molecule. The quenching is recorded by a decrease in the concentration of excited states by EPR methods and from a decrease in the intensity $I_{0(ph)}$ and the phosphorescence lifetime τ_p* and from the enhancement of luminescence of the acceptor by optical methods.

2. From the amount of secondary delayed fluorescence, which in turn can be induced by two mechanisms. In one mechanism, fluorescence is produced by the thermal activation of triplet states to the excited singlet state (the $T_1 \rightarrow S_1$ transition shown in Fig. 3.6). In the second case, delayed fluorescence is produced by the annihilation of the triplet states of two chromophore groups, one being transferred to the ground singlet state and the other to an excited singlet state, i.e. $2T_1 \rightarrow S_0 + S_1$ (Fig. 3.6).

The relaxation of a chromophore from the triplet excited state occurs by the following variants of the exchange mechanism: (1) the exchange-resonance transfer of energy from the triplet level of the donor to the triplet (multiplet) level of the acceptor; (2) the catalysis of the interconversion $T_1 \rightarrow S_0$ under the influence of the external paramagnetic species due to a partial removal of the spin selection rule for the transition; (3) quenching by the mechanism of charge transfer, which may be either reversible or irreversible.

The triplet labels and probes used are phosphorescing chromophores, both the hydrophilic derivatives of fluorescein, eosin, rhodamine, etc., and the hydrophobic derivatives (pyrene, anthracene, dimethyl benzanthracene, etc.). The intrinsic chromophores of the protein (tryptophan, tyrosine, chlorophyll, the heme group after removal of the iron) may sèrve as internal labels. The quenchers of triplet states include organic chromophores, metal complexes and paramagnetic materials. Nitroxide radicals have been found to be effective quenchers [24, 125, 182, 192–196].

The main advantage of the triplet-label technique is the possibility of recording weak exchange interactions and studying slow and very slow dynamic processes. This possibility is associated with the relatively long lifetime of excited triplet states, $\tau_p{}^* = 10$ to 10^{-6} s.

According to Dexter [186], the rate constant of triplet-triplet transfer of energy between the donor and the acceptor brought about by an exchange interaction is determined by the following expression:

$$k_{DA}^{T} = k_0^{T} \exp(-2r/a_b) , \tag{3.23}$$

where a_B is the effective Bohr radius; r is the distance between the donor and the acceptor;

$$k_0^{T} = L_T \int I_D(\nu) \cdot \varepsilon_A(\nu) d\nu , \tag{3.24}$$

where L_T is a coefficient; and $\int$ is the overlap integral of the phosphorescence spectrum of the donor and the absorption spectrum of the acceptor.

If the absorption of the donor and the acceptor is fixed in space (a static exchange), formula (3.23) can be used to determine r in the range of 0.6–1.5 from the experimentally known values of a_B and k_0^T and from the value of k_{DA}^T being measured. The solution of the inverse problem is also possible: the value of the quantitiy a_B, which characterizes the conductivity of the medium to exchange processes can be found from the known values of k_0^T and r and from the measured value of k_{DA}^T. Experimentation has shown that $k_0^T \approx 10^{16}\ s^{-1}$ for non-conducting media with $a_B = 0.07$ nm and for conducting media with $a_B = 0.62$ nm. This topic will be discussed in more detail in Section 11.1.2.

In the case of dynamic exchange, the interaction occurs at the moment of collision between the donor and the acceptor. The efficiency of the exchange, which is characterized by the rate constant k_{ex}^T, is determined by the number of encounters k_d, the steric factor f, the effective interaction radius R_{eff}, the lifetime of the collision complex, τ_{col} and by the rate constant of exchange in the collision complex, k_{ex}^0 (see Sects. 5.4 and 11.1.2).

The long lifetimes of triplet labels in the excited state ($\tau_p{}^* = 10^{-5}$ to 10 s) offer interesting possibilities for the measurement of the rates of slow dynamic processes ($k^T_{ex} = 10^4$ to $10^{10}\,M^{-1}\,s^{-1}$), the determination of the depth of immersion of triplet or quenching centres into proteins and membranes and for the study of very weak exchange interactions. Relatively slow relaxation processes with relaxation times τ_r comparable with $\tau_p{}^*$ are studied on the basis of the magnitude of relaxation shift in the phosphorescence spectrum. As has been reported [197], the use of the phenomenon of triplet-triplet annihilation upon collision recorded from the intensity of delayed fluorescence is very promising.

The above indicated properties of triplet labels offer certain advantages in the study of biological systems as compared with spin and fluorescent labels and probes [99, 129–132, 179–181]. First, the concentration of the labels introduced is lower. For example, it has been shown [194–197] that it is possible to determine the diffusion parameters of spin labels and other quenchers from the extent of quenching of triplet molecules in membranes upon introduction of only one quencher molecule (a spin probe or ferrocene) per 10^4 lipid molecules. The sensitivity of the method appears to be sufficient for the measurement of the number of collisions between macromolecules (myoglobin, cytochrome *c*) in media with a viscosity of up to 100 cp, for the probing of labels in the bulk portions of proteins (Ca^{2+}-ATPase, serum albumins, caseins), in which the accessibility to quenchers is lower by more than three orders of magnitude than in the surface layers [125, 182, 195]. Finally, comparing the efficiency of triplet-triplet transfer betwen labels in proteins and in model systems has made it possible to explore tha ability of the protein matrix to conduct exchange interactions and also the accessibility of the active sites of metalloenzymes to exchange interactions (see Sect. 11.1).

Among other promising trends in the application of triplet labels, mention should be made of the study of collisions between proteins and the localization of the active sites of metalloenzymes in membranes, the dynamics of rigid membranes and the change of the concentration of O_2 in biological systems.

Some of these possibilities have been realized in a new approach known as the "chromophore label-triplet probe" method [198]. The method is based on the phenomenon of the irreversible cis-trans-isomerization of a photochromic label (for example, a stilbene derivative) from the triplet state. The excitation of the label occurs in collision with the triplet probe; the rate of isomerization is controlled by a change in the absorption or fluorescence of the label. An advantage of the approach in question is that the effective characteristic time of the method attains the highest value possible (1–10 s) due to the large lifetime of the triplet state and to the "memorizing" of the act of triplet-triplet transfer upon label-probe collision. The incorporation of the corresponding triplet and photochromic labels into protein macromolecules makes it possible to measure the frequency of protein collisions in biomembranes. The approach described has been demonstrated for the collision of two chymotrypsinogen molecules, one of which was labelled by a stilbene derivative (a photochromic label) and the other by a derivative of erythrosine (a triplet label) [198]. In this case, it was possible to record the collisions at concentrations of protein reagents of the order of 10^{-7} M.

3.3 Electron-Scattering Labels

3.3.1 General

The structural problems of the determination of distances between the chosen portions of the enzyme protein macromolecule, the localization of these portions on the macromolecule and the distribution of the enzyme macromolecules and ligands in a complex biological system, e.g. in the membrane, can be solved by using new variants of the electron-scattering (electron-dense) label technique proposed by a number of investigators [135–142, 872].

A number of cases have been described in the literature illustrating the application of heavy-atom groupings for the study of the structure of nucleic acids [199–200] and biomembranes. The distinctive features of the new approach include the use of polymercurial compounds (mercarbides), whose molecules contain a cluster of mercury atoms, and also a special modification of certain functional groups of a biological object. Such molecules contain, within a relatively small space of 0.5–0.6 nm, about 500 electrons and are therefore powerful scattering centres for electron and X-radiation. The modification of macromolecules and other complex systems by mercury clusters permits the use of methods of electron microscopy [135–139] and small-angle X-ray scattering [140–141] for the investigation of label localization. The introduction of mercury clusters into single proteins crystals can facilitate the deciphering of X-ray patterns, especially in high molecular materials.

As has been shown [135–139], two variants of the modification of biological materials by electron-scattering labels (ESL) can be used in experiments:

1. By single atoms where the centre under investigation contains a large number of atoms to be modified.
2. By polymercury mercarbides when there is only a single centre.

It is convenient to use as an electron-scattering label a so-called Hoffmann base, prepared e.g. by the reaction of ethanol with mercuric oxide. The soluble part of the Hoffmann base is a mercarbide, whose structure is supposed to be as follows [139, 201]:

$$\begin{array}{l} \quad\quad\quad\ \ \mathrm{Hg{-}An} \\ \quad\quad\quad\ \ \ | \\ \mathrm{AnHgC{-}Hg{-}An} \\ \quad\quad\quad\ \ \ | \\ \quad\quad\quad\ \ \mathrm{Hg{-}An} \end{array} \quad (X)\ ,$$

where An are anions. Compounds X with An being acetates have been prepared by chemical synthesis [202].

The functional groups of proteins and other biological objects are modified by taking advantage of the high affinity of mercury compounds for mercaptides [137–139]:

$$\sim CH_2SH + X \rightarrow \sim CH_2SX \quad (I)$$

The binding of electron-scattering labels to other functional groups and their introduction into the hydrophobic portions of biological objects is accomplished with the aid of mediators, for example, according to the following schemes:

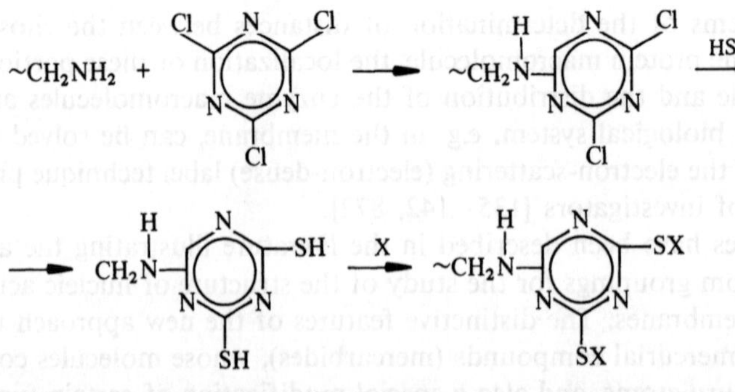

(II)

Hydrophobic portion + X + RSH → Hydrophobic portion. . . RSX (III)

where −RSH is dodecyl sulphide or mercaptoundecanoic acid (MUA).

Reaction (II) can be used for the "multiplication" of −SH− groups by introducing the stage of the intermediate dithiol with trichlorotriazine. This makes it possible, if necessary, to increase the scattering power of the centre. Reaction (III) is used to incorporate labels into the surface and internal layers of the membranes. In the latter case, the carboxyl groups MUA, anchored at the membrane surface, hold the label in the middle of the membrane [138].

3.3.2 Electron Microscopy

The possibility of observing microbiological objects under the electron microscope is associated with their ability to scatter fast electrons. This ability is different for different atoms and depends, among other factors, on the atomic number of atoms in the scattering particle. Under simplifying assumptions, this dependence on heavy nuclei corresponds approximately to the square of the atomic number of the element (Z_A). Thus, in going from the elements that make up the protein macromolecule with atomic numbers of 6−8 to mercury with $Z_A = 80$, the scattering probability is increased by two orders of magnitude. An estimate shows that a cluster of four to six mercury atoms in a space 0.6−0.7 nm in diameter must scatter electrons ten times more intensively than the equivalent portion of the protein macromolecule with a molecular mass of 150000−200000. Experimentation confirms this estimate. Protein molecules are practically transparent for electron emission and merge with the background on the micrographs. In order to reveal them, a special contrasting of the protein surface or the substrate is required, which can be done means of a layer of heavy metal salts. At the same time, molecules containing four to six mercury atoms appear on electron micrographs in the form of individual spots.

Examples of electron micrographs of labelled molecules are given in Fig. 3.7.

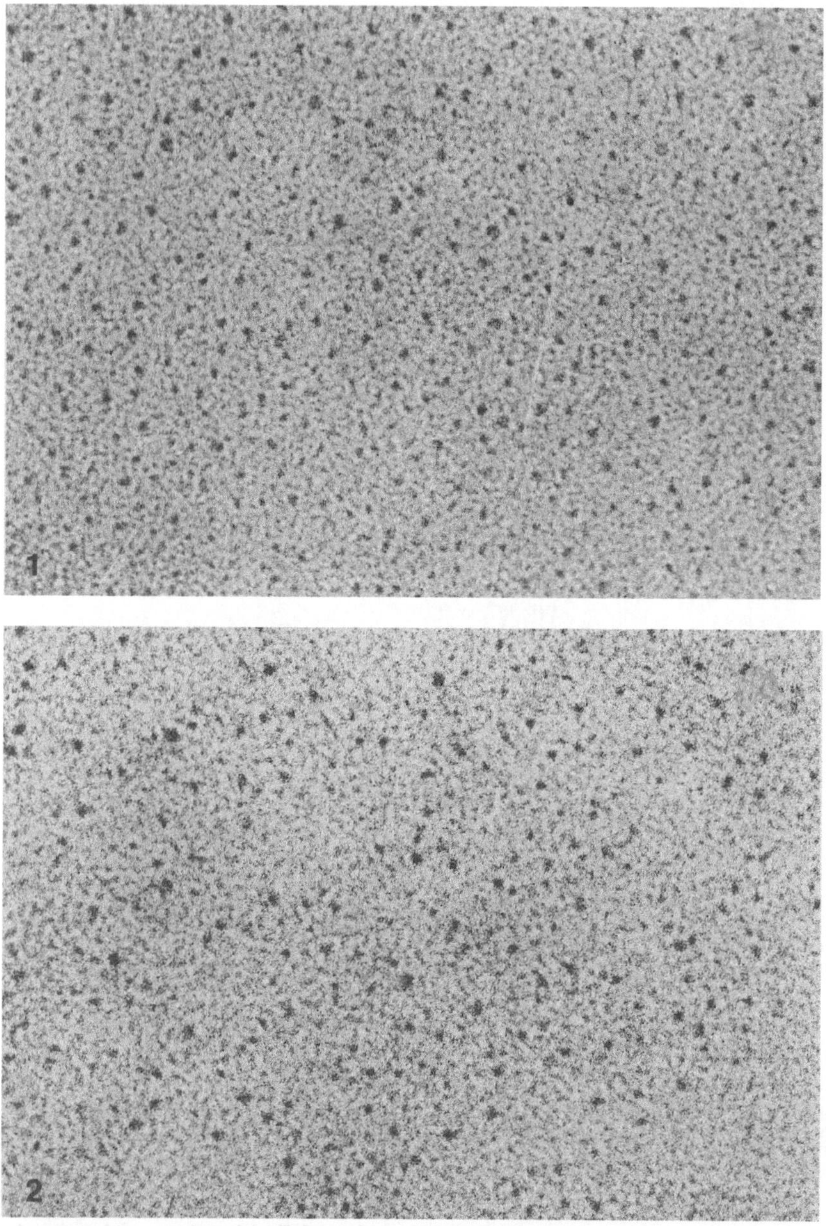

a

Fig. 3.7 a, b. The electron-scattering (electron-dense) label method. **a** Electron micrographs of biological objects labelled with electron-scattering labels (ESL). Enclosed in *parentheses* are the numbers of ESL per molecule (the magnification is indicated in *square brackets*): *1* trichlorotriazine (2–3) [700000]; *2* lysozyme (2) [700000]; *3* human serum albumin (4–5) [700000]; *4* bacteriorhodopsin in membrane in the ligth (6) [300000].
b The possibilities of the method. The *filled circles* denote ESL. *1, 2* Lipids in membranes; *3* proteins in membranes; *4* hydrophobic portions in membranes; *5, 6* distance between the groups in proteins and protein complexes; *7, 8* arrangement of subunits in protein complexes [139]

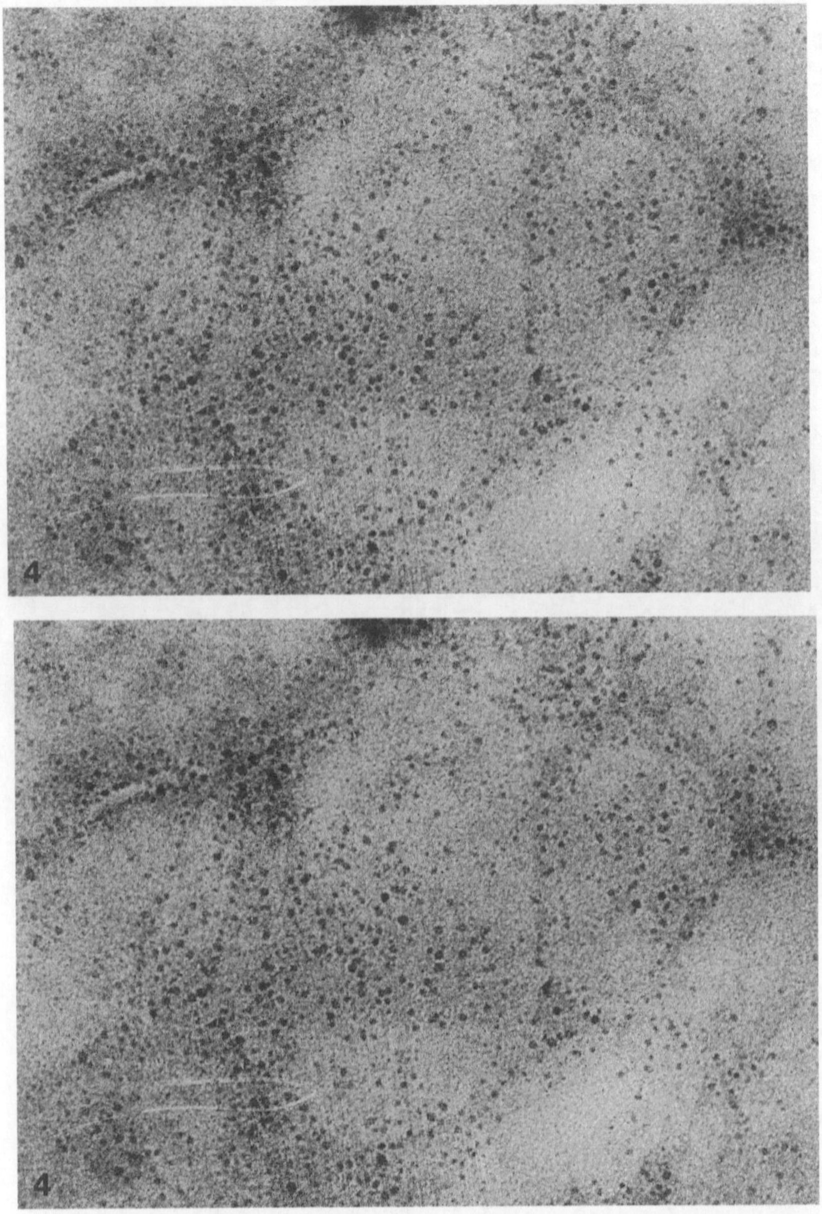

a

Figure 3.7 shows the electron micrographs of modified lysozyme, whose only group was modified according to scheme (II). Apart from the single 6-Å granules on the electron micrographs of electron-scattering labels (Fig. 3.7), electron-scattering granules 1.4 – 1.6 nm in size can also be seen, which corresponds to the size of two closely spaced molecules of the electron-scattering label. The contours of the lysozyme protein globule can hardly be seen on these micrographs. A similar distribution pattern of the labels is observed on the electron micrographs of modified trichlorotriazine (Fig. 3.7).

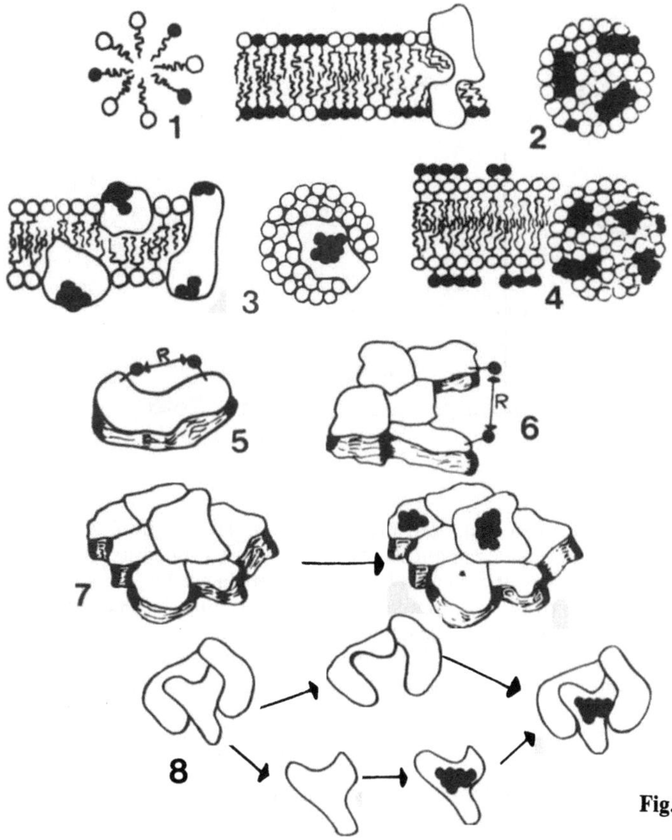

Fig. 3.7b

Experiments with serum albumin have clearly confirmed the possibility of modifying a considerable number of lysine and histidine groups according to scheme (II) (Fig. 3.7). The data available in the literature [135, 139, 142, 203] illustrate the possibilities of modification of the hydrophobic and lipid portions of membranes and also of individual proteins in membranes with subsequent observation of the distribution of the labelled regions.

Examples of the use of electron-scattering labels for studying biological multisite redox systems will be given in subsequent sections.

3.3.3 Small-Angle X-Ray Scattering

When a solution of macromolecules is irradiated with X-rays, the scattering intensity becomes dependent on the angle between the incident and the reflected ray because of the different reflections of the rays from the different portions of the molecule. This dependence is different for macromolecules of different sizes, densities and shapes. The experimental measurement of this dependence and the elucidation of its relationship to the molecular parameters of micro-objects is the concern of a special method, which has already demonstrated its efficiency [204–205].

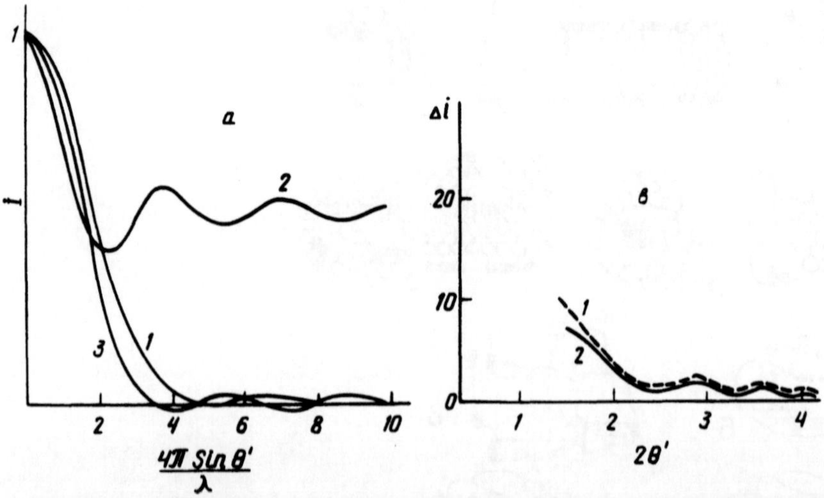

Fig. 3.8 a, b. The curves of small-angle X-ray scattering by protein solutions labelled by electron-dense labels [140, 141]. **a** The normalized components of the scattering intensity for a homogeneous sphere of radius R (a macromolecule) with "point" scatters (labels) added along the diameter; *1* geometric characteristics of the macromolecule; *2* determined by the label-label interaction; *3* depends on the label-macromolecule interaction. **b** Experimental dependencies of the scattering intensity difference (Δi) for solutions of histidine carboxylase with three labels and without labels on the scattering angle θ': *1* experimental curve; *2* calculated curve

Here, we shall consider only the problem of the determination of the distance (r) between heavy atoms residing on the macromolecule. According to the theoretical treatment reported in the literature [206], for molecules of spherical shape randomly distributed in solution, each of the molecules being modified by two heavy atoms situated at the opposite ends of the molecule at a distance r, the function of the dependence of the scattering intensity on the angle θ_x contains three terms.

The first term of the equation describes the scattering of the macromolecule and decreases sharply with increasing θ_x (Fig. 3.8); the third term represents the interference between the molecule and the heavy atoms and also falls with increasing θ_x. An important property of the second term, which reflects the participation of heavy atoms in a scattering event, is that it is described by a function with a number of clear-cut maxima (Fig. 3.8), whose position and intensity are associated with the distance between the heavy atoms. Starting from a certain value of θ_x, this term becomes dominant. For the second term to be experimentally separated, it will suffice that the inequality $Z_e > N_a/100$ is satisfied, where Z_e is the number of electrons in a heavy atom; N_a is the number of light metals of the protein. For a protein with a molecular mass of 50000 the electron-scattering label must contain at least six or seven mercury atoms. This number must be increased roughly in proportion to the molecular mass of the macromolecule.

The relationships given above have been tested on molecules of known structure and it has been found that this approach is completely reliable [141, 206]. For example, the distances between the iodine atoms in a solution of diiodoanthraquinone as determined by this method is 0.9 nm, which is close to the value of 0.9 nm determined by X-ray diffraction analysis. For gramicidin, modified by mercury atoms at the oriented residues, the value of r is 1.0 nm, which is in good agreement with the data obtained by other methods.

Investigations on human hemoglobin, two β-93 SH groups of which were modified by electron-scattering labels, have shown that the distances between the intercluster distances determined by their additional contribution to the scattering curves are equal to 4.2 nm, this being consistent with the X-ray diffraction model of the protein [141]. Experiments with histidine decarboxylase have enabled the determination of the distance between three SH groups of the enzyme localized at different subunits [140].

It may be expected that the electron-scattering label technique will find wide application in experiments using small-angle X-ray scattering for the determination of changes in the distances between certain portions of protein macromolecules during their functioning and also for the study of membrane systems modified at different portions of electron-scattering labels.

3.4 Mössbauer Labels

The Mössbauer-label technique [143–146] makes use of the sensitivity of the nuclear gamma resonance spectrum to the dynamic state of Mössbauer nuclei and to spin-spin interactions with other paramagnetic particles (see Sects. 2.6 and 3.7.3). Mössbauer labels, [^{57}Fe] or [^{119}Sn] nuclei, can be introduced into protein, membrane and other structures by means of chemical synthesis or adsorption in biosynthesis and also by way of replacement of the natural isotope [^{56}Fe] in iron-containing systems.

The NGR (nuclear gamma resonance) spectra are, under certain circumstances, extremely sensitive to the dynamics of Mössbauer labels. For instance, the intensity of resonance absorption (f′) is related to the amplitude of the mean displacement of the nuclei in the direction of recoil (⟨x⟩) during the lifetime of the excited state, $\tau_n^* = 1.3\text{x}10^{-7}$ s, which may be written in numerical form for [^{57}Fe] as follows (⟨x⟩ is expressed in nm):

$$f' = 10^{-2300} \langle x \rangle^2 \ . \tag{3.25}$$

If the [^{57}Fe] nucleus participates in a Brownian diffusion process with a diffusion coefficient D, then with the Stokes' law being fulfilled, the magnitude of the relative broadening of the spectrum is given by:

$$\frac{\Delta\Gamma}{\Gamma} = 2D\tau_n^*/\lambda^2 = 10T/\eta R \ , \tag{3.26}$$

where η is the viscosity in poise and R is the radius of the particle which is rigidly linked to the nucleus; R is expressed in nm.

In the case of jumpwise motion with a jump time τ_{tr} we have:

$$\Delta\Gamma/\Gamma \sim 2\tau_n^*/\tau_{tr} \ . \tag{3.27}$$

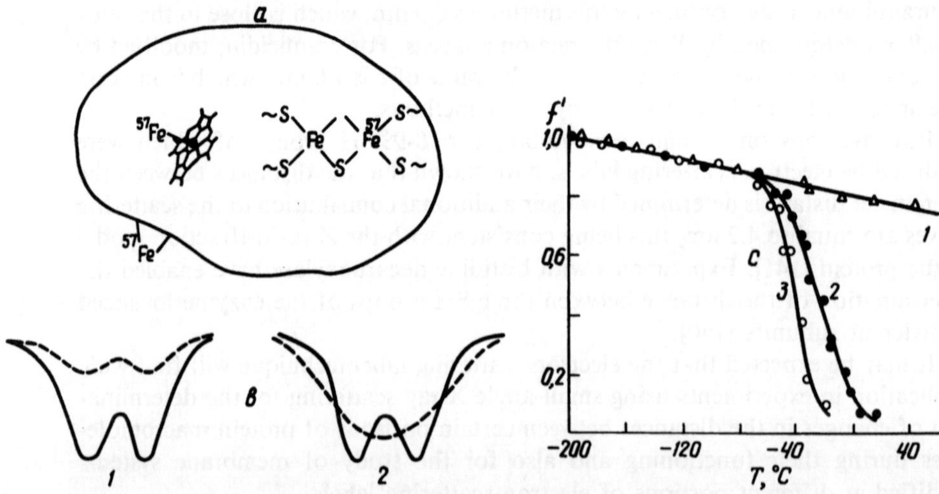

Fig. 3.9 a – c. The Mössbauer label method. **a** Possible methods of introducing labels (^{57}Fe) into proteins. **b** Expected changes in the Mössbauer spectrum parameters for label mobilities with $\tau_c \leqslant 10^{-7}$ s and $\langle x \rangle > 0.01-0.02$ nm: the *solid lines* show the absence of mobility and the *broken lines,* their presence: *1* vibrational motion, f′ decreases, $\Delta\Gamma$ does not change; *2* diffusional motion, $\Delta\Gamma$ increases, f′ does not change. **c** Dependence of the relative value of the parameter f′ for labels on proteins: *1* iron-sulphur protein, in the dry state; *2* iron-sulphur protein in aqueous solution; *3* human serum albumin with labels at the surface [143]

The above formulas given a spectacular representation of the scope of the effect of the dynamics of the medium on the nuclear γ-resonance spectra. For example, the change of the amplitude of vibration of [^{57}Fe] by 0.014 nm results in a decrease of f′ by a factor of 5; if the amplitude changes by 0.05 nm, the value of f′ drops by more than 10^5 times.

The dependence of the parameters of the magnetic hyperfine structure of the spectra (the line width and the magnitude of splitting) of the rate of relaxation of the electron spin on [^{57}Fe] (τ_s) (see Sect. 2.7.1) may be used to determine the distance to the paramagnetic centres that affect τ_s. For this purpose, use may be made of the relations be described in Sections 2.7.1 and 3.1.3 [Eqs. (3.10) and (3.11)]. Estimates based on these formulas show that the MHS of the spectra will be sensitive to the action of paramagnetic compounds at a distance of 0.6 – 1 nm because of the effect of dipolar and exchange interactions on the quantity τ_s.

Mössbauer labels were first used for the study of the intramolecular dynamics of proteins in investigations involving serum albumins [143 – 145]. The [^{57}Fe] ions were bound to the protein surface or were introduced into the bulk portions by way of the reaction with sulphhydride groups obtained by the reduction of the disulphide bridges in the presence of an inorganic sulphide. In the latter case, a polyatomic iron-sulphur cluster, linked covalently to the protein macromolecule at many points, was formed. As can be seen from Fig. 3.9, as the temperature of the solution of the labelled protein increases, the mobility of the Mössbauer labels both in the bulk and on the surface of human serum albumin, recorded by the fall of the parameter f′, increases sharply at a temperature of T ≈ – 50 °C, i.e. long before the environment melts. At $T > -20\,°C$ an additional dynamic effect is observed, a diffusion broadening ($\Delta\Gamma$) of the γ-resonance line. The behaviour of the parameters f' and $\Delta\Gamma$ differs qualitatively from their behaviour for solid particles of iron oxides whose dimensions are close to those proteins (globular and fibrillar proteins, hemoglobin, myoglobin, ferredoxin, etc.) [207].

It has also been shown on a large number of examples [207–213] that the dynamic state of polypeptide chains of proteins, characterized by the vibrational amplitude $\langle x \rangle \gtrsim 0.02$ nm, arises at a certain temperature and at a certain humidity content of the representative samples of a particular object. This type of molecular dynamics is apparently a property common to all protein macromolecules. The results of the application of Mössbauer labels to the study of the dynamic problems of enzyme catalysis will be discussed in greater detail in Chapter 10.

Further application of the method requires widening the range of Mössbauer labels by making use of a series of hydrophobic and hydrophilic compounds with functional groups containing [^{57}Fe] and [^{119}Sn]. This will open up new possibilities for studying the dynamics of various portions of protein and membrane structures. In our opinion, another promising trend in the use of Mössbauer spectroscopy should be based on the principle of double-label and label-probe techniques. The effect of paramagnetic particles on the MHS of the nuclear γ-resonance spectra of [^{57}Fe] at the active sites of enzymes may be used for measuring the distances and for evaluating the location of the active sites. New prospects are offered by the application of other Mössbauer atoms, such as [^{127}I], [^{129}I], [^{67}Zn], [^{73}Ge], [^{181}Ta] and [^{133}Cs] which have different lifetimes τ_n^*.

Thus, the physical label methods enable the solution of a large number of structural and dynamic problems in complex biological systems which cannot usually be studied by traditional methods.

CHAPTER 4

Kinetic Methods in Enzyme Catalysis

One of the key areas of chemical physics is chemical kinetics, a science dealing with the chemical reaction rates. Modern kinetics is equipped with an armamentarium of methodological concepts, which allow one deep insight into the mechanism of a chemical process and, in a number of cases, to predict the course of the process. Chemical kinetics is divided into two main parts: (1) formal kinetics which enables one to split a complex process into individual steps and to measure their rate constants; and (2) a section concerned with the study of the elementary acts of a chemical transformation.

Metalloenzymes catalyze processes involving rather complex kinetics. As a rule, several substrates participate in the reactions, a multi-step electron transfer inside the enzyme often takes place, and cooperative interactions between the active sites may be encountered.

This chapter deals with some problems of formal kinetics of enzymatic reactions. Since a large number of works have been published, which are devoted to the kinetics of chemical and enzymatic reactions [214–222], we shall only briefly consider the general problems of kinetics. Greater emphasis will be placed on the fields that have a more direct relation to the basic topic of this monograph. The theory of the elementary acts of chemical and enzymatic processes will be expounded in Chapter 5.

4.1 Kinetics of Cooperative Processes

A cooperative system may be defined as a system consisting of interacting components. In such a system, the thermodynamic state of each of the components and their rate of change of this state at any moment of time depend, in a general case, on the states of all the other components. As applied to enzyme catalysis the cooperativity concept is used mainly in two ways. In the first case, we deal with enzymes having several catalytic or regulatory sites. Here, it is assumed that the thermodynamic and kinetic parameters of the chemical transformations at each of the sites depend on the state (e.g. the degree of occupancy by the reactants) of the other sites [223, 224, 225]. In the second case, the protein macromolecule is thought to be a cooperative assembly of components, e.g. amino acid fragments and water molecules [227–231]. In either case, taking into account the cooperativity of the system implies that the equilibrium constants K or the rate constants k of the individual steps are considered to be dependent on the state of the other components of the system. Examples illustrating the specific features of the kinetics of cooperative processes are given below.

In enzymology, the phenomenon of allosterism is very widespread; allosterism denotes that the reagents or effectors (substrates, inhibitors or activators) interact on the enzyme while being bound by various spatially separated sites. Let us consider two such systems.

1. Homotropic, allosteric systems with direct cooperativity (the Volkenshtein model or the Koshland-Nemeti-Filmer model) [223, 224]. The enzyme macromolecule contains n identical sites (subunits). The thermodynamic and kinetic properties of each subunit depend on whether the neighbouring subunit is free or occupied by substrate molecules. For a two-site system, in which the square denotes the state of the subunit not occupied by the substrate and the circle represents the state occupied by the substrate, the system may be represented by the following scheme:

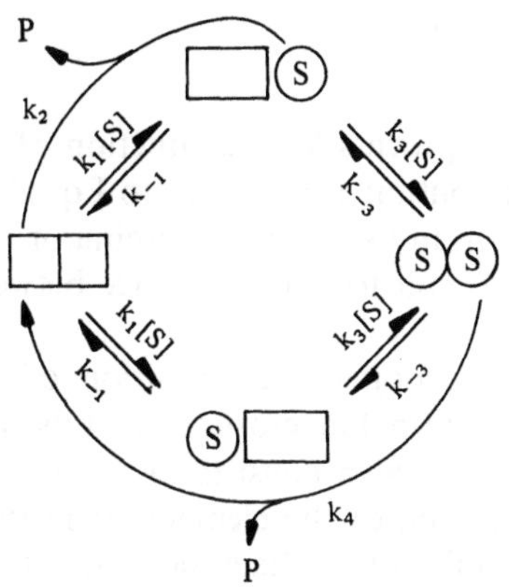

and the velocity of the reaction is given by:

$$V = 2k_2[E]_0[S](\beta K + \alpha[S])/(\beta K^2 + 2\beta K[S] + [S]^2) \ , \tag{4.1}$$

where

$$\alpha = k_4/k_2,\ \beta = K'K^{-1},\ K = (k_{-1}+k_2)/k_1 \text{ and } K' = (k_{-3}+k_4)/k_3 \ .$$

Cooperativity manifests itself at $\alpha \neq 1$ and $\beta \neq 1$. The conditions for the point of inflection on the curve of V vs S are as follows:

(1) $\alpha < 0.5$ and any β;
(2) $1 > \alpha > 0.5$ and $\alpha/(2\alpha-1) > \beta > \alpha^2/(2\alpha-1)$;
(3) $\alpha > 1$ and $\alpha^2/(2\alpha-1) > \beta > \alpha(2\alpha-1)$.

The maximum is possible at $\alpha < 0.5$ and any β.

The value of the Hill coefficient, which characterizes the sharpness of the change of the function V(S), is given by:

$$n = \frac{d \log \{V_{max}/(V_{max}-V)\}}{d \log [S]} \ , \tag{4.2}$$

at small [S]; in this case, n = 2. For a system with m identical sites, n = m. This may be used for the evaluation of the number of sites in a system.

2. A homotropic system with indirect cooperativity (the Monod-Wyman-Changeux model [225]). A dimer of subunits may, as a whole, be either in two or more symmetric conformational states, R and T (R for relaxed and T for tense); the affinity for ligand, characterized by the equilibrium constants K_R and K_T, varies with the change of the state of the oligomer:

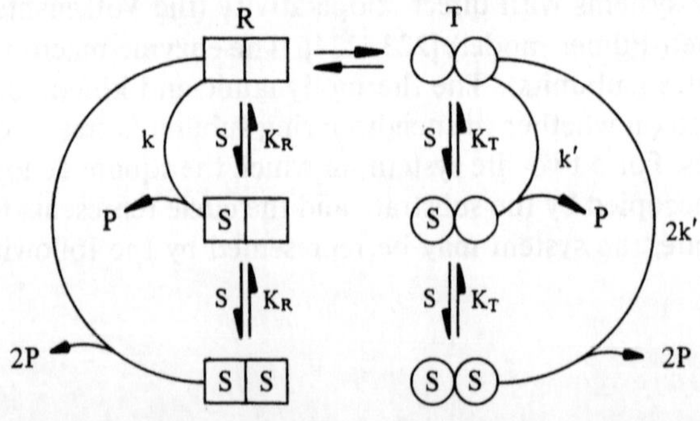

The mathematical expression for the rate of the process V as a function of [S] has been given in the literature [225]. In general outline it is similar to Eq. (4.1). The difference is in the absence of the linear term in [S] in the denominator and its presence in the numerator. It is therefore difficult to make a choice between the models on the basis of kinetic data alone.

As an example of the simplest model of a cooperative transition of the macromolecule from the initial macrostate I to the final macrostate F, let us consider a large ensemble of linear systems, each of which consists of n identical elements [24, 99, 227–231]. In the initial state each of the elements is in the i microstate and in the final state it is in the f microstate. Each element of the system has neighbours, the interaction with which does not allow it to undergo the i→f transition independently. The process starts with such a transition in the weakest link, e.g. from the extreme element with the rate constant of induction k_1. After this transition the cooperative bonds in the neighbouring link are weakened and the i→f transition is facilitated. The subsequent transitions occur in an analogous manner. The process may be described by the following scheme:

$$\underset{\text{I}}{\text{i–i}\cdots\text{i–i}} \underset{k_{-1}}{\overset{k_1}{\rightleftharpoons}} \text{f–i}\cdots\text{i–i} \underset{k_{-2}}{\overset{k_2}{\rightleftharpoons}} \cdots \underset{k_{-n}}{\overset{k_n}{\rightleftharpoons}} \text{f–f}\cdots\text{f–i} \xrightarrow{k_{n+1}} \underset{\text{F}}{\text{ff}\cdots\text{f–f}} .$$

If the induction step ocurs at a much slower rate than the preceding steps, steady-state conditions are set up in the system. In the simplest case, where $k_2 = k_3 = \ldots = k_+$ and $k_{-1} = k_{-2} = \ldots k_-$ and the subsequent step (k_{n+1}) is irreversible, the reaction involving the I→F transition proceeds by a unimolecualr mechanism and the apparent rate constant is given by:

$$k_{app} = k_1(K-1)/(K^n-1) , \tag{4.3}$$

where $K = k_-/k_+$. Equation (4.3) describes the main features of cooperative processes.

At $k_- > k_+$ and $K > 1$, then: $k_{app} = k_1/K^{n-1}$

In this region the overall rate constant k_{app} at sufficiently large n depends sharply on the equilibrium constants of the elementary transitions, K, and will be rather sensitive to all the factors that have a bearing on K (temperature, pH, substances added, etc.). In the other extreme region, $k_+ > k_-$, $K < 1$ and $k_{app} = k_i$.

For an irreversible reaction involving n steps, each of which proceeds with the rate constant k_+, the buildup of the product concentration ([F]) at initial times $t < k_+^{-1}$ is described by:

$$[F] = [I]_0 \frac{(k_+ t)^n}{n!}, \tag{4.4}$$

where $[I]_0$ is the initial concentration of the original state; the appearance of the product is associated with the induction period and the rate of its appearance increases the more sharply, the larger the number of elements in the cooperative ensemble.

Thus, despite their obvious simplicity, the models described above account well for the characteristic feature of cooperative processes [232]: (1) the sigmoidal dependence of the rate of enzymatic transitions on the reagent concentration [Eq. (4.1)]; (2) the presence of lag periods and other non steady-state phenomena [Eq. (4.4)]; (3) the non-linear dependence of the rate of the process on the enzyme concentration (in the case of dissociating enzymes); (4) the sharp, extreme dependence of the rate of cooperative processes (e.g. the denaturation of proteins) on temperature and other factors (Eq. (4.3)].

4.2 Experimental Methods of Studying the Kinetics of Enzymatic Reactions

The rate constants of steady-state enzymatic reactions are measured by recording the rate of disappearance of substrates (S) or the rate of accumulation of the product (P) as a function of time or initial concentration of the enzyme and substrates. The methods used to treat the data have been reviewed extensively [214–223]. Of primary interest are non-stationary methods, which permit the study of fast intermediate steps. These methods may be divided into the following groups: (1) flow methods; (2) relaxation methods; (3) pulse methods; and (4) combination methods [233–235].

Three types of flow methods according to the mixing rate of the reagents can be differentiated: the constant flow method, the accelerated flow method and the stopped flow method. The stopped flow technique has found the widest application; it permits one to study bimolecular and polymolecular reactions with characteristic times of $\tau \gtrsim 10^{-4}$ s. This limit is due to the possibility of rapid mixing of the reagents.

Relaxation methods are based on a rapid interaction, which brings the system under study out of the equilibrium or stationary state. The relaxation to the initial state is recorded by methods that enable rapid recording made, as a rule, by optical methods. The effect most widely used is the sharp increase of temperature by means of electric discharge ($\tau \approx 10^{-5}$ s) or by laser action ($\tau \approx 10^{-8}$ s).

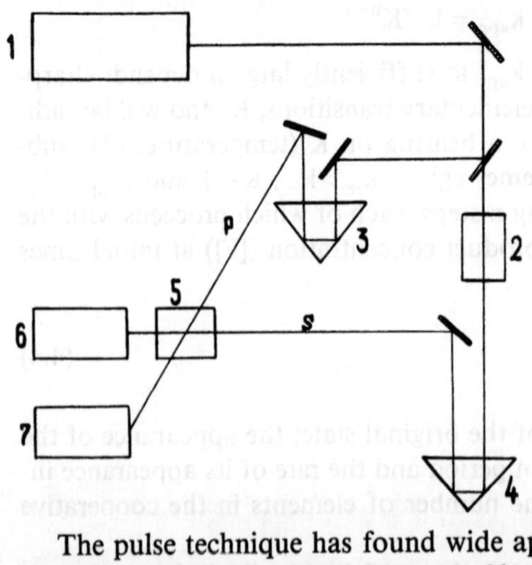

Fig. 4.1. Block diagram of the setup for measuring the kinetics in the picosecond regime [235]: *1* laser; *2* self-phase modulation cell; *3, 4* optical delay line; *5* sample cell; *6* spectrograph; *7* excitation intensity monitor; *P* pump pulse path; *S* probe continuum pulse path

The pulse technique has found wide application. The process is started by a light flash (flash photolysis). The use of laser action combined with fast recording methods has made it possible to study processes in nanosecond and, later, in picosecond ranges [220, 226, 236, 237]. Figure 4.1 shows a simplified scheme of the setup used to study the kinetics of reactions in the subpicosecond range [237]. A laser ray of duration 0.4 ns is split, after being amplified, into two rays: an exciting ray and a probing ray. The exciting ray impinges on the sample earlier than the probing ray (by $\Delta\tau$). The lag time, which is several picoseconds, is specified by the differences in the paths travelled by the rays be means of a mirror system or by the differences in the refractive indices of optical media. Special photosensitive devices equipped with computers record the dependence of the change of optical density at a certain wavelength on the time $\Delta\tau$.

Combination methods may include the flow, pulse and relaxation procedures. Methods have been lately proposed, which combine the relaxation and pulse principles (the periodic-action or the "kinetic-resonance" method) [238–240]. We shall consider these techniques in more detail.

The photosensitive system is periodically irradiated with light. The result of this action depends, under certain conditions, on the extent to which the system being studied has relaxed to the initial state during the illumination. To a first approximation, the condition for the response of the system to the outside influence is the closeness of the characteristic relaxation times of the system and the duration of the external influence ("kinetic resonance"). The main advantage of this approach is the possibility of obtaining information on the fast equilibrium steps of a complex process by way of measuring the reagent concentration or reaction rates averaged over many time periods, which depends on the average catalyst concentration. The random measurement errors are averaged simultaneously.

For a system which dissociates irreversibly under the action of light:

$$ES \xrightarrow{I,\, h\nu} E+S \ .$$

Upon periodic change of the light intensity by the law $I = I_0\ (I+\theta \cos \omega t)$, theory gives at small θ the following expression for the concentration of the enzyme-substrate (ES) complex [238]:

$$[ES] \approx \left\{ \frac{[E]_0 [S]_0}{[S]+K} - \theta^2 B \right\} - A\theta \cos(\omega t + \varphi) \ , \tag{4.5}$$

where B is a constant which is independent of time; A is the amplitude of the first harmonic of vibration; φ is the phase shift between the action and the response. All these quantities are functions of the reagent concentrations and the rate constants of the individual steps. By measuring the dependence of the concentration ES on the pulse frequency at various initial concentrations, one can check the validity of the scheme chosen and determine the characteristic times of the process, τ, in the range 10^{+2} to 10^{-6} s [238].

In another variant of the kinetic resonance technique [239, 240], use is made of short (10^{-5} s) periodic pulses with a period between pulses of $T' = 10^2$ to 10^{-3} s. After each pulse the system departs from the equilibrium position, which substantially facilitates the measurements and increases the accuracy of recording the differences in average concentrations and, especially, in average rates of the processes. A further advantage of this version is the simplicity of the mathematical apparatus. Figure 4.2 shows various responses of the system under study of the external influence as a function of the ratio of the rate constant of a unimolecular process (k) and of the period T'. For a reaction proceeding according to the following scheme:

$$A \underset{k}{\overset{I,\, h\nu}{\rightleftharpoons}} B \ ,$$

theory gives the following value for the relative concentration:

$$f(T') = [A(T')]/[A]_0 = [1 - \exp(-\ k \ln T')]/kT' \ . \tag{4.6}$$

Expression (4.6) predicts the presence of an inflection point on the curves of f vs T'; at this point $f_1(T') = 0.30$ and $k = 1.8 \times T'^{-1}$. If the reverse reaction B→A

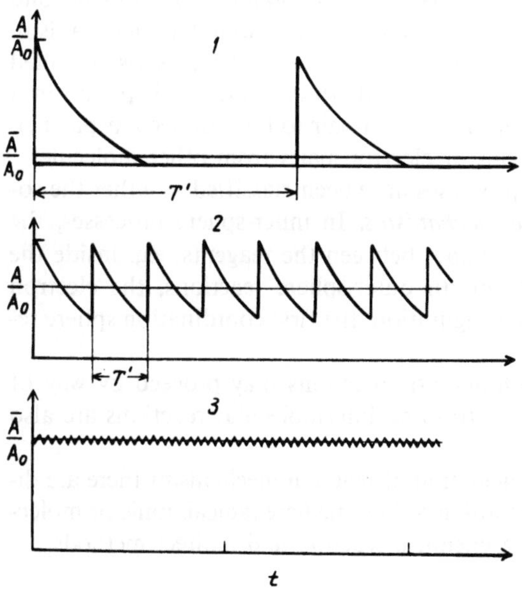

Fig. 4.2. Dependence of the relative and average concentrations (A/A_0 and $\bar{A}/A_0$) on time at different relations between the pulse duration (T') and the reaction rate constant (k): *1* $k^{-1} \ll T'$; *2* $k^{-1} \approx T'$; *3* $k^{-1} \ll T'$ [239, 240]

is second-order, then $f_1(T') = 0.21$ and $k = 2.16 \times [B]_0^{-1} T^{-1}$ The method makes it possible to determine simultaneously the order of the reaction and the value of the rate constants. In the coordinates of df/fT′ vs T′ a maximum must be observed corresponding to certain characteristic times of the reaction. Analysis of such a "kinetic-resonance" spectrum must provide information on the mechanism of the process under investigation.

The two variants of the kinetic-resonance technique have been tested for the photodissociation of carboxyhemoglobin and carboxymyoglobin [238, 240].

The method can be further improved by developing a periodic pH shock technique through the use of a system that changes reversibly the pH of the medium upon illumination by the mechanism of the Bohr effect, for example, the pH of carboxyhemoglobin (Sect. 6.1.1). In principle, it is possible to develop a periodic temperature shock technique. In this method, use may be made of reactions in closed vesicules, in which heating is done with the aid of a laser pulse and cooling occurs rapidly due to the small size of such systems.

Thus, modern methods allow one to study the kinetics and mechanism of enzymatic processes in a wide range of characteristic times, $\tau = 10^4$ to 10^{-12} s.

4.3 Methods of Studying the Mechanism of Enzymatic Reactions

As has been pointed out above, the methods of formal kinetics, combined with methods of measuring fast processes, permit the establishment of the sequence of the individual steps and their rate constants. The next step is the elucidation of the mechanism of each of the steps. At present, the following classification of reactions and reagents has been adopted [241].

Oxidation-reduction (redox) reactions include the transfer of electrons from the oxidizing to the reducing agent. If all the electrons are transferred, we speak of an electron-transfer reaction. *Acid-base processes* involve the participation or formation of real or generalized acids or bases. The breaking of a covalent bond into radicals and the reverse reaction are regarded as the homolytic mechanism. If the bond breaking results in the formation of ions, the process is called heterolytic. *Nucleophilic reagents* are compounds that, while participating in a reaction, donate all their electrons or an electron pair to their molecular partner. *Electrophilic reagents* accept electrons or electron pairs from other molecules.

Oxidation-reduction, or redox, processes have been classified as either the so-called *outer-sphere* or *inner-sphere mechanisms.* In inner-sphere processes, the conversion takes place upon direct contact between the reagents, e.g. inside the coordination sphere of a metal atom. In outer-sphere reactions, the electron transfer is effected with the ligand configuration, the first coordination sphere remaining unaffected.

Nucleophilic, electrophilic and homolytic reactions may proceed by way of substitution or addition. Concerted intra- and intermolecular reactions are also discussed in the literature.

Among the approaches to the elucidation of reaction mechanisms there are direct methods which involve direct recording of intermediate radical, ionic or molecular particles during the course of a chemical reaction and indirect methods.

Indirect methods make use of the experimental dependencies of the rates of a process or individual steps on the steric and electronic properties of substrates and inhibitors, the polarity of the donor-acceptor ability of the solvent, the ionic strength, various substances added, which react with intermediate particles (e.g. radicals), etc. The methods used to take into account some of these effects will be described in Chapter 5, which deals with factors that have an influence on the rate of chemical reactions. At this point, we confine ourselves to two approaches which are of importance for the elucidation of the mechanism: the isotopic substitution (or replacement) method [242–245] and reaction rate measurements in magnetic fields [246–253]. The application of the first method has become traditional in enzyme catalysis. The second is only beginning to be used.

Thermodynamic and kinetic isotope effects are caused chiefly by the differences in the vibration frequencies of isotopes in the initial, transient and final states [242–245]. These differences are reflected in the bond energies and activation energies. Other mechanisms of the effect of the isotopic composition of molecules on the conversion rate are as follows: the change of the mobility of particles and the change of the probability of tunnelling of nuclei (see Sect. 5.3.5). All these effects are the stronger, the larger the differences in masses of atoms or groups of atoms that participate directly in a chemical transformation conversion. The most frequently encountered case is one where a heavier atom is bound more strongly in the initial than in the transition state and the replacement of the light isotope by a heavy isotope results in the slowing down of the reaction. For example, for the replacement of hydrogen by deuterium in the absence of nuclear tunnelling theory predicts the following values of the maximum kinetic isotope effect: $a_{max} = k_H/k_D$ for the breaking of the X–H and X–D bonds (a_{max} is given in parentheses): X=C (6.9), N (8.5), O (10.6). In practice, the isotope effects are, as a rule, weaker ($a = 1.2-5$); cases are also encountered when $a<1$. The latter anomalies are accounted for by the non-elementary nature of the step observed, the presence of intermediate equilibria shifted to the side of the reactant upon deuteration, etc. The transfer of D^+ in uncomplicated cases occurs usually five to six times more slowly than the transfer of H^+. The isotope effects of the solvent are not strong when it has a non-specific effect. For instance, upon replacement of H_2O by D_2O in isomerization reactions of molecules the rate is decreased by 15–20%.

According to theoretical estimates [244], the corrections to the tunnelling effects become significant with the vibrational frequency near the maximum of the potential barrier, which is equal to $\nu^{\neq} > 1000\,cm^{-1}$, and with the value of $E/RT > 20$. Under these conditions, hydrogen, for example, will tunnel about two times more effectively than deuterium. The correction to the tunnelling process increases with decreasing temperature. The effect of the replacement of H by D on the dielectric constant of water leads, according to the estimates made on the basis of the Marcus theory (see Sect. 11.1.2), to an isotope effect in the electron-transfer reaction at room temperature, this effect being equal to 1.5–2. The dependence of the reaction rate on the ratio of the light and heavy isotopes (for example, in water) may be used to establish to participation of one or several protons in the transient state and to determine the rate of the limiting step [244].

A new type of isotope effect has recently been discovered, the magnetic isotope effect, the nature of which will be discussed below [248].

The theory of the effect of the magnetic field on the rate of chemical reactions is based on the fundamental law of conservation of momentum [246–248]. This law naturally extends to the intrinsic momentum of electrons and nuclei (spin). Therefore, in systems in which no interaction takes place between electron spins and orbital momenta or nuclear spins, any changes in the total spin are forbidden. This forbiddenness is partially removed in the presence of the interactions indicated, since channels are opened for the transfer of the momentum to other electrons and nuclei.

The efficiency of electron-nuclear, spin-spin and spin-orbital interactions depends on the magnetic field strength. This is eventually responsible for the effect of the field on chemical processes, the rate of which depends on the population of certain spin states of the system. This population, in turn, depends on electron-nuclear, spin-spin interactions, this being responsible for magnetic isotope effects.

Let us consider the mechanism of the effect of the magnetic field using, as an example, the recombination of the radical pair, $A\uparrow + B\downarrow \rightarrow A\uparrow\downarrow B$. Such a situation is encountered in free-radical reactions and also in the photoseparation of charges upon photosynthesis (see Sect. 9.2.1). According to the principle of spin conservation, the formation of a singlet state in the product $A\uparrow\downarrow B$ is possible only upon interaction of the two particles in the singlet state (S = 0), $A\uparrow + B\downarrow$ (S). The triplet pairs $A\uparrow + B\uparrow$ (T) with spin S = 1 do not react and are converted in other channels. However, as a result of spin-nuclear and spin-orbital interactions, when the magnetic field H_0 is switched on, there is a probability of the triplet-singlet conversion with a characteristic time given by [253]:

$$\tau_{TS} = [(\Delta g \beta H_0 \mp a)]^{-1}, \tag{4.7}$$

where Δg is the difference in g-factors for particles A and B; a is the energy of the contact electron-nuclear magnetic interaction. According to formula (4.7) at $H_0 \approx 50-100$ mT the time $\tau_{TS} = 10^{-7}$ to 10^{-8} s. Thus, the magnetic field has an effect on the T→S conversion in the pair and, hence, on the population of these levels. The magnetic isotope manifests itself through the quantity *a* [253].

Figure 4.3 illustrates the effect of the external magnetic field strength on a process, the rate of which is proportional to the population of the singlet level of a radical pair. As an example of such a process we may cite the reverse recombination of charge photoseparation in photosynthesis recorded by delayed fluorescence [250–251]. In the absence of a magnetic field ($H_0 = 0$), the triplet level T is degenerate. The population of the singlet level S formed upon photodissociation of the pair decreases with time due to the S–T transition with a rate constant k^0_{ST}.

When the magnetic field is turned on, the level T splits into three levels: T_{+1}, T_0 and T_{-1}. The efficiency of the S→T transition is changed because of the differences in the values of k_{ST} for the transition to various triplet levels. The transition $S_0 \rightarrow T_0$ is more effective than the transition to other levels. As the field strength H_0 increases, the T_+ and T_- levels move away from the initial state more and more and the T_0 level remains where it is and, therefore, its contribu-

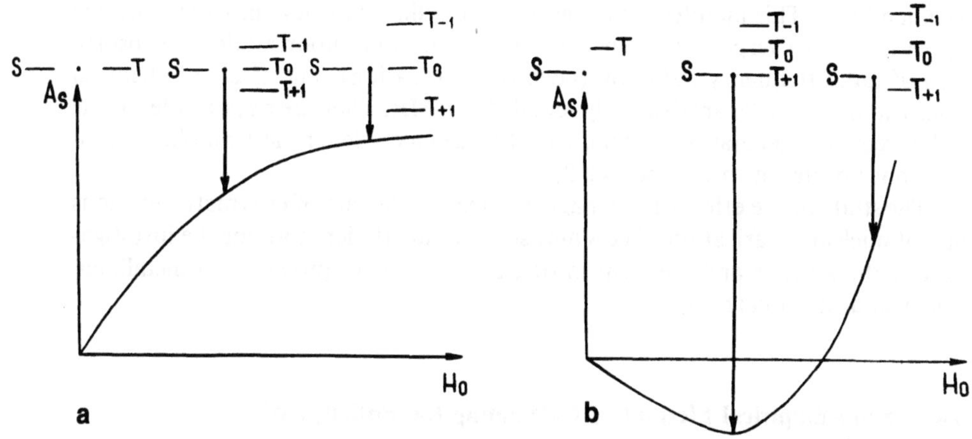

Fig. 4.3a,b. The effect of the linear field strength (H_0) on the population (A_s) at the singlet level, which depends on the rate constant of the S→T transition ($k_{S\rightarrow T}$) for a radical pair: **a** without exchange interaction (J = 0) and **b** with exchange interaction (J > 0). The *vertical arrows* indicate qualitatively the correspondence between the triplet level splitting and certain portions of the $A_s(H_0)$ curve

tion to the overall rate of the S→T transition becomes less and less. The relative population of the S level increases and the rate of recombination becomes faster and faster (Fig. 4.3). As the field strength H^0 continues to increase, the T_+ and T_- levels are found to be at such a large distance that the $S\rightarrow T_\mp$ transition becomes insignificant and the rate of recombination becomes insensitive to the change of H_0.

A qualitatively different picture is observed if there is a sufficiently strong exchange interaction between the components of a radical pair. In such a case, in the absence of a magnetic field the energy levels of the system split into two: a singlet and a triplet level with a difference equal to the doubled value of the exchange integral (J). When the magnetic field is switched on, the triplet levels are split and the T_+ level is brought close to the S level. This contributes to the increase of the overall rate of the S→T transition, to the decrease in the relative population of S and, hence, to the decrease of the rate of recombination (V) (Fig. 4.3). However, when H_0 continues to increase, the S and T_+ levels come closer together and then begin to move away from each other; the dependence $V(H_0)$ will pass through a minimum and will then be saturated. At the minimum point the following equality is fulfilled:

$$J = h^{-1} g \beta H_{min} , \tag{4.8}$$

which may be used for evaluating the integral J.

A further consequence of the S→T conversion in radical pairs, as a result of the transfer of the electron spin moment to the nuclear spins, is the polarization of these nuclear spins [253]. The concerted tumbling of electron and nuclear spins upon energy exchange contributes to the anomalous population of the nuclear

Zeeman levels. This manifests itself either in a radio frequency emission when the upper levels are overpopulated (negative polarization) or upon additional absorption of radio frequency radiation (positive polarization). The effect of chemical polarization of nuclei are distinctly revealed in NMR. They are responsible for the radio frequency generation that has recently been recorded in electron phototransfer from porphyrin to quinone [252].

The study of the effect of the magnetic field on the rate of enzymatic reactions and of nuclear polarization effects may serve as an efficient tool for the investigation of the kinetics and mechanism of electron-transfer processes in metalloenzymes and metal carriers.

4.4 A Semi-Empirical Method of Calculating the Enthalpy of Chemical Reactions

The thermodynamic feasibility of chemical reactions which is characterized, in particular, by the standard enthalpy change ΔH_0 of the system can be estimated from the difference in the bond-making and bond-breaking energies (D). For these energies to be calculated use can be made of a semi-empirical method described in the literature [254].

Numerous experimental data indicate that the energy required to break the bond between two atoms depends substantially on the nature of the surrounding atoms and radicals. According to the concepts developed by Semenov [255], the value of the energy of the x−y bond (D_{x-y}), defined as the energy required to split the molecule into radicals $x^{\cdot}$ and $y^{\cdot}$, may be represented in the following form:

$$D_{x-y} = E^0_{xy} - B_x - B_y \;, \tag{4.9}$$

where E^0_{xy} are the average bond energies of a given type in the molecule calculated by the additivity rule; B_x and B_y are the energies of the conjugation of radicals $x^{\cdot}$ and $y^{\cdot}$; for atoms it is assumed that $B_{x,y} = 0$.

The values of B_x and B_y calculated from the known values of D_{xy} and E^0_{xy} may be used to calculate unknown values of D_{xy}. The accuracy of such calculations is usually 6−8 kJ mol^{-1}.

The method under consideration may be illustrated by an example of the calculation of the energy parameters of nitrogen-hydrogen compounds, which will be used in Chapter 11 for the energy estimation of various N_2 reduction mechanisms.

According to the generalized additivity rule, the heat effects of the sequential addition of a hydrogen atom to nitrogen may be split into the components:

$$D_1 = D_{N_2-H} = E^0_{N-H} - B_1 \;; \tag{4.10}$$

$$D_2 = D_{HN_2-H} = E^0_{N-H} - B_2 \;; \tag{4.11}$$

$$D_3 = D_{H_2N_2-H} = E^0_{N-H} - B_3 \;; \tag{4.12}$$

$$D_4 = D_{H_3N_2-H} = E^0_{N-H} - B_4 \;; \tag{4.13}$$

where E^0_{N-H} is the average energy of the N–H bond; B_1, B_2, B_3 and B_4 are the energies of stabilization (conjugation) of the radicals.

According the the law of energy conservation, the following equalities must be satisfied:

$$B_1+B_2 = E_3, \quad B_3+B_4 = E_2, \quad E_1+E_2+E_3 = D_{N_2} ,$$

where E_1, E_2 and E_3 are the energies of the first, second and third bonds respectively, in the nitrogen molecule; D_{N_2} is the energy of N_2 dissociation into its atoms.

The calculation of the quantities of interest, D_{N_2-H} and D_{HN_2-H}, includes the following stages:

1. The calculation of E_1 from the following formulas (the energy is expressed in $kJ\,mol^{-1}$ and the values of D are taken from a handbook [256]):

$$E^0_{N-H} = (D_{NH_2-H}+D_{NH-H}+D_{N-H})/3 = 386 \;;$$

$$B_{NH_2} = E^0_{N-H}-D_{NH_2-H} = -50 \;;$$

$$E_1 = D_{H_2N-NH_2}+2B_{NH_2} = 153 \;.$$

2. The calculation of E_2 from the formulas:

$$B_3 = E^0_{N-H}-D_{NH_2NH-H} = 158 \;;$$

$$B_4 = E^0-D_{H_2N_2-H} = 72 \;;$$

$$E_2 = B_3+B_4 = 230 \;.$$

3. The calculation of E_3 from the formula:

$$E_3 = D_{N_2}-(E_1+E_2) = 560 \;.$$

4. The calculation of B_2. First, the quantity B_2' is calculated:

$$B_2 = E^0_{C-N}-D_{H_3CN_2-CH_3} = D_{H_3C-NH_2}-D_{H_3CN_2-CH_3} = 97 \;.$$

Then, the correction is made for the replacement of CH_3 by H:

$$B_2 = B_2'-5 = 70 \;.$$

5. The calculation of D_2:

$$D_2 = E^0_{N-H}-B_2 = 318 \;.$$

6. The calculation of D_1 using the following formulas:

$$B_1 = E_3-B_2 = 465 \;;$$

$$D_1 = E^0_{N-H}-B_1 = -93 \;.$$

The bond energies of the molecules are given below [254–257]:

Compound, bond	Dissociation energy of bond energy, kJ mol^{-1}
H_2N_2H-H	$D_4 = 330 \pm 20$
H_2N_2-H	$D_3 = 380 \pm 20$
HN_2-H	$D_2 = 330 \pm 20$
N_2-H	$D_1 = 93 \pm 20$
N_s-first	$E_1 = 160 \pm 13$
N_2-second	$E_2 = 230 \pm 20$
N_3-third	$E_3 = 530 \pm 20$

The material presented in this chapter indicates that modern kinetic methods, when combined, make it possible to establish the sequence of the steps of enzymatic processes, to measure the rate constants of the individual steps in the range from 10^{+12} to 10^{-4} s^{-1}, to study the mechanisms of elementary steps and to estimate the thermodynamic feasibility of such processes.

CHAPTER 5

Factors Affecting the Efficiency of Catalysis

Enzymatic process reactions proceed at a high velocity under mild conditions (normal temperature, pressure, aqueous medium close to neutral, weak oxidizing and reducing agents). At the same time, such reactions are characterized by high selectivity, specificity and the ability to be controlled, depending on the biological need for a given enzymatic process. Why do enzymes exhibit such unique, outstanding catalytic properties? The answer to this question would be of decisive importance both for biochemistry and for the theory and practice of catalysis as a whole.

At present, there is no unified viewpoint on the nature of enzyme catalysis. A large group of scientists, including the present author, believe that an enzymatic reaction, like any chemical catalytic process, is a certain sequence of elementary chemical steps. Each of these steps proceeds by "ordinary" laws of chemical kinetics. The accelerating action of a catalyst is accounted for by the fact that its active centres become involved in such chemical reactions with substrate molecules which lead to an increase in the velocity of the process as a whole. In other words, the catalyst involved in a chemical process alters the path of the reaction in a more favourable direction. Within the framework of these concepts, enzymes and other catalysts do not possess a special "catalytic power" or other properties that distinguish them from other reagents. However, enzymes ar characterized by a set of certain specific properties which have been "polished off" in the course of biological evolution. These properties, taken together, allow them to effect chemical reactions under conditions close to optimal.

In an analysis of the mechanisms of enzymatic reactions it seems to be reasonable to employ a simple methodological rule: no new concepts should be introduced if the phenomena under consideration can be described unambiguously within the framework of traditional concepts, i.e. the basic conceptions of modern chemical kinetics.

The present chapter offers a brief outline of the modern concepts of the elementary acts of chemical reactions. The specific properties of redox enzymes of various classes and the concrete mechanisms of the reactions catalyzed by these enzymes will be considered in subsequent chapters.

5.1 General Conceptions of Chemical Kinetics

The theory of a chemical process rests on the concept of the thermal activation of molecules involved directly in the reaction [214–222, 258–269]. The reaction involves only those "hot" molecules whose energy is greater than a certain minimal value called the energy of activation E_a. Other factors that affect the rate of the reaction between two molecules are: the number of collisions (k_d) and the steric factor (f_s).

The value of the rate constant of a bimolecular reaction is given by the Arrhenius equation:

$$k = k_d f_s \exp(-E_a/RT) \ . \tag{5.1}$$

For a unimolecular reaction we have:

$$k = k_0 \exp(-E_a/RT) \ . \tag{5.2}$$

It should be emphasized that a simple physical interpretation of the activation parameters and the feasibility of Eq. (5.1) for elementary steps are valid only under certain conditions. These conditions may be formulated in the following manner: (1) the system must be statistic, i.e. it must consist of a very large number of particles, whose behaviour is governed by the laws of random thermal motion; (2) during the course of the reaction the Maxwell-Boltzmann equilibrium energy distribution must not be violated. In application to enzymes, this means, in particular, that at the active sites and in other parts of the system super-equilibrium "hot" states must not arise, whose temperature differs from the average temperature and which have time to interact before being relaxed to an equilibrium state; (3) the displacement of nuclei during a given elementary act must be classical, i.e. the nuclei must move with an average thermal verlocity (50–500 m s^{-1}); (4) the adiabaticity condition must be realized in the system, according to which the electron shell of the reacting molecules must have time to be adjusted, at any instant of time, to the corresponding nuclear configuration. Since the characteristic time of electronic motion over short distances τ is of the order of 10^{-15} to 10^{-14} s and that of nuclear motion is 10^{-13} to 10^{-12} s, it follows that the latter condition is fulfilled for many reactions. There are, however, cases in which the rate of electron transfer is limited by quantum-mechanical selection rules or electrons or protons must be displaced over large distances, e.g. by a tunnelling mechanism. These cases should be discussed separately [261–268].

There is no ground for believing that any of the above four postulates, which form the basis of the concepts of chemical kinetics, is violated in the elementary steps of chemical reactions involving the reconstruction of heavy nuclei at the active sites of enzymes, in the acts of conformational displacements of protein groups, in the reorientation and movement of solvent particles, etc. The totality of enzyme and substrate molecules, which are randomly moving in solution and exchanging their energies with one another and with the solvent, is undoubtedly a statistic ensemble. The macromolecules of enzymes that are in contact with the mobile aqueous environment or with lipids undergo rather effective processes of thermal relaxation and the lifetimes of the excited vibrational and rotational states are of the order of 10^{-11} to 10^{-14} s [24, 259, 262]. The possible non-equilibrium nature of the enzyme conformation in the initial state or in the intermediate steps of the process (see Chap. 11) does not imply the inapplicability of Eq. (5.1) for steps that take a shorter period of time than the lifetime of these conformationally non-equilibrium states.

According to modern concepts, the occurrence of the elementary step of chemical processes, including catalytic reactions, proceeds at a sufficient rate characteristic of enzymatic catalysis ($k = 10^1$ to $10^7\ s^{-1}$) provided the following factors are operating in concert.

1. *The Thermodynamic Feasibility of the Step.* According to the empirical correlation equations of chemical kinetics, the change of the positive standard free energy (ΔF_0) must not be greater than about 20–30 kJ mol^{-1}. In certain cases, the rule, the better is the thermodynamics of the step, the better is the kinetics applied, a quantitative expression of which is, for example, the Polanyi-Semenov and Bronsted equations [214, 255].

2. *The Proximity and Orientation Effects of the Substrate Molecule and the Catalytic Site.* The requisite condition for the occurrence of a chemical reaction is the approach of the reacting particles to a certain distance and the presence of an orientation favourable for a given reaction. The preliminary approach of the two reagents during a complex-formation process, resulting from the interaction of the groups that do not participate in the subsequent reaction, increases the effective value of the rate constant of the reaction (k) by about 10^2 times as compared with the value of k under standard conditions with unit reactant concentrations. The precise orientation of the substrate may provide an acceleration of 10^2 to 10^6 times, depending on the type of the reaction. For reactions involving three and more molecules, the acceleration due to these effects may be considerably greater.

3. *The Frequency Factor.* The quantity ν_0 may be close to the frequency of the bond being broken, $\nu_0 = 10^{12}$ to 10^{13} s (e.g. in a diatomic molecule) or to the oscillation frequency of an electron in a potential box, $\nu_0 = 10^{15}$ s (in the simplest tunnelling model) of an electron through a potential barrier). In other cases, ν_0 is a composite quantity which may include, in particular, the frequency of solvent reorientation.

4. *Quantum-Mechanical Factors.* According to theory, the rate constants of an elementary step of a chemical process (k) depend significantly on the degree of positive overlap of the electron wave functions in the final and initial states. Quantitatively, this phenomenon is characterized by the resonance integral V. Under certain conditions, $k \sim |V|^2 \sim S^2$, where S is the overlap integral.

Over short distances within a single coordination sphere the overlap is, as a rule, very significant and the frequencies of electronic motion exceed the frequencies of nuclear motion with characteristic times $\tau = 10^{-13}$ to 10^{-12} s. In this case, the so-called adiabatic approximation is valid. If the overlap is slight, i.e. the centres are separated by a large distance or electronic transitions are symmetry-forbidden, then

$$k \sim |V|^2 \sim \varkappa < 1 , \tag{5.3}$$

where $\varkappa$ is the transmission coefficient of the reaction. The quantum effects associated with spin selection rules can also lower the value of $\varkappa$ (see below).

5. *Factors Governing the Matching of the Nuclear Environment to the Electronic Configuration.* The transition of electrons in a chemical reaction requires that the law of energy conservation be satisfied. In many cases, such a transition is impossible without a simultaneous change in the energy of the nuclear core. The vibrational (including collective vibrations, phonons), rotational and translational degrees of freedom of the nuclear environment serve as a sink for the excess electron energy during the transition. The electron transfer leads to a change in the balance of the interaction forces between electrons and nuclei and, hence, in the internuclear distances. In other words, for each system a fit between electrons and nuclei must be realized.

The rate constant of electron-transfer processes depends substantially on the free activation energy of the reorganization of the reagents and the medium resulting from a given elementary step ($\Delta F_r^{\neq}$):

$$k \sim \exp(-\Delta F_r^{\neq}/RT) . \tag{5.4}$$

For inner-sphere reactions $\Delta F_r^{\neq}$ depends on the free energy change during the breakage and formation of chemical bonds in the coordination sphere of the complex and on the reorganization energy of the medium. Outer-sphere reactions are accompanied by the reorganization of the medium alone.

It is expedient to distinguish between two types of electron-nuclear complementarity: an electronic-vibrational [269] and an electronic-orientational complementarity [223].

Electronic-vibronic complementarity is achieved upon synchronization of nuclei as a result of the rapid vibrational relaxation during the time comparable in order of magnitude with the duration of the elementary step of an adiabatic chemical reaction (10^{-13} s). The efficiency of the fit is sharply diminished during this time with increasing number of nuclei participating in the process.

Another type of fit of the nuclear environment is associated with the relatively slow reorientation of nuclei. A special case of the electronic-orientational fitting is the electronic-conformational complementarity between the configuration of the protein moiety of a metalloenzyme and the electronic state of the active site [223]. If the forward and reverse conformational changes occur faster than the chemical reaction proper (k_r), the apparent rate constant of the process may be written as:

$$k_{app} = K_{conf} k_r , \tag{5.5}$$

where K_{conf} is the equilibrium constant of the corresponding conformational transition.

5.2 Thermodynamic Feasibility of the Chemical Step

The fundamental possibility of a chemical process taking place in closed systems (in which no exchange of substances with the surroundings occurs) at constant temperature and constant volume is described by the standard free energy ΔF_0. In chemical thermodynamics, this quantity is defined as the chemical work performed upon transition of a given system from a certain standard state, in which the concentrations (in a general case, the activities) of the solutes are equal to unity (e.g. moles per litre) and all gases are characterized by a partial pressure (or fugacity) equal to 1 atm, to an equilibrium state.

The quantity ΔF_0, which is usually expressed in $kJ\,mol^{-1}$, is also a quantitative measure of the extent of reaction, since it is uniquely connected with the equilibrium constant K by the well-known relation:

$$\Delta F_0 = -2.3\,RT \log K = \Delta H_0 - T\Delta S_0 \ . \tag{5.6}$$

For real processes, in which the concentrations are not necessarily equal to unity in the state before the reaction is started, the free energy of the process is given by:

$$\Delta F = \Delta F_0 - 2.3\,RT \log \{[C]^{m_1}\,[D]^{m_2}/[A]^{n_1}\,[B]^{n_2}\} \ , \tag{5.7}$$

where [A] and [B] are the concentrations (activities) of the reactants; [C] and [D] are the concentrations (activities) of the products; n_i and m_j are the stoichiometric coefficients $(i,j = 1,2)$.

Oxidation-reduction (redox) processes can conveniently be described by using the concept of the standard potential (E_0^0), which is defined as the standard free energy of the reaction at pH = 0 and is expressed in electron volts:

$$X + \tfrac{1}{2}\,H_2 = H^+ + X^- \ , \tag{5.8}$$

where X and X^- are the oxidized and reduced forms of a given substance respectively. Apparently, the potential difference for the two redox couples is $\Delta E_0^0 = \Delta F_0/96.3$ eV.

Simple estimates show that at physiological concentrations (e.g. at 10^{-3} M) and with the extent of reaction being equal to 10%, the values of ΔF_0 will be approximately equal to 5.5, 11 and 16.5 $kJ\,mol^{-1}$ for unimolecular, bimolecular and trimolecular reactions respectively.

The above applies to biochemical systems only in those cases where these systems may be regarded as closed systems. This condition is apparently not always fulfilled for separate portions of electron-transfer chains and thier coupling with other reactions during the course of the process. If the process as a whole is sharply shifted to the side of the product formation or if the reaction products are rapidly withdrawn, a situation may arise in which one of the intermediates present in small concentrations under equilibrium conditions will nervertheless be formed at a sufficiently high rate contrary to the thermodynamic disadvantage of the process at this particular step. It is only important that this intermediate state be converted even more rapidly to the end products.

The foregoing does not mean that in open systems thermodynamically unfavourable steps are allowed kinetically. Modern kinetics permits the estimation of the limits of positive values of ΔF_0, above which the step will proceed at negligible rates. Such estimations can be made using the well-known correlation equations that relate kinetic parameters, such as the rate constant (k) and the energy of activation (E_a), to thermodynamic parameters: the equilibrium constant (or ΔF_0) and the enthalpy (ΔH_0).

These equations are represented by the Polanyi-Semenov equation [255]:

$$E_a = \alpha_c + \beta_c |\Delta H_0| \ , \tag{5.9}$$

and the Bronsted equation:

$$\log k = \log k_0 - \beta_0' \Delta F_0 / 2.3\,RT \ , \tag{5.10}$$

where α_c, β_c and β_c' are the experimentally obtained coefficients, whose values can be predicted theoretically in a number of cases.

Considering that enzymatic reactions proceed at high rates ($k \gtrsim 1\ s^{-1}$) and the numerical values of the coefficients for endothermic reactions $\alpha_c \approx 45$ kJ mol^{-1}, $\log k_0 = 10-12$ and β_c, $\beta_c' \approx 0.5-0.8$, it can be shown that the individual unimolecular steps can proceed at sufficiently high rates at $\Delta F_0 \leqslant 20-30$ kJ mol^{-1} and $\Delta E_0^0 \leqslant 0.2-0.25$ eV.

The correlation Eqs. (5.9) and (5.10) may also prove useful in the evaluation of the rates of individual steps in those cases where the thermodynamic parameters of the steps are known from experimentation or can be calculated. In such cases, however, one should consider a number of factors that may shift these equations either to an increase or a decrease in the coefficients: (1) tunnelling transitions of electrons and protons; (2) the presence of rate-limiting, conformational steps; (3) an appreciable smoothing-out of the energy profile because of vibrational interactions [269]; (4) low values of the orientational and synchronization factors (see Sect. 5.5.5), etc.

The energy profile of an elementary chemical process, which is taken here to mean a sequence of the thermodynamic energy levels of the reactants, intermediates and products, depends in large measure on the reaction mechanism. In calculating the energy profile of alternative mechanisms, we can distinguish the most preferred ones and exclude some mechanisms as being undoubtedly infeasible. The next step in this approach is to use data on enzymatic reactions.

The possibility of the occurrence of a chemical reaction is most completely described by the free energy of the process. However, if there are no data on ΔF_0 or E_0 in estimating calculations, one may use data on the enthalpy or the bond energy. The chemical changes at an enzyme proceed, as a rule, unimolecularly and, hence, must not involve considerable changes in entropy, while the conformational and solvation rearrangements are often accompanied by the compensation of the energy and entropy parameters and therefore make no significant contribution to the quanity ΔF_0 (see Sect. 10.1).

In order to estimate the thermodynamics of a reaction that occurs in the coordination sphere of the metal (M), it is expedient to conditionally divide the reaction into steps that represent the process taking place under certain idealized conditions (e.g. in the gas phase) and the interaction of the reactants (A) and products (B) with the metal complex:

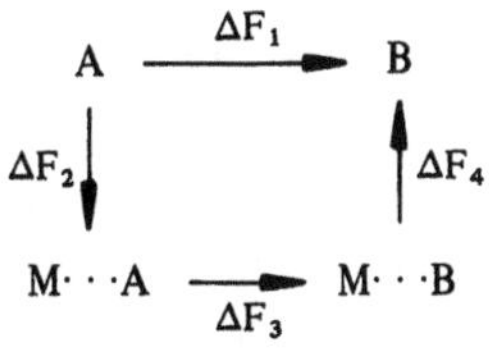

It is obvious that $\Delta F_3 = \Delta F_1 - \Delta F_2 - \Delta F_4$. The most difficult problem here is to calculate the free energies of complex formation (ΔF_2 and ΔF_4).

It should be kept in mind that for many enzymatic reactions, the principle of the stepwise equilibrium occurrence of the process is observed (otherwise, the reaction cannot be characterized by a smooth thermodynamic relief required for a catalytic process). Therefore, to a first approximation, $\Delta F_2 \approx \Delta F_4 \simeq 0$ and $\Delta F_3 \approx \Delta F_1$. This means that for a crude estimate of the energetics of the step in the coordination sphere of M, one may use data on analogous reactions taking place under idealized conditions. Obviously, such estimates are valid if the alternative mechanisms of the reaction, one of which is to be chosen, differ in energetics by an amount greater than the possible differences in the values of ΔF_2 and ΔF_4.

5.3 Collision of Reactants. Steric and Electrostatic Factors

The intensity of the translational motion of particles in the condensed phase is characterized by the diffusion coefficient D_{tr}, i.e. the number of particles that cross per unit time the unit surface area at unit particle concentration. If the size of the particle (R) is larger than that of the molecules of the medium, the latter may be considered continuous and isotropic. For such systems, Stokes' law is valid:

$$D_{tr} = k_B T / 6\pi R \eta \ , \tag{5.11}$$

where η is the viscosity of the medium. The following equation has been suggested for the estimation of the diffusion coefficient (D_{tr} in $cm^2\,s^{-1}$) [273].

$$D_{tr} = 11.6 \times 10^{-6} (M^n - 1.89) \ , \tag{5.12}$$

where n = 1/3 for spherical particles; for other particles, n = 0.3 – 0.5, depending on their shape; M is the molecular mass.

For two particles, A and B with radii R_A and R_B, the rate constant of collisions (first encounters) [270–272] is given by:

$$k_d = 4\pi R D_{tr} \ , \tag{5.13}$$

where $R = R_A + R_B$ and $D_{tr} = D_A + D_B$.

In biological systems, many enzymatic processes are localized in membrane structures and the diffusion, in fact, takes place in two-dimensional space (lateral diffusion). In this case, the dependence of the rate constant of the collision between two particles on the lateral diffusion coefficient D_l is described by the following formula [274]:

$$k_d = \frac{8}{\pi} R^2 \int_0^{\infty} \frac{\exp-(D_1 t u^2/R^2)}{U[I_0^2(u)+Y_0^2(u)]} du \ , \tag{5.14}$$

where I_0 and Y_0 are the zero-order Bessel functions of the first and the second kind respectively; u is an auxiliary variable; t is the reaction time. There are tables of the integrals contained in formula (5.14) which can be used to calculate k_d from the particular values of D_e and t [275]. The formula for estimation of D in membranes has been given in elsewhere [272].

A bimolecular reaction in the liquid phase is usually considered a two-step process:

$$A+B \underset{k_{-1}}{\overset{k_d}{\rightleftharpoons}} AB \xrightarrow{k_r} \text{Products} \ .$$

The encounter of particles as a result of diffusion with the rate constant k_d means that they are entrapped in a single liquid "cage", a kind of microreactor. In this cage, the particles undergo secondary collisions and react at the rate constant k_r. Emergence of the particles from the cage is possible at the rate constant k_{-1}. The apparent rate constant for a steady-state process may be written in the form:

$$k = 4\pi D_{tr} R k_r/(k_r + k_{-1}) = 4\pi D_{tr} R_{app} \ . \tag{5.15}$$

This dependence forms the basis for the classification of reactions as kinetic and diffusion-controlled reactions. With $k_r \gg k_{-1}$ the reaction upsets the equilibrium, statistical distribution and the entire process is controlled by the number of first encounters: $k = k_d$. With $k_r \ll k_{-1}$ the concentration of diffusion pairs and the rate of the chemical conversion become steady state:

$$k = k_d k_r/k_{-1} = K_{AB} k_r = 4\pi R^2 \Delta_r k_r = V_r k_r \ , \tag{5.16}$$

where K_{AB} is the equilibrium constant of the formation of a diffusion pair; V_r and Δ_r are the volume and thickness of the so-called reaction layer respectively.

Within the framework of the reaction layer model, it is assumed that the entry of particles into this layer is the prerequisite for the occurrence of the reaction at a rate constant k_r. No reaction takes place outside this layer. An approximate estimate shows that at R = 0.3 nm, the value of Δ_r is about 0.01 nm. For processes in which the reaction occurs with the particles being separated by a considerable distance (long-range processes), the value of R_{app} depends on D_{tr} and on the factors that characterize the effectiveness of the long-range interaction. For example, if the quantity k_r depends exponentially on the distance $R-R_{AB}$ with an exponential factor a_B [Eq. (5.35)], the quantity R_{app} in Eq. (5.13) will be given by:

$$R_{app} = R_{AB} + [\ln (k^0/a_B^2 D_{tr}) + 1.15]/a_B \ , \tag{5.17}$$

where k^0 is the rate constant at a distance R_{AB}. The physical significance of Eq. (5.17) is that at small values of D_{tr} (rate encounters), the effective radius of interaction increases due to the reaction between the remote particles.

For cases where on collision only those particles that are oriented in some special way become involved in the reaction, the concept of the apparent steric factor f_{app} is introduced, which characterizes the ratio of the number of effective collisions to the total number of first encounters. As has been shown in the literature [270, 277], under diffusion-controlled conditions $f_{app} \approx \sqrt{f_{AB}} = \sqrt{f_A f_B}$, where f_A is a geometric factor equal to the ratio of the area of the reaction spot of the particle A to the total area of the particle; f_B is the geometric factor of the second particle. For the kinetic regime, $f_{app} = f_{AB}$.

The differences in the values of f_{app} for different regimes are qualitatively accounted for as follows. Upon diffusion approach in the condensed phase, the particles, before being separated by large distances, undergo multiple collisions. As a result, for particles that have initially collided with no contact between their reaction spots, such a contact is possible upon repeated collisions. As a consequence, the effective geometric factor increases. For reactions that proceed under kinetic conditions, the particles which have entered the reaction volume as a result of repeated collisions have the possibility of leaving it unreacted. This possibility is taken into account by introducing the equilibrium constant K_{AB} [Eq. (5.16)]. In such a case, the apparent steric factor is equal to the geometric factor, i.e. it has a lower value than in the case of the diffusion-controlled regime.

The electrostatic factors for the collision of charged particles under diffusion-controlled conditions are taken into account, to a first approximation, by the Debye equation. This equation relates the ratio of the rate constants of the collisions of charged ($k_d^{\pm}$) und uncharged (k_d^0) particles to their charges z_A and z_B, the separation of the charges r_z and the dielectric constant of the medium ε:

$$\alpha_d = k_d^{\pm}/k_d^0 = -\delta/(1-\exp\delta) \,, \tag{5.18}$$

where $\delta = N z_A z_B / 2RT\varepsilon r_Z$. As can be seen from Fig. 5.1, which shows the dependencies of α_d on $z_A z_B$ at different r_z for the spin exchange between nitroxide radicals of different structures and charges and between these radicals and paramagnetic ions, the experimental data are in satisfactory agreement with the theoretical dependence (5.18) at $\varepsilon = 80$. Therefore, for diffusion processes, this

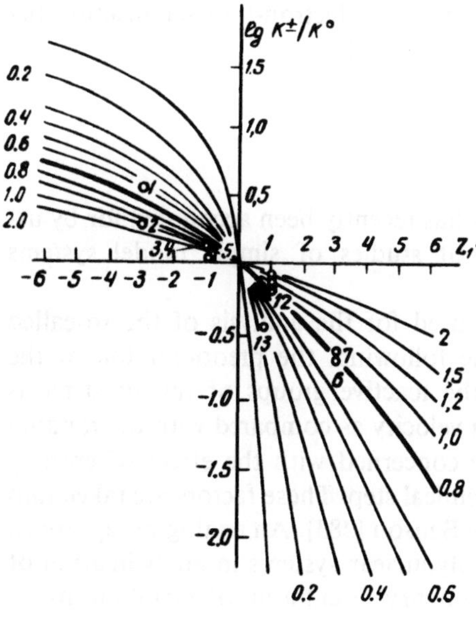

Fig. 5.1. Dependence of the relative rate constant on the charge (Z_i) and the distance r_z between the charges for a diffusion process. The theoretical curves have been calculated according to Eq. (5.18); the *points* represent experimental data on the spin exchange between nitroxide radicals [99]; the *numbers* on the curves correspond to the charge separation in nm

equation and the plots in Fig. 5.1 can be used for the approximate evaluation of r_z with z_A and z_B being known and for the determination of z_A and z_B with r_z being known.

For processes occurring under kinetic conditions it would be more correct to use the equilibrium collision complex model. Taking into account the effect of the electrostatic factors on K_{AB} leads to the expression:

$$a_r = k^{\pm}/k^0 = \exp(-\delta) . \tag{5.19}$$

Comparison of Eq. (5.18) with Eq. (5.19) shows that at $\delta > 0$ diffusion-controlled processes are more sensitive to the effect of the electrostatic factors than kinetic processes. At large, positive δ the value is $a_d/a_r \approx \delta$; at $\delta < 0$ the relationships are reversed.

Additional information on the charge state of particles is obtained by studying the dependencies of the rate constant of the reaction on the ionic strength, $I' = 1/2 \sum_i c_i^2 z_i^2$, where c_i is the concentration of ions. According to Wherland and Gray [278]:

$$\ln \frac{k}{k_\infty} = 3.576 \left[\frac{\exp(-\varkappa_1 R_A)}{1+\varkappa_1 R_B} + \frac{\exp(-\varkappa_1 R_B)}{1+\varkappa_1 R_A} \right] \frac{z_A z_B}{R_A + R_B} , \tag{5.20}$$

where k_∞ is the rate constant at $I' = \infty$ and $\varkappa_1 = 3.3\,(I')^{1/2}$ (R in nm). The physical mechanism of the action of neutral ions on the reaction rate is that the charged particles of the reactants attract ions of opposite sign and repel ions of the same sign. As a result, an ion atmosphere with ions of opposite sign is formed around the particles, which shields the electrostatic interactions.

The electrostatic effects in reactions involving enzymes may be very strong because of the presence of local accumulations of charges in the complex mosaic of charges on proteins. This will lead to a large value of $z_A z_B$ and, hence, to a sharp change in the reaction rate as compared with uncharged particles. This effect has been demonstrated, for example, for the electron-transfer reaction between deoxymyoglobin and ferricytochrome *c* [279].

5.4 Orientational Effects

The high efficiency of enzymatic catalysis has recently been accounted for by using the concepts based on the results of studies of simple model systems [280–282].

Such an approach has recently been used for the analysis of the so-called orientational effects which consist in the following: the preorientation of the catalytic groups of an enzyme and of the reactive groups of the substrate is responsible for an increase in the reaction velocity as compared with the random arrangement. In this section we shall be concerned with the effect of entropy orientational factors on the rate of the chemical step. These factors are taken into account by using the method developed by Benson [283]. An analogous approach may be used to analyze data on model polynuclear systems in an estimation of the orientational entropy gain upon preliminary orientation of metal atoms.

A routine procedure for evaluation of the contribution of orientational effects is to compare the rates of three types of reactions. In reactions of the first type, the reacting groups diffuse freely in solution; in those of the second type they are bound by flexible "legs" within a single molecule; and in reactions of the third type the structure of the molecules is such that even in the initial state the reacting groups are oriented in the direction of formation of reaction products. A rather precise comparative estimation is possible for the orientational effects operating in reactions of the second and third types. A detailed analysis of the available data on the entropies of formation of various cyclic and acyclic compounds has led Benson to the formulation of the following quantitative rules.

Rule 1. Upon cyclization of aliphatic molecules the decrease of entropy per bond is 19.5 (here and further in the text, the entropy ΔS_{or} is expressed in $J K^{-1} mol^{-1}$) for all rings, except the C_4, C_5 and C_8 hydrocarbons, for which the corresponding values are: 16.5, 15.2 and 15.2.

Rule 2. When a double bond is introduced into C_4 and C_5 rings, their entropy mobility is lowered by $\Delta S_{or} = 13.5$ and 22.5; the mobility of the C_6 ring is changed only slightly ($\Delta S_{or} = -2.5$). The introduction of three double bonds into cyclohexane, which makes the ring practically rigid, is accompanied by a decrease in entropy by a factor of 21. A close value must be observed when the internal mobility of the six-membered ring is arrested, for example, by transverse bridges.

Rule 3. The entropy of free rotators lies in the range of 19 to 25 for small groups (for example, for CH_3, CH_2 and OH groups, the values of ΔS_{or} are equal to 24, 22.5 and 19 respectively) and in the range of 33 to 46 for more bulky radicals. The presence of rotational barriers between single bonds (which amount to 9–21 $kJ mol^{-1}$) can lower the value of ΔS_{or} by a factor of two to eight.

The Benson rules can be used for comparative evaluation of the contribution of orientation entropy effects ΔS_{or} for reaction types 2 and 3 [282]. While using the Benson rules, it must, however, be kept in mind that the entropy values refer to stable molecules in the ideal gas state at 1 atm and 25 °C.

The reactions of interest to enzymologists and chemists proceed in the liquid phase. The mobility of real systems in the initial states may be somewhat lower and the mobility of cyclic transition states higher than the calculated values. Therefore, calculations of ΔS_{or} give only the upper limit of the effect.

A comparison of the experimental data on the reactivity of a number of compounds [282] with theoretical values calculated with account taken of the entropy factors has led to the following conclusions: (1) The preorientation of the reactive groups in a cyclic molecule leads to an increase in the reaction rate; the amount by which the reaction rate is speeded up either approaches the maximum possible entropy factor or exceeds it by one to eight orders of magnitude (the "orbital overlap" effect) [281]. (2) The highest excess of experimental values over theoretical values is observed for cases in which the structure of the cyclic complex has the least mobility. (3) The experimental value of the acceleration effect depends on the catalyst used and, hence, on the reaction mechanism.

It is of interest to carry out kinetic experiments on model synthetic systems, in which the centres of metal atoms are also oriented relative to one another and to the substrate molecule. Such experiments could help one to elucidate the possible role of preorientation and "orbital overlap" [280, 281] in redox multi-electron processes.

5.5 Elementary Acts of Chemical Reactions

5.5.1 General Principles

A treatment of the elementary act of a chemical reaction is based on the following fundamental laws: (1) the law of energy conservation; (2) the law of momentum conservation and (3) quantum-mechanical properties of electrons and nuclei. In accordance with these laws, the elementary act of a chemical reaction, i.e. the transition of a certain molecular system from one state to another, may be regarded as the result of three events.

In the first event, intramolecular or intermolecular interactions, light absorption or emission or other energy-exchange processes result in the equilibration of the energy levels of the initial and the final state. The probability of this event is denoted P_F (the "energy conformity" factor). The second event is the equilibration of the momentum, including the intrinsic momentum of an electron (spin) (the probability P_M). The probability P_q of the third event, i.e. the transition proper, is determined by the quantum-mechanical properties of the system.

The value of the rate constant of the process may be given by:

$$k = \nu_0 P_F P_M P_q \ , \tag{5.21}$$

where ν_0 is a factor which has the dimensions of the frequency; depending on a particular situation, the value of this factor can be determined by the electron oscillation in the potential box, the vibrational frequency of the nuclei in the molecule, by the rate of reorganization of solvent particles, or it may depend on a number of various factors (see Sect. 5.5.4). In a general case, the quantities figuring in Eq. (5.21) may be mutually dependent.

The role of the "energy conformity" factor (P_F) and the factor P_M which determines, in particular, spin-forbidden transitions, will be considered in the subsequent sections. Here, we shall examine the general quantum-mechanical features that determine the value of P_q.

The solution of the Schrodinger equation, which takes into account the electric and wave properties of electrons and nuclei, permits, in principle, the theoretical calculation of any statistic properties of molecules and their aggregates. In order to describe dynamic processes, use is made of a more general equation:

$$\partial\Psi/\partial t = i\hat{H}\Psi/h \ . \tag{5.22}$$

From a quantum-mechanical viewpoint, the elementary act of a chemical reaction is treated as the transition of a molecule or atom from the initial state (i), which is described by the wave function, ψ_i, to the final state (f) with the wave function ψ_f under the action of a certain perturbation, characterized by the Hamiltonian $\hat{H}_{if} = \hat{V}$. Since in many cases the energy of the perturbing interaction is much lower than the total energy of the molecule, recourse is made to perturbation theory.

In the case of two states, the total wave function may be represented as:

$$\Psi = a_i \psi_i + a_f \psi_f \ . \tag{5.23}$$

According to the theory [284–286], the rate of the transition is proportional to the probability of finding the system in the final state (P_{if}), which is given by:

$$P_q = P_{if} = a_f^2 = (1/\hbar^2) \left| \int_0^t V(t) \exp(-i E_{if} t/h)\, dt \right|^2 , \tag{5.24}$$

where E_{if} is the energy of the transition between the final and the initial level.

From Eq. (5.24) it follows that the time dependence of the transition probability will be determined by both the time dependence of the perturbing action, V (t), and by the oscillation of the exponential term with period $\tau_e = h/E_{if}$.

When the perturbation is applied instantaneously, e.g. when a hydrogen atom is placed in a constant electric field, the probability of finding the system in the first excited state will first increase, reach the maximal value P_{max} during the time $\tau_e \approx 10^{-15}$ s, and then oscillate from 0 to P_{max} with frequency $\nu_e \approx \tau_e^{-1}$.

Of greatest interest is the case in which the electric field is gradually increased. This case simulates the situation that arises upon approach of the reacting molecules as a result of their vibrational or translational motion. If the rate of increase of the electric field strength (E_e) is linear:

$$v = dE/dt = \alpha_e E_e \ , \tag{5.25}$$

and the field is increased slowly, i.e. if $\alpha_e \ll \tau_e^{-1} \approx 10^{15}\ s^{-1}$, the transition probability will increase linearly with time. If the electric field strength ceases to increase, the probability of finding the particles in various states does not depend on time. This case corresponds to the adiabatic course of the reaction when the motion of electrons has the chance to follow the slower motion of the nuclei.

Let us now consider the situation involving the transition of the system from one state to another using the concept of energy terms. The dependence of the total energy of the system in the initial (i) and final (f) states on the generalized nuclear coordinates Q are shown in Fig. 5.2. With a certain value of the coordinates Q_{tr} the energy of the initial and final states is the same and the law of energy conservation allows the term-to-term transition. The probability of such a transition, however, depends essentially on the extent of adiabaticity of the process. The extent of adiabaticity is determined by the magnitude of the resonance integral $V = \int \psi_i \hat{V} \psi_f$, which characterizes the efficiency of the interaction in the reacting system. If the value of V is sufficiently high, the terms in the region of Q_{tr} are split (Fig. 5.2) with a decreasing activation barrier.

In fact, two new adiabatic terms are formed: the lower and the upper energy term. In this case, which corresponds to the adiabatic regime, the attainment of the maximum energy region implies, in fact, the transition of the system from

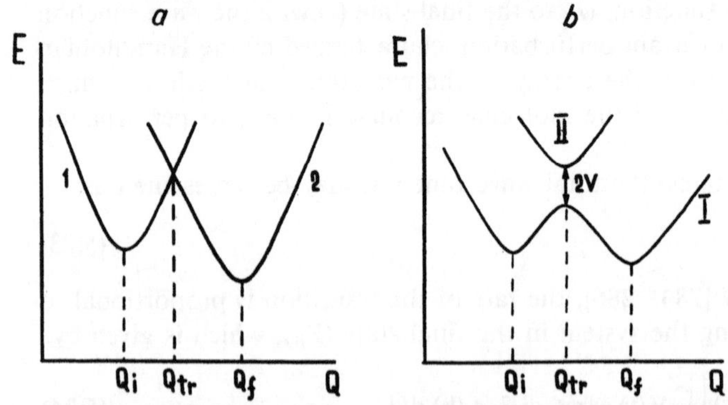

Fig. 5.2 a, b. Variation in the energy of the system along the reaction coordinate: **a** diabatic terms of the reactants (*1*) and products (*2*); **b** adiabatic terms of the ground state (*I*) and excited state (*II*); *V* resonance integral

state i to state f. In another extreme case, where the interaction of the terms in the region of the coordinate Q_{tr} is close to zero, the terms practically do not split, and the probability of the transition from the initial to the final term is very low. Such basically non-adiabatic reactions proceed very slowly.

In the general case, the probability of the transition i→f is described by the Landau-Zener formula [285, 286]:

$$P_q = P_{if} = 1 - \exp(-4\pi^2 V^2/hv|S_1 - S_2|) , \tag{5.26}$$

where S_1 and S_2 are the slopes of the terms in the Q_{tr} region. If the exponent of the exponential function is small, then:

$$P_q \approx 4\pi^2 V^2/hv|S_1 - S_2| . \tag{5.27}$$

Thus, the probability of the process is the higher, the greater the magnitude of the resonance integral, the lower the velocity of nuclear motion and the smaller the curvature of the terms in the region of their intersection. At $P_q = 1$ the process is adiabatic. Equations (5.21), (5.26) and (5.27) may be generalized in the form of the Fermi golden rule [see Eq. (5.37) in Table 5.1, see pp. 110 and 111].

5.5.2 Adiabatic Reactions. Activation Energy and Entropy

The elementary acts of chemical reactions are examined by using the concept of the potential surface. For the simplest reaction of a diatomic molecule AB with an atom C, the potential surface is the dependence of the energy of the system on the distances r_{AB} and r_{BC} (Fig. 5.3). In the initial state, the B-C distance is large and the system is in the left energy valley. The chemical reaction is described as the movement of the system to the right valley along the reaction coordinate, i.e. along the most favourable path through the saddle between the "mountain" and a high "plateau". The "plateau" corresponds to the atomization of molecules AB and BC.

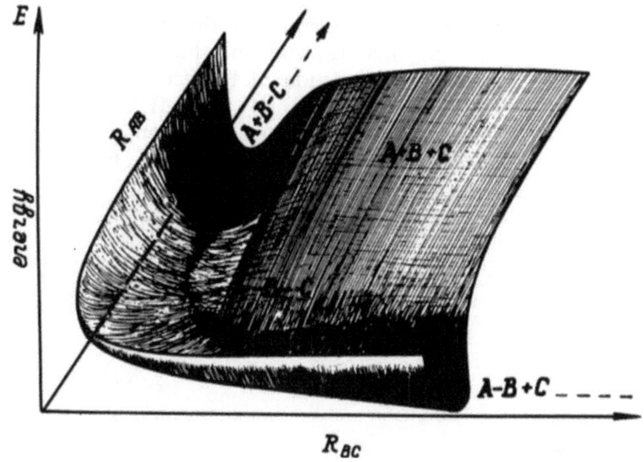

Fig. 5.3. The potential energy surface for the reaction AB+C→A+BC. The *broken line* indicates the reaction path. R_{AB} and R_{BC} are the coordinates of the initial and final states

The difference in the energy of the system at the saddle point (the transition state or the activated complex) and in the initial state, $\Delta H_0^{\neq}$, is close to the energy of activation E_a, and the difference between the energies of the initial and final states corresponds to the reaction enthalpy ΔH_0.

To describe reactions in which the configurations of a large number of nuclei are changed, one has to make of the concept of multi-dimensional space. In such case, the energy profile of the reaction is tentatively represented as the dependence of the energy of the system on the change of the generalized coordinates Q along the reaction coordinate (Fig. 5.3). The rate of reaction is equal to the frequency of the passage through the potential barrier.

As has been shown by Eyring, if the four conditions described above in considering the Arrhenius equation (see Sect. 5.1), namely the statistical nature of the system, the fulfillment of the Maxwell-Boltzmann distribution law during the course of the reaction, the classical nature of nuclear motion and the adiabaticity of the process, are satisfied, the value of the rate constant will be given by:

$$k = \frac{k_B T}{h} \frac{Z_1^{\neq} \dots Z_i^{\neq}}{Z_1 \dots Z_j} \exp(-\Delta H^{\neq}/RT)$$

$$= \frac{k_B T}{h} \exp(\Delta S^{\neq}/R) \exp(-\Delta H^{\neq}/RT) \; , \tag{5.28}$$

where $Z_i^{\neq}$ are the translational, rotational and vibrational partition functions in the transition state, with the exception of the vibrational partition function for the reacting bond, e.g. the bond is broken; Z_j are the corresponding partition functions in the initial state; $\Delta S^{\neq}$ is the entropy change for the formation of the transition state. Note that although the values of $\Delta S^{\neq}$ and $\Delta H^{\neq}$ formally have resemblance to thermodynamic functions, the transition-state theory does not presuppose a thermodynamic equilibrium between the transition state and the reactants. Within the framework of this theory, the formation of the transition state automatically implies that the system descends into the product valley.

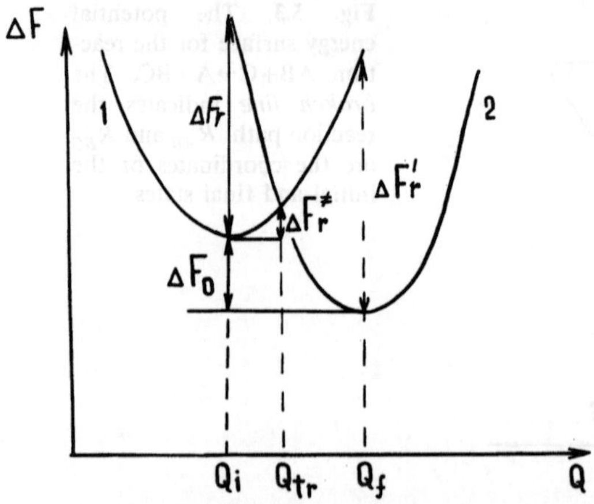

Fig. 5.4. The free energy of the initial (*1*) and the final (*2*) state of the reacting system as a function of the generalized classical coordinates. ΔF_r and $\Delta F_r^{\neq}$ are the free energy of reorganization and activation respectively; ΔF_0 is the free energy of the process; $\Delta F_r'$ is the energy of reorganization of the reverse process

The transition-state theory (also known as the absolute reaction-rate theory or the activated-complex theory) has played an outstanding role in chemical kinetics. Though quantitative calculations based on the apparatus of the theory may be made only for the simplest systems in the gas phase, it has been successfully used to account for a large body of experimental data, including the "anomalous" values of the pre-exponential factors, and to predict qualitatively such effects as the effect of the solvent of the rate of chemical reactions.

For example, the acceleration of the reaction involving the formation of ions in a polar solvent is treated as the result of the solvation of the transition state. Since the enthalpy and entropy of the system diminish upon solvation, such reactions are characterized by negative values of $\Delta S^{\neq}$ and low values of $\Delta H^{\neq}$. The accelerating effect of donor and acceptor reagents on the rate of nucleophilic and electrophilic reactions is accounted for by a decrease in the free energy of activation $\Delta F^{\neq}$ upon their incorporation into the transition state. The change of the microstructure of the solvent in reactions involving non-polar molecules may also affect the activation parameters of the process [287, 288].

The transition-state concept as applied to the adiabatic reorganization of the solvent particles, which accompanies the electron transfer, forms the basis of the Marcus theory formulated in his earlier works [260].

In the Marcus theory it is assumed that the efficiency of the electron transfer is determined by the rate of the transition from the initial potential curve, which characterizes the harmonic dependence of the energy on the generalized coordinate of the solvent, Q, to the final potential curve (Fig. 5.4). The corresponding calculation leads to the following dependence of the rate constant of the electron transfer on the standard free energy of the process (ΔF_0) and on the free energy of reorganization (ΔF_r is the difference in the energy of the system before and after the electron transfer at fixed positions of the solvent molecule) (Fig. 5.4):

$$k_{tr} = \frac{k_B T}{h} \exp\left(-\frac{\Delta F_r^{\neq}}{RT}\right) = \frac{k_B T}{h} \exp\left\{-\frac{(\Delta F_r + \Delta F_0)^2}{4RT\Delta F_r}\right\} . \tag{5.29}$$

For particular systems the value of ΔF_r can be calculated by proceeding from the values of the radii of the reactants, the vibrational frequencies of the ligands in the coordination sphere, the dielectric properties of the medium and the given transition-state configuration (see Sect. 11.1.3). One of the interesting corollaries of the theory is the prediction of the Bronsted correlation between the free energy of activation and the free energy of the process. Indeed, it is easy to show that:

$$\log k_{tr} \sim \Delta F_r^{\neq} = \Delta F_r^{0\neq} + (0.5 + \Delta F_0/2\Delta F_r)\,\Delta F_0 \ . \tag{5.30}$$

In terms of potential curves (Fig. 5.2), the presence of such correlations can be explained by the fact that with the thermodynamics being more favourable, a decrease in the potential curve occurs corresponding to the final state of the system (i). If the curvature of the curve is changed insignificantly, this automatically leads to a decrease in the height of the energy barrier. Another possible cause of the decrease of the activation energy may be the decrease of the curvature of the potential curves, due, for example, to vibronic effects of the second kind [269]. Note that the increase of the reaction coordinate, the relative shift of the curves in the horizontal direction, leads to an increase in the activation energy and, hence, to the slowing down of the process. A physical factor responsible for such a shift may be an increase in the number of bonds being broken in molecules during the course of concerted processes. Other factors that limit the concerted mechanisms will be considered in Sections 5.5.5 and 11.2.

The transition-state theory remains an important tool of modern chemical and enzymatic catalysis. There are, however, a considerable number of processes associated mainly with the electron and proton transfer when the conditions of classical nuclear motion and adiabaticity are known to be violated. Such cases require special study (see Sects. 5.5.3 and 5.5.4). For concrete reactions involving many nuclei it is necessary to take into account the synchronization of nuclei during movement along the reaction coordinate (see Sect. 5.5.5).

5.5.3 Inner-Sphere Non-Adiabatic Reactions. Quantum-Mechanical Selection Rules

Many non-adiabatic chemical reactions proceed by an inner-sphere mechanism with the reactants being in direct contact with one another, for example, in the coordination sphere of transition-metal or enzyme-substrate complexes. Under these circumstances the rate of change of the perturbation determined by the vibration frequency of the nuclear system (10^{12} to $10^{13}\ s^{-1}$) is high and the quantum-mechanical probability of the reaction, P_q, depends on the magnitude of the resonance integral V [Eq. 5.27)]. The resonance integral, in turn, depends on the value of the overlap integral $S = \langle \psi_i | \psi_f \rangle$ and:

$$k \sim P_q \sim |V|^2 \sim |S|^2 \ . \tag{5.31}$$

Thus, the rate of the chemical reaction is determined to a considerable extent by the positive overlap of the wave functions of the initial and final states of the reaction system. The overlap differs from zero only in those cases where the wave functions ψ_i and ψ_f have the same symmetry [267, 268]. Processes that are accompanied by a change in the symmetry of the wave functions have a zero transition probability P_q and, hence, are forbidden. A system with an energy high

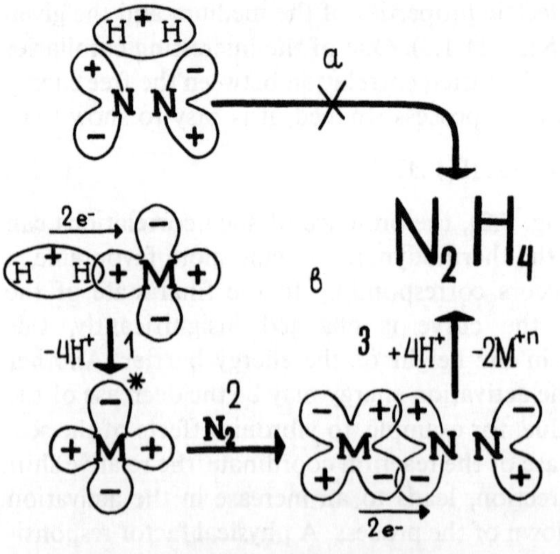

Fig. 5.5 a, b. The "by-passing" of the symmetry-forbidden reaction paths for the reduction of N_2 by molecular hydrogen involving the formation of hydrazine with participation of the orbitals of transition metals (*M*). **a** The forward reaction is symmetry-forbidden; the positive overlap is zero; **b** symmetry-allowed "by-passing" steps: *1* reduction by hydrogen; *2* reduction of N_2 in a complex with M; *3* hydrolysis of the complex. The interaction with only one of the $\pi^*_{N_2}$ orbitals and with one M-atom is shown. The *asterisk* designates the d-orbital of the metall filled with two electrons.

enough to surmount the energy barrier cannot pass to the final state and the reaction is not realized. The forbiddenness may be partially removed if other states of corresponding symmetry and combined with ψ_i and ψ_f. As has been shown by Woodward and Hoffmann, who formulated the selection rules for the conservation of orbtial symmetry, for the solution of many problems it is sufficient to take into account the so-called boundary orbitals (the lowest, filled orbitals of the donor and the highest, empty orbitals of the acceptor).

Let us illustrate the selection rules for the conservation of orbital symmetry (also known as the Woodward-Hoffmann rules) [268] and possible ways of circumventing the forbidden reaction paths upon introduction of a catalyst into a chemical process, taking as an example the hypothetical reaction of N_2 reduction by molecular hydrogen. The reaction for the preparation of N_2H_2 is thermodynamically, substantially unfavourable (Sect. 11.2.4); we shall therefore consider a more real process:

$$N_2 + 2H_2 = N_2H_4 \;, \tag{5.32}$$

in which H_2 acts as a donor and N_2 as an acceptor. Figure 5.5 shows the orbitals of the initial and final states of the reaction. For simplicity, only one of the two planes is shown. As can be seen in Fig. 5.5, the symmetry of the combination of the bonding σ_{H_2} orbital of H_2 and the anti-bonding $\pi^*_{N_2}$ orbital of N_2 does not correspond to the symmetry of the two bonds being formed in the N_2H_4 molecule. Figure 5.5 also demonstrates the absence of a positive overlap between σ_{H_2} and $\pi^*_{N_2}$. Thus, the forward reaction of formation of N_2H_4 is symmetry-forbidden.

Let us now consider reactions involving a catalyst, i.e. a transition-metal complex:

$$H_2 + M^{+n} \rightarrow M^{+n-2} + 2H^+ \;,$$

$$N_2 + 2M^{+n-2} \rightarrow M^{+n} = N - N = M^{+n} \;;$$

$$M^{+n} = N - N = M^{+n} + 4H^+ \rightarrow 2M^{+n} + N_2H_4 \;;$$

which lead eventually to the formation of N_2H_4. At the first stage, H_2 reduces the complex and the d-orbital of each metal atom is filled with two electrons. As seen from Fig. 5.5, this process is symmetry-allowed. At the second stage, an allowed interaction occurs between two filled d-orbitals of the metal and the two anti-bonding π^*-orbitals of N_2 with the formation of a binuclear complex (Fig. 5.5 shows only the interaction with one $\pi^*_{N_2}$ orbital). The hydrolysis of the binuclear complex, resulting in the formation of free hydrazine, is also symmetry-allowed. On the whole, the process can proceed at a sufficiently high rate at elevated temperatures and pressures when the reaction is thermodynamically favourable.

The application of the conservation principle of orbital symmetry allows a choice to be made between the possible reaction mechanisms without performing complicated quantum-mechanical calculations.

Another quantum-mechanical selection rule, the conservation principle of total spin, follows from the law of momentum conservation (Sects. 4.5 and 5.1). We shall illustrate this rule the example of the electron-transfer reaction between a Cu^{1+} complex and an Fe^{3+} complex, the latter being in high and low-spin states (the magnitude of the spin is given in parentheses):

$$Cu^{1+}(0) + Fe^{3+}(1/2) = Cu^{2+}(1/2) + Fe^{2+}(0) \ ;$$

$$Cu^{1+}(0) + Fe^{3+}(5/2) = Cu^{2+}(1/2) + Fe^{2+}(2) \ ;$$

$$Cu^{1+}(0) + Fe^{3+}(1/2) = Cu^{2+}(1/2) + Fe^{2+}(2) \ ;$$

$$Cu^{1+}(0) + Fe^{3+}(5/2) = Cu^{2+}(1/2) + Fe^{2+}(0) \ ;$$

Of the four possible reactions only the first two are allowed; the last two reactions are spin-forbidden and cannot be realized.

The spin-orbital and spin-spin electron-nuclear interactions can only partially remove the spin selection rule (see Sect. 4.5).

5.5.4 Outer-Sphere Non-Adiabatic Reactions. Long-Range Tunnelling Processes

The concepts of tunnelling mechanisms were formulated by Gamow and his classical work, in which he was concerned with the problem of the passage of a quantum particle through the potential barrier [289]. We shall examine this process by taking, as an example, the emergence of an electron from the potential well. This model is a crude approximation of the process of molecule ionization. It is assumed that the electron is uniformly delocalized on the molecule, the nuclear configuration remains unaffected upon electron transition and that after the transition through the barrier, the electronic energy may assume any value. If the electron was initially localized in the potential well and was oscillating there with a frequency of $\nu_e \approx E^2/h \approx 10^{16}\,s^{-1}$, where E is the energy of the particle, the rate constant of the transition (k_{tr}) of the electron through a unidimensional potential barrier of height E_0 and width L (Fig. 5.6) will be given by:

$$k_{tr} = \nu_e P_{tn} = \nu_e \exp[-a_{tn} L (2M\Delta E_0)^{1/2}/\hbar] \ , \qquad (5.33)$$

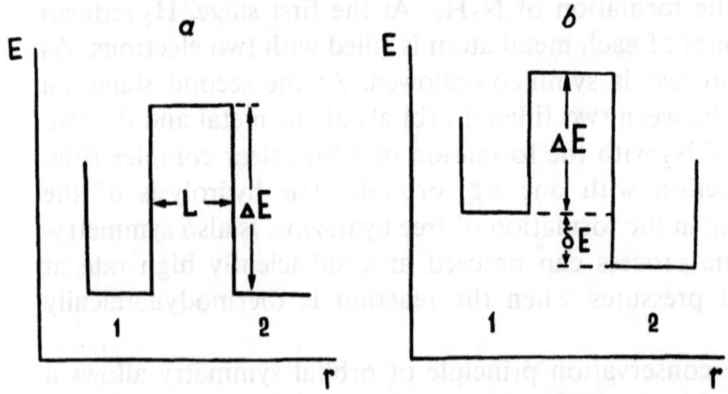

Fig. 5.6a, b. The electron energy of a molecular system upon energy transfer versus the distance between the centres at time-fixed nuclear coordinates; *1* and *2* initial and final states; *L* and *ΔE* are the width and height of the barrier respectively; *δE* is the energy change upon transition. **a** Energy of the final state may assume any value; **b** energy of the final state has a fixed value

where P_{tn} is the tunnelling probability; M is the mass of the particle; a_{tn} is a coefficient which depends on the shape of the barrier (for a rectangular barrier, $a_{tn} = 2$). Formula (5.33) is also valid for other quantum particles at a fixed barrier shape. As can be seen from formula (5.33), the rate of tunnelling depends substantially on the barrier width and, to a lesser extent, on its height. In going from an electron to a proton and further to heavier nuclei the rate of the process falls sharply.

For many cases the value of k_{tr}, calculated by formula (5.33), is much higher than the rate constant of the activation transfer of a light particle due to thermal fluctuations. In such cases, on speaks of the tunnelling of the particle through the barrier, of the tunnelling mechanism of the transfer process. The Gamow model describes well such processes as the alpha-decay of nuclei and the cold emission of electrons from metals. This model, however, cannot be applied directly to the process of electron transfer between donor and acceptor molecules.

In real systems, an electron is localized and has fixed energy not only in the initial but in the final state as well (Fig. 5.6) [290–297]. The electron-transfer process involves the reorganization of the nuclear environment and the shape of the barrier is continually changed. The rate of electron transfer is determined not only be the tunnelling probability (P_{tn}), but also by the probability of equilibration of the energy levels of the initial and final states. In the system, the excess of electron energy δE must be transferred, upon transition from the left to the right potential well, to the donor and acceptor molecules and to the environment. The mechanism of energy transfer may consist in the excitation of normal vibrations of molecules, the excitation of collective vibrations of the medium (phonons), the increase of the kinetic energy of the surrounding groups, etc. These nuclear relaxation processes complete the irreversible electron transfer process.

The outer-sphere electron-transfer processes have lately been studied using the Landau-Zener model (Sect. 5.1). Within the framework of this model, a chemical reaction is regarded as the result of the transition from the initial electronic term to the final term (Fig. 5.2). At the point of intersection of the terms, the tunnelling probability P_{tn} is equal to the probability of a quantum-mechanical transition from term-to-term, P_q (Eq. (5.27)]. Under the assumption that the process is fully spin-allowed, the rate constant of the transition may be written in the following form [see formulas (5.26) and (5.27)]:

$$k_{tr} = \nu_0 4\pi^2 V^2 P_F/(hv|S_1 - S_2|) \ . \qquad (5.34)$$

According to formula (5.34), the rate of the process is determined, first, by the frequency of the appearance of a transition state, in which the nuclear configurations of the reactants and products are identical and the energy levels are equilibrated ($\nu_0 P_F$) and, second, by the probability that in this transition state a transfer will take place at fixed nuclear configuration (P_q). Note that in formula (5.34) the quantities characterizing the intensity of nuclear motion are contained both in the numerator (ν_0) and in the denominator (v) and may be cancelled after averaging over ν and performing certain transformations with account taken of the dimensions of the quantities. The resonance integral (V) is related to the value of the overlap integral in the wave functions of the initial and final states:

$$V \sim |S| = S_0 \exp[-(r-r_0)/a_B] \ , \qquad (5.35)$$

where r and r_0 are the distances between the centres and the closest-approach distance between the centres (e.g. upon van der Waals contact) respectively; S_0 is the overlap integral on closest approach.

It is not difficult to see that in the transition state the dependence of the system on the electron coordinates is described by the model of two potential wells with equal energy levels in the initial and final states (Fig. 5.6); the frequency of electron oscillation between these wells is given (at $a_B = 0.07\ nm^{-1}$, for example) by:

$$\nu_e = V/h \simeq 5 \times 10^{15} \exp[-1.4(r-r_0)] \ , \qquad (5.36)$$

where r is given in nm. For instance, at a distance of $r - r_0 \simeq 1.8$ nm, $\nu_e \sim 10^4\ s^{-1}$ and at a distance of 1 nm the value is $\nu_e \sim 10^9\ s^{-1}$; at the van der Waals contact distances, $\nu_e \sim 10^{15}\ s^{-1}$.

Thus, at distances of up to about 1.8 nm the frequency of electron-density oscillation in the transition state is sufficiently high. However, as can be seen from formula (5.34), the value of the rate constant of electronic transitions (k_{tr}) is substantially influenced by the efficiency of the energy adjustment of the nuclear environment and by the physical mechanism of the reorganization; calculation of P_F is the task of the theory.

We shall briefly characterize the existing conceptions [the corresponding Eqs. (5.37–5.54) are given in Table 5.1]. Chernavsky and Grigorov [291] have considered a model in which it is assumed that the long-range tunnelling, electronic transition is accompanied by the excitation of normal modes of vibration. The probability of this event depends on the ratio of the excess electron-transfer energy to the energy of the vibrating quantum, on the width of the resonance line, Γ_k, and also on the "coupling strength" between the electronic and vibrational systems (Table 5.1); the "coupling strength" is characterized by the ratio $\Delta' = \Delta\varphi_e/\Delta\varphi_0$, where $\Delta\varphi_0$ and $\Delta\varphi_\varepsilon$ are the amplitude of zero vibrations in

Table 5.1. Formulas for rate constants

Expression for the rate constant	Equation No.	Remarks	Reference
A. Elementary Reactions			
$2\pi h^{-1}\lvert V_{fi}\rvert^2 FC$ where $V_{fi} = \frac{\int \psi_f^* V \psi_i d\tau - (\int \psi_i^* \psi_f d\tau)(\int \psi_f^* \psi_i d\tau)}{(1 - \lvert\int \psi_i^* \psi_f d\tau\rvert^2)}$	(5.37)	ψ_i and ψ_f = Wave functions of the initial and final states; FC = Franck-Condon factor	[260]
$\sigma = \exp\{(E_r^i/h\nu_i)\,[\tanh(h\nu_i/4kT) - h\nu_i/4kT]\}$	(5.38)	σ = Nuclear tunnelling factor E_r^i = Reorganization energy of the ith vibration ν_i = Frequency of ith vibration	[260]
$2\pi h^{-1}\lvert V\rvert^2 \frac{e^{-S}S^v \exp - (\Delta F_0 + vh\nu + \Delta F_r^s)/4\Delta F_r k_B T}{h\nu\Gamma(v+1)}$ $S = \Delta F_r/h\nu_i$ $v = -[\Delta F_r^s - \Delta F_0 - 2\Delta F_r^s k T_B (h\nu)^{-1} \ln(\Delta F_0 + \Delta F_r)/h\nu]/(h\nu)^2$	(5.39)	ΔF_0 = Free energy ΔF_r = Reorganization energy of medium Γ = Gamma function	[260]
B. Non-Adiabatic Reactions			
At $k_B T \gg h\omega_0$: $(\pi \cdot \hbar^{-2} E_r^{-1})^{1/2} \lvert V\rvert^2 \exp[-(\Delta E + E_r)^2/4E_r k_B T]$	(5.40)	ω_0 = Frequency factor of solvent	[261]
At $k_B T \ll h\omega_0$: $2\pi\lvert V\rvert^2 \hbar^{-2}\omega_0^{-1} \frac{(E_r/h\omega_0)^{m_0}}{m_0!} \exp-(E_r/h\omega_0)\exp(\Delta E/k_B T)$	(5.41)	E_r = Reorganization energy of the system ΔE = Energy level difference m_0 = $\Delta E/\hbar\omega_0$	
$(\pi\hbar)^{-1} V_0^2 g^2 \cdot \exp(-2R/a_B) \cdot \Gamma_v/[\Gamma_v^2 + (\Delta E - \Delta E_v)^2]$	(5.42)	Γ_v = Width of vibrational level a_B = Wave function attenuation parameter ΔE_v = Change of vibrational quantum	[291]
$2\pi\hbar^{-1}\lvert V\rvert^2 \left[2\pi k_B T \coth\left(\frac{T_a}{2T}\right)\right]^{-1/2}$ $\cdot \exp[-(\Delta E - \Delta)^2/2k_B T\Delta \coth(T_a/2T)]$	(5.43)	Δ = Vibronic interaction parameter T_a = Characteristic temperature	[265]
At $k_B T \gg \hbar\langle\omega\rangle \gg \hbar\langle\omega_s\rangle$: $2\pi\hbar^{-1}\lvert V\rvert^2 [2\pi E_r k_B T]^{-1/2} \exp -(\Delta E + E_r)^2/4E_r k_B T$	(5.44)	S = $\sum_i \mu_i \omega_i h^{-1} \Delta q_i^2/2$ μ_i = Reduced mass	[266]

At $k_B T \ll \hbar\langle\omega\rangle \ll \hbar\langle\omega\rangle_s$: $2\pi(\hbar^2\langle\omega_s\rangle)^{-1}\lvert V\rvert^2 \exp(S_s - S) \sum_{m=0}^{\infty} S_s^{p(m)} S^m/p(m)!m!$	(5.45)	$p(m=)$ $\langle\omega\rangle_s$	$= \frac{\Delta E - m\hbar\langle\omega\rangle}{\hbar\langle\omega_s\rangle}$ = Vibrational frequency of solvent	
At $k_B T \gg \hbar\langle\omega\rangle \gg \hbar\langle\omega_s\rangle$: $(2\pi)^{-1}\langle\omega\rangle\varkappa \exp[-(\Delta F_r + \Delta F_0)^2/4\Delta F_r k_B T$ $\varkappa = 2\cdot\lvert V/2\rvert^2/(h^2\langle\omega\rangle^2 \Delta F_r k_B T\tau^{-3})^{1/2}$	(5.46) (5.47)	ΔF_0 ΔF_r	= Free energy of reaction = Free energy of reorganization	[293]
At $T \to 0$: $\hbar^{-1}\lvert V\rvert^2 (E_r E_k)^{-1/2} \exp[-(\Delta E - E_r)^2/2E_r\hbar\omega]$	(5.48)	E_k	= Kinetic energy of translational degrees of freedom	[290]
At $k_B T \ll \hbar\omega$: $\nu_e \exp[-E_0^2(R_0)/F^2(\delta_0^2+\delta_1^2]$ At $k_B T \gg h\omega$: $\nu_e \exp[-E_0^2(R_0)/(F^2\delta_0^2 + 4E_r k_B T)]$	(5.49) (5.50)	F δ_0 δ_1	= Curvature of the term in the intersection region = Amplitude of zero vibrations = Mean vibrational amplitude	[296]
At $Z^1 \gg \omega^2(E_a/\lvert V\rvert^2)$: $\omega_2 [Z'(E_a/\pi k_B T)^{1/2}]^{-1} \exp(-E_a/k_B T)$	(5.51)	Z' E_a	= Rate of vibrational relaxation = Energy of activation	[316]
$\omega_A[(\beta^2/4+\omega_c^2)^{1/2} - \beta/2] \exp(-E/RT)2n\omega_c$ $\beta = 6\pi r_0 \eta/m$	(5.52)	ω_A and ω_c r_0 and m η	= Vibration frequencies in initial and transition states = Radius and mass of particles = Viscosity of the medium	[317]
$\nu_{in}(E_r'/E_r) + DE_r''T^{1/2}/2\pi\delta r^2(E_r V)^{1/2} \exp(-E/RT)$ $E_r = E_r' + E_r''$	(5.53)	E_r' and E_r'' ν_{in} δ_r D	= Reorganization energies inside and outside the reaction complex = Vibration frequency inside the complex = Change in the radius of the complex = Diffusion coefficient	[317]
$\omega = (D_0/Ll) \exp[-(E_A + E_D)/RT]$	(5.54)	D_0 E_A E_D L l	= Pre-exponential factor = Weight of conformational barrier = Activation energy of diffusion to the top of barrier = Barrier length = Free mean path	[318]

the molecule and the mean displacement of the nuclei upon electronic transition respectively. If $\Delta' \gg 1$, the coupling is strong; if $\Delta' \ll 1$, the coupling is weak. Chernavsky and Grigorov have examined the effect of vibrations with strong coupling. A similar model has been considered by Hopfield (Table 5.1). Jortner has examined a more general case for the interaction of electronic and vibrational transitions in a solid. He takes into account the possibility of the excitation not only of normal modes with strong coupling, but also of the collective vibrations of the nuclear environment (phonons) with weak coupling. The corresponding expression is given in Table 5.1.

Aleksandrov, Zamaraev and Khairutdinov [290] presume that, apart from the electronic-vibrational interactions, there may be excited classical kinetic and vibrational degrees of freedom (Table 5.1).

A theory of electronic transitions in polar media has been developed in a series of fundamental investigations carried out by Levich, Dogonadze and Kuznetsov and co-workers [261, 293]. These authors have extended the Marcus theory (see Sect. 5.5.2), taking into account, apart from the classical motions during the reorganization of the medium, the quantum-mechanical effects of both electronic and nuclear tunnelling (Table 5.1). The tunnelling processes are described by these authors with the aid of the Landau-Zener theory. Expressions have been derived [261, 293] which relate the reorganization energy to the parameter of the dielectric properties of the medium. A recent work by Dogonadze et al. [293] takes into account the possibility of a contribution from the reorganization of the protein environment upon electron transfer.

A theoretical model of chemical reactions, which takes into account that the electron transfer is accompanied by both the outer-sphere reorganization of the polar medium and the motion of heavy nuclei, e.g. a proton, inside the sphere has been put forward by Bendersky and Ovchinnikov [295]. In another work of these authors carried out in collaboration with Goldansky [296] attention is drawn to the possibility of the preliminary approach of the reacting molecules to a distance smaller than the van der Waals contact distance. This approach, invoked by the easy contraction of the van der Waals shells as compared to the extension of rigid chemical bonds, reduces the width of the tunnel barrier (Table 5.1). As has been shown by Bendersky et al., this model is capable of explaining, under certain circumstances, the presence of a low temperature limit in reactions involving heavy nuclei.

In the majority of the above mentioned theoretical works, it is assumed that the electron-exchange interaction and the nuclear reorganization occur independently (the adiabatic approximation) and perturbation theory is used to describe the process. Kozhushner and Ivanov [297] have recently examined a more general case of electron transfer. Taking into account the electronic-vibrational interaction in donor and acceptor molecules and also the electrostatic effects has led the authors to a general expression for the rate of electron transfer. Theory predicts a number of interesting phenomena, in particular the appearance of vibrationally excited states of the donor and the acceptor, the deviation from an exponential type dependence [Eq. (5.35)] at large distances etc.

In a number of works the reactions in viscous media are described by using the conceptions developed by Kramers [317]. According to these concepts, the potential barrier of the reaction is overcome under the action of a random force. In contrast to gas-phase reactions, in which the system moves along the reaction coordinate with a thermal velocity after a sufficient amount of energy has been accumulated in order to surmount a smooth potential barrier, in this particular case the system experiences a "friction" against the resisting medium. The magnitude of the "friction" coefficient, in turn, depends on the effective mass and radius of the reactant particles and on the viscosity of the medium [see Eqs. (5.52–5.54) in Table 5.1]. In low viscosity media, in which the "friction" during the motion to the point of intersection of the terms is insignificant, the corresponding expressions lead to ordinary kinetic expressions similar to the Arrhenius equation. In viscous media the viscosity parameters of the medium enter into the expression for the pre-exponential factor [317] and for the apparent activation energy [318].

Conceptions of the diffusion over the conformational states [785] and of the diffusive motion along the coordinate of the medium [786] have been developed recently, which depend on the viscosity of the medium under certain conditions, in contrast to the motium along the internal coordinates of the reaction system. Note that the conceptions that the thermodynamic and kinetic parameters of the conformational and solvation changes of the protein and the medium are included in the apparent kinetic parameters of chemical reactions were worked out already in the 1960s [726, 729]. The dynamic properties of protein and their effect on the rate of enzymatic reactions will be considered in more detail in Chapters 10 and 11.

Thus, at present, a large number of theoretical concepts have been accumulated on the various mechanisms of electronic transitions, including long-range electron transfers. The principal task of experimentalists is to elucidate the relationship between these theoretical conceptions and the actual photochemical and dark redox reactions in biological systems. Some applications of the theory to particular experimental data will be described in Chapter 11.

5.5.5 Synchronization of Nuclei in an Electronic Concerted Reaction

In order to explain the high efficiency of enzymatic processes, wide use is made of the concepts of energetically favourable, concerted mechanisms [216–219]. Multi-electron redox processes can also be accompanied by a change in the configuration of a large number of nuclei (Sects. 11.1 and 11.2). It may, however, be presumed that [24, 99, 298–301] that certain kinetic limitations exist on the realization of polynuclear concerted mechanisms, in which the process occurs upon simultaneous time- and space-concerted motion (taking 10^{-13} s) of nuclei along the reaction coordinate. In a concerted process the transition from the initial to the final state occurs upon motion of nuclei in a certain direction, which is the only possible path that leads to reaction products. Obviously, the statistic thermal nature of chemical processes limits the number of nuclei (the principle of least motion) [302].

In this section we shall estimate the synchronization factor (α_{syn}) for a number of simple model processes. The available values are used for the analysis of possible alternative mechanisms of molecular nitrogen fixation and other enzymatic and chemical reactions (Sect. 11.2). The purpose of such calculations is to estimate the values of pre-exponential factors for a number of concrete typical systems and to formulate some quantitative conceptions of probability factors that limit the energetically preferred concerted processes. We shall now consider models which we conditionally term the "diffusion", "oscillatory" and "normal vibration" models.

1. *The Diffusion Model.* Let us assume that the process of interest occurs due to the thermal motion of a certain number of particles (molecules) of the reactants or solvent lumped into a collision complex. It is presumed here that the particles have been preliminarily reoriented, this reorientation being taken into account by the entropy factor ΔS_{or}. In such a case, for the transition to the reaction products, apart from thermal flucutations, which are taken into account by the Boltzmann energy term [in Eqs. (5.1) and (5.29)], it is necessary that the particles move in the direction leading to the reaction product. If it is assumed that the motion of the particles is thermo diffusional in character and all the directions of motion are equally probable, then we may suppose that for two neighbouring particles the probability of motion in a given direction, limited by a certain solid angle, will be close in order of magnitude to the value of the steric factor for two colliding molecules in the gas phase ($f_s = 10^{-2}$ to 10^{-3}).

The probability of occurrence of n such motions will, in fact, be equal to the synchronization factor α_{syn}; in the simplest case, where all the particles are identical:

$$\alpha_{syn} = f_s^{n-1} = (10^{-2} \text{ to } 10^{-3})^{n-1}$$

Inspection of this formula shows that even in the most favourable hypothetical case, where the particles are properly pre-oriented and the energy of activation is equal to zero, no more than six to eight particles can participate simultaneously in the elementary process. Considering that real processes have an energy of activation and require a pre-orientation of the particles, we may assume that in the diffusion model one elementary act involves no more than four or five particles simultaneously. In principle, however, there can be systems in which the particles are packed in a special matrix, in which each of the particles can move either along the reaction coordinate or in the opposite direction. In such cases, $f_s = 1/2$ and $\alpha_{syn} = (1/2)^{n-1}$.

2. *The Oscillatory Model.* This model has been considered by Aleksandrov [303]. According to this model, a chemical reaction occurs as a result of the simultaneous transition (taking approximately 10^{-13} s) of a system of independent oscillators from the state, in which the mean displacements of nuclei correspond to the amplitude of thermal vibrations φ_0, to the state in which this displacement exceeds a certain critical value (φ_{cr}). Then, if $\varphi_{cr} > \varphi_0$ and the activation energy of the concerted process is $E_{(n_1)} \gg RT$, theory gives the following expression for the ratio of the pre-exponential factors of the synchronous and direct processes:

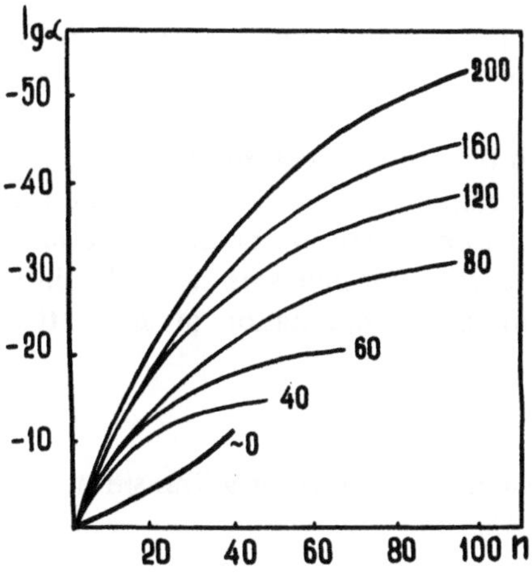

Fig. 5.7. Theoretical dependencies of the synchronization factor (a_{syn}) on the number of degrees of freedom (n) of the nuclei involved which participate in a concerted reaction [24, 299–301, 303]. The curves have been constructed in accordance with Eqs. (5.55) and (5.56); the *numbers* are the activation energies [$E_{(n_1)}$] in kJ mol^{-1}

$$a_{syn} = \frac{n_1}{2^{n-1}} \left(\frac{n_1 RT}{\pi E_{(n_1)}}\right)^{(n_1-1)/2} \tag{5.55}$$

At $\varphi_{cr} < \varphi_0$ and $E_{(n_1)} < RT$,

$$a_{syn} = n_1/2^{n_1-1} \ , \tag{5.56}$$

where n_1 is the number of vibrational degrees of freedom of the nuclei participating in the concerted transition.

It is not too difficult to see that the oscillatory model corresponds to a case where the reacting bonds in the reaction complex have been properly oriented and that the concerted movement of the particles along the reaction coordinate is sufficient for the reaction to be realized. Analysis of Eqs. (5.55) and (5.56) provides a clear idea of the scale of the synchronization factor and of the dependence of this factor on the number of bonds and the activation energy (Fig. 5.7). In the case of the concerted mechanism, the decrease of the synchronization probability (a_{syn}) with increasing n_1 must be compensated for by an appreciable decrease in the activation energy. Constrained, concerted processes at moderate energies (20–40 kJ mol^{-1}), which are characteristic, for example, of enzymatic processes, are most probable. Here, the incorporation of each new nucleus into the transition state leads to a ten-fold decrease in the rate of the process.

3. *The Normal Vibration Model.* The transformations in complex molecules which are accompanied by an appreciable rearrangement of the structure, by a change in internuclear distances and valence (bond) angles, and by the rearrangement of the ligands and solvent particles, etc., can be described with the aid of the concept of normal modes of vibration (Sect. 2.6). For instance, the transition of a system consisting of a central atom in an octahedral environment of six ligands to a state with an increased distance between the central atom and the ligands corresponds to a normal, totally symmetric mode. The value of the statistical weight of this mode will in this case be equal to $a_{syn} = (1/15)$.

In a general case, the reaction may be accompanied by the motion, which is described by using a set of n_1 normal modes and, in the limit, of $n_1 = 3\,m-6$

modes, where m is the number of nuclei in the system under consideration. In order to evaluate the synchronization factor in such cases, one can use Eqs. (5.55) and (5.56).

It is evident that the models of concerted processes discussed above are only a crude approximation to the motion of a complex system of nuclei along the reaction coordinate. However, such a crude approximation apparently permits one to make a choice between the possible reaction mechanisms. The reliability of such a choice increases through a comparative examination of alternative mechanisms.

5.6 Activation of the Substrate in the Coordination Sphere of a Transition Metal. Specific Feature of the Structure of Clusters

The entry of the substrate into the coordination sphere of a metal creates, in many cases, favourable conditions for their further transformation. We shall briefly summarize these conditions [305–308].

The binding of two substrate molecules by metal atoms contributes to their approach and orientation. As a result of the interaction with the metal atoms, the energy profile of the process is changed and conditions are created for its optimization (see, for example, Sect. 5.5.3).

The displacement of the electron density brought about by a partial transfer of electrons from the metal atom to the substrate and from the substrate to the metal atom can improve the energetics of the interaction with electrophilic and nucleophilic reagents and with external oxidizing and reducing agents respectively. If the reaction is accompanied by the rupture or loosening of the bonds in the substrate, the activation energy of the process may be lowered substantially as a result of the vibronic effect of the first and second kinds [269, 304]. As can be seen from Fig. 5.8, the shift of the potential curves brought about by the effect

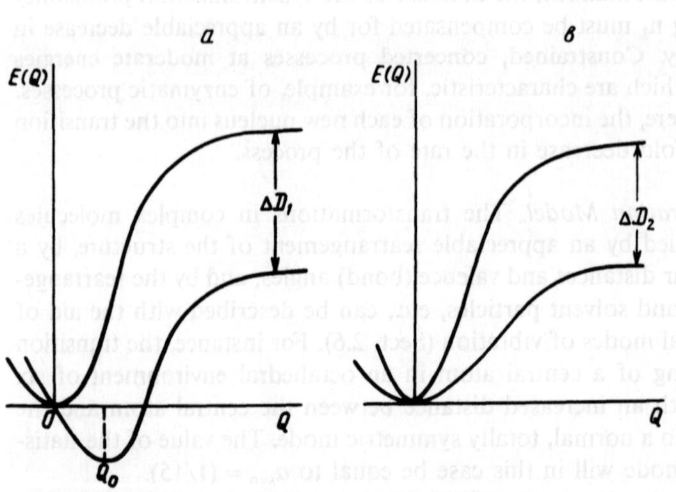

Fig. 5.8 a, b. The vibronic effects in molecules [304]. **a** The effect of the first kind: the shift of the potential curves; **b** the effect of the second kind: the curvature change of the potential curves

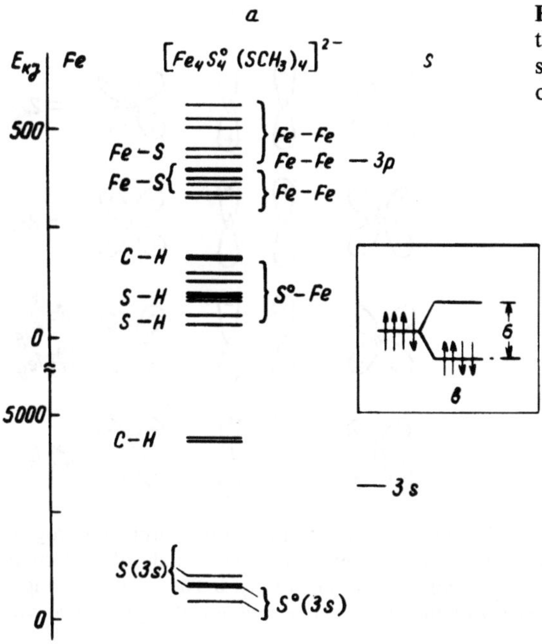

Fig. 5.9. Energy diagram of orbitals in the complex $[Fe_4S_4(SCH_3)_4]$ and the splitting of the uppermost occupied orbital (inset) [315]

of the first kind and the change of their curvature as a result of the effects of the second kind may lead to a considerable decrease in the energy barrier of the reaction. The factors that contribute to the effects of the first kind are the filling of the anti-bonding orbitals of the substrate by electrons and the emptying of the bonding orbitals. The effects of the second kind are assisted by the presence of low lying, unfilled orbitals of the transition metal.

Note that polynuclear complexes capable of undergoing a multi-orbital interaction with the substrate and characterized by the presence of a system of low lying, unfilled orbitals are most suitable for the realization of all types of energy and electronic-vibronic effects. The atoms in clusters, which are linked together by chemical forces, are at such distances from one another that the wave functions of their valence shells can be effectively overlapped [24, 311]. This gives rise to a system of molecular delocalized states (Fig. 5.9). The energies of these states are close to the ground-state energies of the atoms. The difference between them does not exceed the interatomic interaction energy in order of magnitude.

The atomic orbitals of the atoms that make up a cluster can form molecular orbitals of different symmetry. The levels of each type of symmetry are almost equally distributed over the energy interval E. As a result, among the group of the closely spaced levels of the nearest excited states, there are different symmetry levels.

When the charge of the cluster is changed as a result of the addition or subtraction of electrons, the system of one-electron levels can lose or gain one, two or more electrons and is therefore a system of high electron capacity.

Slight local changes in electron density cause small changes in the forces that act on the nuclei in the cluster, for which reason one may expect that the diplacement of the equilibrium positions of the nuclei in redox processes is also small.

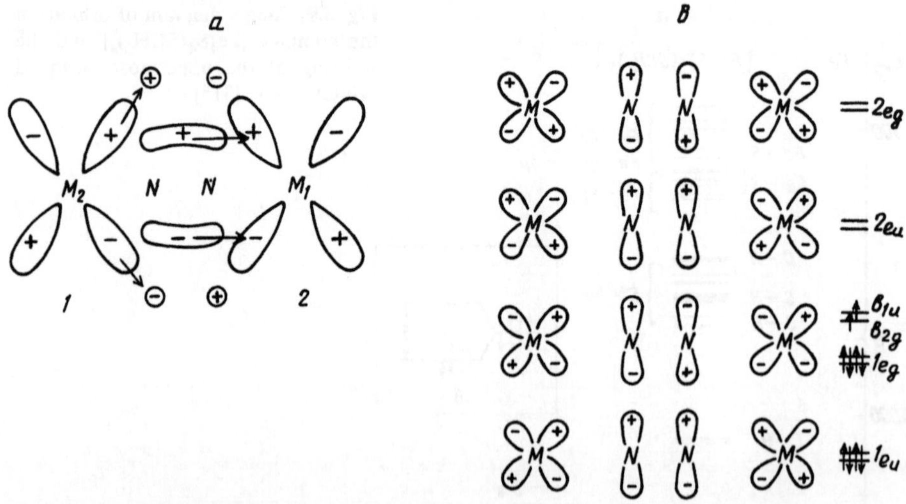

Fig. 5.10 a, b. Multi-orbital binding in a binuclear complex of Mo with molecular nitrogen. **a** The possibility of overlap of the d-orbitals of the metal with the orbitals of molecular nitrogen: *1* anti-bonding π^* orbitals of N_2; *2* bonding orbitals of N_2. **b** The relative arrangement of the energy levels and the symmetry of the orbitals in a complex of N_2 with two Mo atoms [855]

A slight reconstitution of the nuclear system during electronic transitions contributes to multi-electron transfers that involve clusters, since the synchronized motion in transfer processes is facilitated due to small nuclear displacements in the system.

The clusters display a strongly pronounced tendency towards the multi-orbital binding of the ligands when, for example, a transfer of electrons occurs from the filled d-orbitals of the metal to the anti-bonding orbitals of the ligand accompanied by the reverse transfer of electrons from the bonding orbitals of the ligand to the empty d-orbitals of the metal. The process of multi-orbital binding is illustrated in Fig. 5.10 for the case of the formation of a diatomic N_2 complex. Figure 5.10 also shows a system of molecular orbitals formed.

In chemical terms, multi-orbital binding may be defined as a parallel donor-acceptor and dative interaction of a metal complex with the substrate. An important distinction of a multi-orbital binding from a one-orbital binding is the possibility of a much stronger change in the electronic structure of the substrate upon coordination. The point is that in a single-orbital binding the electron-density transfer is appreciably hindered by the charging effect, as a result of which an electric field is produced directed against the transfer. In multi-orbital binding, because of the compensating effect of the oppositely directed transfers, a much stronger change may occur in the population of the substrate bonds during coordination.

Furthermore, the charging charge-exchange effects are of little importance for clusters because of the distribution of the transferred charge among the atoms in the cluster due to the delocalization of the d-states of the cluster. This delocalization is also important in another respect [307]. The atomic d-orbitals are sufficiently strongly localized. Therefore, two electrons in one d-orbital undergo a strong coulombic repulsion. In going to the cluster, the formation of a molecular orbital, delocalized over many centres, from atomic d-orbitals leads to a decrease in the coulombic repulsion because of an increase in the interelectronic distance. As a result, an additional effect appears which facilitates the transfer of several electrons to the d-levels of the cluster as compared with mononuclear complexes because of a decrease in the energy required to overcome the coulombic electron-electron repulsion.

As noted above, the clusters have many closely spaced levels of different symmetry types. Therefore, upon interaction with other molecules or upon alteration of the electronic state of the complexes there is a greater probability that orbitals of the corresponding symmetry will be involved in this process. This leads to the possibility of by-passing symmetry- and multiplicity-forbidden reaction paths.

The above mentioned properties of the clusters makes them promising catalysts for redox processes. It is not surprising that nature makes wide use of this type of complex in solving its chemical problems (see Chaps. 6–11).

5.7 The Principle of "Optimal Motion" in Elementary Acts of Chemical and Enzymatic Processes

The energy preferableness of concerted elementary processes and the simultaneous decrease of the probability in the case of the participation of a considerable number of nuclei have led to the formulation of the principle of "optimal motion" [24, 99, 298–301]. According to this principle, the number of nuclei, whose configuration is changed in the elementary act of a chemical reaction must be sufficiently large to provide the favourable energetics of the step and, at the same time, sufficiently small for the high value of the nuclear synchronization probability to be maintained during motion along the path to the reaction products.

The optimal dependence of the rate constant of the concerted reaction (k_{syn}) on the number of nuclei may be given by the following expression:

$$k_{syn} = \alpha_{syn}(n)\, k_0 \exp\left(-E_{(n)}/RT\right) , \qquad (5.57)$$

where n is the number of chemical and physical bonds which are broken or formed in a given reaction; k_0 is the pre-exponential factor of an analogous, non-concerted (direct) reaction.

The condition for the preferableness of the concerted reaction over the direct reaction with a rate constant k_{dir} and an activation energy E_{dir} is the inequality;

$$k_{syn}/k_{dir} > \alpha_{syn}(n) \exp\left\{(E_{dir} - E_{syn}(n))/RT\right\} . \qquad (5.58)$$

Though the estimates that illustrate the principle of optimal motion are based on simplified models and approximation formulas (see Sect. 5.5.5), they have, nevertheless, made it possible to draw some conclusions which are apparently useful in taking into account in the treatment of complex elementary acts of chemical and enzymatic processes.

1. In polymolecular reactions in the gas and liquid phases, the formation of a preliminary complex along the reaction coordinate must contribute to the ac-

celeration of the process by two or more orders of magnitude as compared with the collision mechanism. The cause of the acceleration is the constraint imposed in the preliminary complex on the possibly useless (from the standpoint of the process under consideration), diffusional motions that do not lead to reaction products. This circumstance may appear to be one of the factors responsible for the substantial acceleration of reactions in the case of the rigidly forced orientation of the reacting particles.

2. In liquid-phase processes, accompanied by a change in the state of an assembly of solvent particles or of the water-protein matrix of enzymes, the simultaneous rearrangement of more than four or five particles is hardly probable. In the actual mechanism of the reorganization of large cooperative ensembles the system apparently contains a weak link which is the first to undergo rearrangements. The cooperative binding of particles in this link is weakened, which contributes to the formation of a series of consecutive and parallel transitions by analogy with chain processes. A kinetic analysis of such cooperative processes [227–231] has made it possible to account for a number of experimentally observed regularities of liquid-phase and enzymatic reactions, including the anomalous values of the energy and entropy parameters, their high sensitivity to reaction conditions, compensation effects, etc. (Sects. 4.3 and 10.1).

3. As has been shown for nitrogenase and lysozyme reactions, despite the obvious energy preferableness of certain concerted mechanisms, they may prove inefficient because of the too low value of the synchronization factor [299–301]. In such cases, the sequential transformation of the system through a number of steps is favourable. Here, the role of a multi-functional catalyst is to provide favourable energy and probability factors through the optical use of the corresponding functional groups at each step of the process. A detailed study of a number of enzymes, such as chymotrypsin, ribonuclease, glutamate aspartate, aminotransferase, catalase and aminooxidase, has made it possible to detect experimentally a number of intermediate steps in each of them [309–310, 312].

4. An important condition for the occurrence of catalytic concerted and other polyfunctional processes is the possibility of the dynamic adjustment of the shape of the catalyst matrix to the substrate structure and to the reacting catalyst groups which undergo changes in the course of the reaction [24, 99, 313]. This adjustment results in the stabilization of the charge and dipolar states during the course of the reaction. The absence of such an adjustment would require additional energy for the resistance of the matrix to be overcome during the change of the matrix structure. Rapid, reversible, spontaneous, slight conformational changes in enzymes and the rearrangement of solvent particles can provide the required preliminary adjustment.

Fluctuational changes in the structure of proteins and enzymes have been postulated to account for the kinetic data on the proton exchange between water and protein and have been recently detected by physical methods (Sect. 10.2).

Thus, the principle of optimal motion permits one to explain qualitatively, from a unified viewpoint, a large number of experimental regularities of chemical and enzymatic catalysis. It is evident, however, that further theoretical and experimental studies are required for this principle to be effectively applied to particular reactions.

PART II

CHAPTER 6

Iron-Containing Proteins

Beginning in 1920s, when the classical works of Keilin and Warburg were carried out, up to the mid-1950s, the only class of iron-containing proteins studied had been hemin enzymes and oxygen and electron carriers [46, 47, 317–320]. The prosthetic heme group consisting of a porphyrin ring (Fig. 6.1), at the centre of which there is an iron atom, possesses sharply pronounced spectral properties. Because hemoglobin, myoglobin and cytochromes are contained, as a rule, in the rather large amounts in cells, these properties literally catch the investigator's eye. Heme-containing enzymes, such as catalase and peroxidases, are present in relatively small concentrations, but they are noted for their enormously high enzyme activity and are widespread. As early as the beginning of the last century the ability of blood to decompose hydrogen peroxide into water and O_2 was detected. The oxidation of various substrates by hydrogen peroxide, which is catalyzed by peroxidases, is also easily detected by means of simple, coloured reactions.

The ideas of the presence of other, non-heme forms of iron, which are of great biological value, were advanced more and more often in the 1940s and 1950s. The fact that the total amount of iron in preparation exceeds by two or more times the amount of iron linked to heme was detected by an increasing number of authors. It soon became possible to isolate a few proteins containing non-heme iron, whose function was to transport and store iron in animal tissues [321] and to transport oxygen in some bacteria [322]. The most important event was the

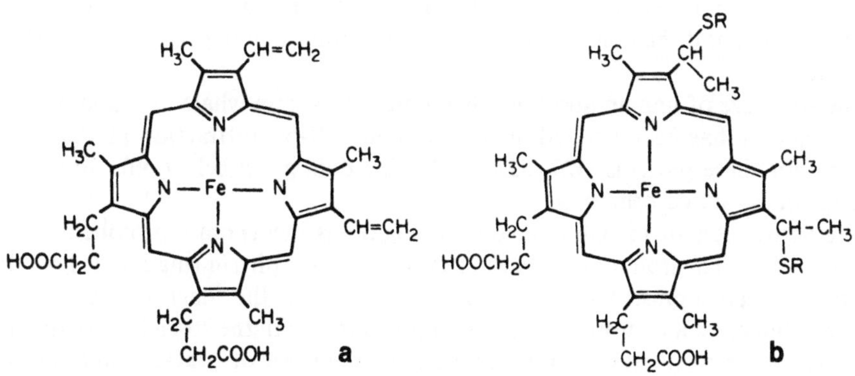

Fig. 6.1. Formulas of **a** protoheme and **b** mesoheme

discovery of a new class of non-heme proteins, i.e. iron-sulphur carriers [323]. The first proteins of this type were isolated from microorganisms by Mortenson and co-workers [323] and from spinach by Tagawa and Arnon in 1962 [324]. Of great importance for the further development of investigations was also the detection and interpretation by Beinert and Sands of a new EPR signal with a g-factor in the region of 1.89 [10].

Heme-containing proteins are classified into the following main groups: (1) oxygen carriers, myoglobin and hemoglobin; (2) electron carriers, cytochromes; (3) oxidoreductases, catalase, peroxidase, cytochrome P-450.

The non-heme proteins known to date may be subdivided into the following groups: (1) iron-sulphur electron carriers (rubredoxins and ferredoxins); (2) enzymes containing a non-heme iron along with other prosthetic groups (nitrogenase, xanthine oxidase, etc. (see Chap. 7)); (3) oxygen carriers (hemerythrin); (4) iron-transport and -storage proteins (transferrin, ferritin).

The present chapter is concerned with the basic principles of the structure and functioning mechanism of iron-containing proteins. We will describe the oxygen carriers, myoglobin und hemoglobin, which are not enzymes, strictly speaking. However, the obvious evolutional and structural relationship between this group of proteins and other heme-containing proteins and a sufficiently deep understanding of the mechanism of their functioning make it expedient to devote a separate section to these proteins.

6.1 Heme Proteins

6.1.1 Myoglobins and Hemoglobins

Myoglobins (Mb) and hemoglobins (Hb) play an important biological role, supplying oxygen for numerous oxidation processes taking place in animals. An oxygen carrier, leg hemoglobin (Lb) has been detected in root nodules, plant tissues and legumes. Extensive literature has been devoted to the various medicinal, physiological and structural aspects associated with this class of proteins [see, for example, 21–24, 223, 320, 325–328, 428].

Here, we will consider briefly the problems that have a direct relation to the basic topic of the present monograph: the active-site structure, the mechanism of oxygen binding and the conformational changes upon binding of ligands to the active site.

The structure of human and horse hemoglobins, sperm whale myoglobin and leg hemoglobin has been studied most thoroughly. X-ray diffraction models are available for these proteins [327, 329–333]. The data given below refer to these proteins unless stated otherwise.

The active sites of myoglobins and hemoglobins incorporate protoheme IX (Figs. 6.1–6.3). The iron atom is in the centre of the porphyrin ring and is linked to four nitrogen atoms. The main ring is attached to the macromolecule via numerous linkages: a coordination bond in position 5 with the so-called proximal histidine, two covalent ethereal bonds and a large number of contacts (about 60) with the hydrophobic groups (Fig. 1.3). The sixth coordination site in the ligand-

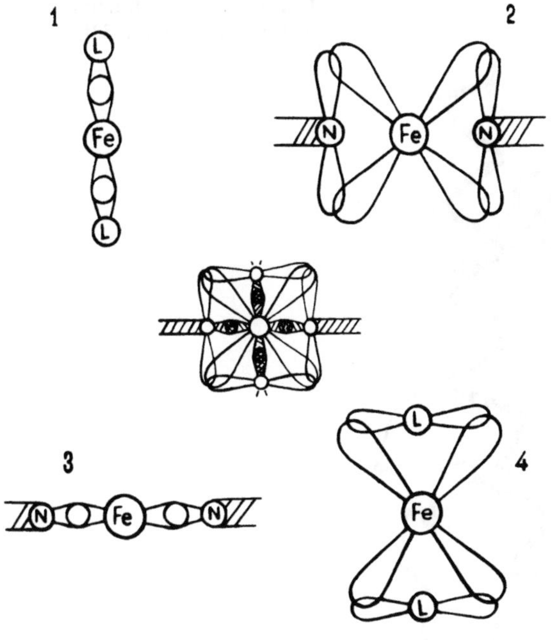

Fig. 6.2. The overlap of the d-orbitals of the central iron atom in the heme with the orbitals of ligands (*L*): *1* p-orbitals of axial ligands (the σ-bond); *2* p-orbitals of the porphyrin ring (the π-bond); *3* p-orbitals of the nitrogen atoms of porphyrin (the σ-bond); *4* p-orbitals of the ligands (the π-bond). Shown at the *centre* is the general overlap scheme

free reduced form of the protein with Fe^{2+} (deoxymyoglobins, deoxyhemoglobins) is free and may be occupied by an oxygen molecule (the oxy-form) or by the corresponding ligands (CO, NO).

In the oxidized meta-form the sixth coordination site in the heme is occupied by a water molecule (neutral and acid media) or by a hydroxyl ion (alkaline media). The diagram of the overlap of the molecular orbitals in heme-ligand complexes is shown in Fig. 6.2, and the arrangement of the energy levels of the orbitals is given in Fig. 6.4. The physico-chemical properties of the complexes are given in Table 6.1. In deoxyheme the lowest unoccupied orbital e_g^3 differs in energy by only 0.2 eV from the nearest occupied orbital; a relevant theoretical treatment [269] shows that in this case favourable conditions appear for the manifestation of the Jahn-Teller pseudoeffect. As a result of this effect, the nuclear system assumes a less symmetric configuration. In the case of heme this effect must manifest itself in the emergence of the iron atom out of the plane with a decrease in the symmetry of the system.

According to X-ray diffraction data [332], the Fe atom in a single crystal of deoxyhemoglobin is brought out of the plane by 0.06 nm to the side of the proximal histidine and by 0.055 nm deoxymyoglobin. In the meta-forms this distance is 0.021 nm (in the subunits of hemoglobin) and 0.04 nm (in myoglobin).

An important role in the binding of ligands is ascribed to the distal histidine (Figs. 1.3 and 6.2), which does not come into direct contact with iron atoms but can interact with O_2 in the coordination sphere, thus forming a hydrogen bond. This circumstance, coupled with the hydrophobic nature of the O_2 environment (valine, phenylalanine), creates favourable conditions for the binding of O_2 and for the prevention of the oxidation of the Fe^{2+} heme with the formation of a free superoxide ion O_2^-. The solvation of this superoxide ion in a polar medium is

Fig. 6.3. X-ray diffraction model of leg hemoglobin [333]

evidently the main cause of the rapid oxidation of heme upon breakdown of the molecular structure of proteins.

The addition of O_2 to the heme group occurs as a result of the interaction of the d_{xz} and d_{yz} orbitals of Fe with the anti-bonding π_g^* orbitals of the oxygen molecule. As a result, a new system of orbitals is formed, which is filled with spin-paired electrons (Fig. 6.4). Experiments with model systems have disclosed the substantial role of the basicity of the proximal ligand and the polarity of the environment and the possibility of rotation of the proximal ligand upon binding of O_2 [334].

Upon binding, the O_2 molecule lies at an angle of 156° relative to the perpendicular to the plane of the heme. An analysis of the hyperfine splitting of the EPR spectrum of hemoglobin, in which the Fe^{2+} atom ion has been replaced by a paramagnetic Co^{2+} atom and the isotope [^{17}O] is used, has shown that about 66% of the electron density in the oxy-forms is transferred to O_2. The frequency of the deformation vibrations of the O–O bond in oxyhemoglobin corresponds to the frequency in the $H-O_2^{\cdot}$ radical. This is also an indication of a significant contribution of the $Fe^{3+}O_2^-$ structure [335, 336].

The binding of O_2 and other ligands is accompanied by a substantial reconstitution of the system of energy levels of the heme group. The differences in the energy levels become small and the system as a whole passes over from the high-spin state in the deoxy-form to the low-spin states in the oxy-form (Fig. 6.4).

Detailed information on the electronic structure of the heme-oxygen complex has been obtained for cobalt-substituted heme proteins by means of EPR [337,

Table 6.1. Selected physico-chemical properties of heme proteins and enzymes

Protein (source)	S	Optical spectra λ_{max} (nm)	MCD spectra[a] λ_{max} (nm) λ_{min} (nm)	EPR (g-factor)	NGR δ(mm s^{-1})	ΔE_Q (mm s^{-1})	Redox potential E_0 (mV)
Sperm whale		[440]	[441]	[441]	[442–443] T = 195 K		
Mb^{+2}	2	434 556	$\overset{-}{420}$ $\overset{+}{437}$ $\overset{+}{535}$ $\overset{+}{550}$ $\overset{-}{595}$	Not observed	0.86 T = 300 K	1.78	[444]
$Mb^{+2}O_2$	0	418 543 581	$\overset{+}{408}$ $\overset{-}{427}$ $\overset{+}{527}$ $\overset{-}{550}$ $\overset{+}{572}$ $\overset{-}{584}$		0.25	2.3	+47
$Mb^{+3}H_2O$	5/2	409 507 635	$\overset{+}{404}$ $\overset{-}{429}$ $\overset{+}{470}$ $\overset{+}{490}$ $\overset{-}{690}$	$g_\parallel = 2$, $g_\perp = 6$	T = 300 K 0.42	1.3	
Man		[445]	[446]	[446] Unobservable	[447] T = 195 K		
Hb^{2+}	2	430 556	$\overset{+}{432}$ $\overset{+}{545}$ $\overset{-}{590}$		0.90	2.4	αHb
$Hb^{2+}O_2$	0	415 542 577	$\overset{+}{410}$ $\overset{-}{425}$ $\overset{+}{524}$ $\overset{-}{555}$ $\overset{+}{568}$		T = 195 K 0.20	1.89	[444] +113
$Hb^{3+}H_2O$	5/2	407 500	Spectrum weakly pronounced	$g_\parallel = 2$ $g_\perp$	T = 195 K 0.20	2.00	βHb +52

[a] MCD spectra were recorded in all cases at room temperature at neutral pH; + and − indicate that the effect is positive or negative.
[b] Cyt c^{3+} is present in solution at neutral pH in two forms: (1) with the ligands Fe^{3+} Met, Im; (2) with the ligands Met, Im (Im is imidazole).
[c] The NGR spectrum is given for P-450 of camphor.
[d] Compound I has a common spin S = 3/2 consisting of the spin of Fe^{2+} (s = 1) and the spin of the radical cation of porphyrin (s = 1/2).
[e] Compound I of peroxidase has a signal from the radical cation near g = 2.004. The signal from Fe (IV) in compound I and compound II is not observed.
[f] The NGR spectrum is given for cyt *c* from yeast.

Table 6.1 (continued)

Protein (source)	S	Optical spectra λ_{max} (nm)	MCD spectra[a] λ_{max} (nm) λ_{min} (nm)	EPR (g-factor)	NGR δ(mm s^{-1})	ΔE_Q (mm s^{-1})	Redox potential E_0 (mV)
Mitochondria		[449]	[448, 449]		[475]		
Cytochrome *c* oxidase							
cyt a^{2+}	0	438 511 594	+412 −420 +440 −450		T = 4.2 K 0.45	1.1	+25
		[449]	[448, 449]	[450]			
cyt a^{3+}	1/2	422 540 592	+402 +420 −434 −582 +606	3.03 2.21			+215
cyt a^{2+}	2		−430 +445.5	Unobservable			
					[451]		
cyt a_3^{3+}	5/2		Spectrum weakly pronounced	Unobservable	T = 195 K 0.3	1.34	[452]
Horse		[453]	[453]		[455]		[456]
cyt c^{2+}	0	413 521 550	−400 +413 −427 +506 −510 +514 −517 +520 −526 +540 +546 −550		T = 195 K[f] pH = 6 0.42	1.2	+260
				[454]			
cyt c^{3+} [b]	1/2	408 530 565	+401 −413 +444 −567	3.06 2.25 1.25 3.2 2.08	0.21	2.14	
Pig liver		[457]	[457]				
cyt b_5^{2+}	0	423 557	−415 +426 −438 +553 −561				
				[356]			[356]
cyt b_5^{3+}	1/2	413 523 562	+406 −419 +553 −570	3.0 2.23 1.43			+25

Microsomes		[409]	[458]	[459]	[460] T = 190 K		[461]
cyt P-450^{2+}	2	412 559	+404 −422 +439 +550 −573	Unobservable	0.78	2.39[c]	−410
cyt P-450^{3+}	1/2	415 535 570	+402 −426 +512 −534 +553 T = 0 °C	2.41 2.24 1.92	T = 200 K 0.31	2.75	pH = 7.0 [442] −20
cyt P-420^{2+}		[409] 427 530 560					
cyt P-420^{3+}	5/2	414		2.25 6.0			
Liver		[462]	[462]	[464] T = 77 K	[463] pH = 6.9 pH = 7.0	T = 195 K	
Catalase (Fe^{3+})	5/2	405 500 622	−403 −429 +530 −550 +620	g = 1.92 g = 6.3	0.55	0.82	
Compound I	1 + (1/2)	400 656	+372 −416 490–630 −668	Unobservable			
Compound II	1	427 534 567	+425 −442 +521 +575	Unobservable			
Horseradish				[465] T = 4.2 K[e]	[466]		[467]
Peroxidase	5/2	[387] 403 499 641		pH = 6.2	0.25	1.96	
Compound I	1 + (1/2)[d]	400 570	[387] +387 −405 +550 −650	Unobservable	0.0	1.2	941
Compound II	1	416 528 546	+408 −422 +528 −548	Unobservable	0.03	1.36	963

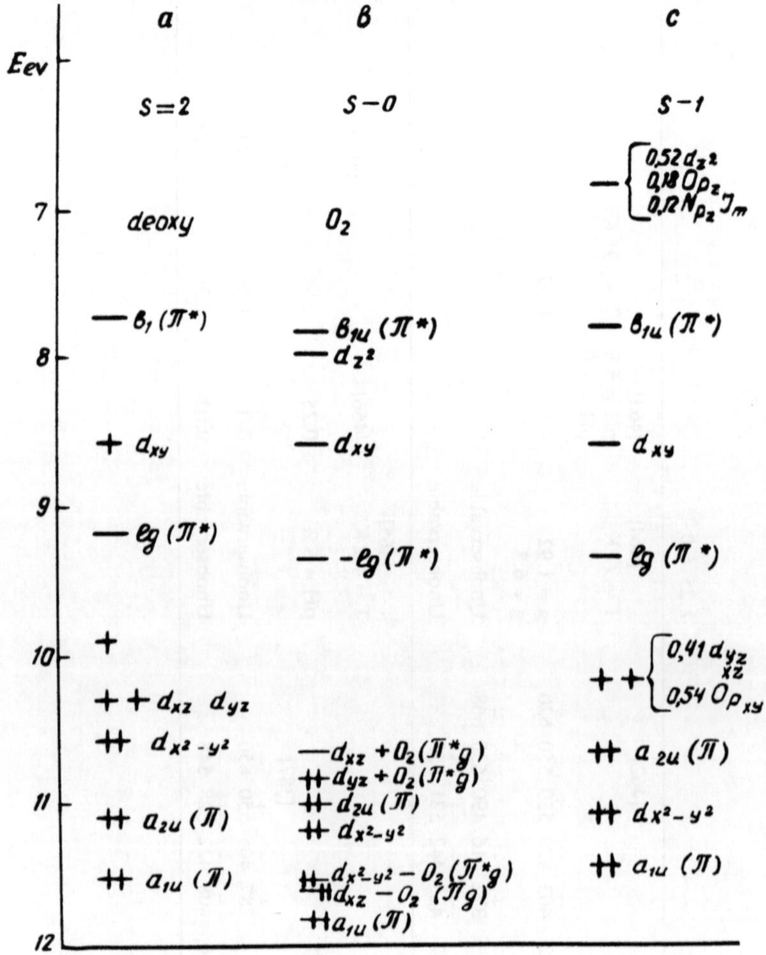

Fig. 6.4a–c. Molecular-orbital diagram of heme derivatives: **a** Fe^{2+}; **b** $Fe^{2+}O_2$ (hemoglobin) [319]; **c** $Fe^{3+}O$ [425]

338]. For example, in leg hemoglobin in the deoxy-form the unpaired electron is localized on the d_{z^2} orbital of Co^{2+}. Upon binding of dioxygen, one unpaired electron stays on the π^* orbitals of the ligand. On the basis of a thorough analysis of the EPR spectra at different pH, it has been concluded by some [337] that the protonation of the distal histidine at pK = 6.4 leads to the formation of a symmetric ethylenelike structure of the complex. In alkaline media oxygen forms an ozonelike structure with a higher, unpaired electron density on the O_2 atoms. The important role of the acid-base state of distal histidine in myoglobin upon binding of dioxygen (the acidic Bohr effect) has been convincingly demonstrated by the hyperfine shift in the NMR spectra [340].

According to X-ray diffraction data [327], the binding of O_2 and CO to Hb and Mb is accompanied by the movement of the iron atom to the plane of the heme ring, i.e. by a displacement of about 0.04 nm. The data on the coordination of the Fe atom in deoxyhemoglobin obtained by the EXAFS method have placed some doubt on the correctness of the X-ray diffraction data [341]. However, additional data have been obtained [342] in favour of the emergence of the Fe atom out of the heme plane. Note that the replacement of the iron in the heme group of hemoglobin by a Co atom alters only slightly the macromolecular and functional properties of Hb. But X-ray diffraction analysis does not reveal a difference in the position of cobalt in the oxy- and deoxy-forms of the proteins. The EXAFS data for carp Hb and the hybrid Hb (Zn_2Fe_2) have also not revealed these differences [338, 339].

The detailed macromolecular structure of Hb, Mb and Lb from various sources has been established by means of X-ray diffraction analysis. For example, Fig. 6.3 shows the latest data on the leg hemoglobin of lupine [333]. As seen in Fig. 6.3, the polypeptide chain is foled into a protein globule. The structure incorporates L-helical segments joined by folds. The proline adjacent to Glu and Ala resides most often at the fold sites. The designations adopted for the segments are given in Fig. 6.3. An analogous structure is exhibited by myoglobins from other sources. The application of neutron diffraction, which enables one to establish the coordinates of the hydrogen atoms, has led to the construction of a myoglobin model with localized water molecules [343]. For example, in carbomonoxymyoglobin the position of 40 hydrogen molecules is localized, 27 molecules are involved in the bonds between the protein atoms and 9 molecules form hydrogen bonds with the main chain. It is interesting that many of the hydrogen bonds are in a strained state and the angle of departure from the equilibrium position reaches 25°

There is a closer relationship between the primary, secondary and tertiary structures of Hb, Mb and Lb. To a first approximation, Hb may be regarded as a complex of four Mb molecules. The molecular masses are 17000 (Mb) and 60000 (Hb). The presence in the polypeptide chain of identical, indispensable amino acids and the closeness of the steric and electronic properties of the replaceable residues are responsible for the closeness of the secondary and tertiary structures. However, noticeable differences are observed between the structures that give rise to functional differences. For instance, the heme "pocket" of the α-subunit of oxyhemoglobin is large compared with the corresponding "pocket" in oxymyoglobin. The subunits of Hb differ appreciably in amino-acid composition and structural details from those of Mb and Lb.

The structural differences described above are responsible for appreciable functional differences. The main differences are in the affinity for O_2. The association constants K_{O_2} of oxycomplexes at pH 7 and 20 °C (in nM^{-1}) are 5.6×10^4 (human Hb), 1.2×10^6 (the α-chain, human Hb), 7.7×10^4 (human Mb) and 2×10^7 (*Askaris* Lb). The increased affinity for O_2 in Lb as compared with other heme proteins is due to the high rate constant of association ($k_1 = 1.5 \times 10^8\ M^{-1}\ s^{-1}$) [344]. The rate constant of dissociation k_{-1} for all proteins is of the same order of magnitude ($5-10\ s^{-1}$).

A qualitatively new property of hemoglobins is the cooperative interaction between the subunits and, as a consequence, the sharply pronounced allosteric properties. The binding of four O_2 molecules is characterized by an S-shaped nonhyperbolic dependence of the fraction of complexed molecules on the O_2 concentration in solution. The corresponding curves are described by equations containing the Hill coefficients $n \approx 2.8$ (Sect. 5.3). The allosteric properties of hemoglobins undergo substantial changes upon binding of regulators, e.g. inosine hexaphosphate (IHP) [327].

The kinetic and structural features of allosteric phenomena in hemoglobin are the subject of a large number of investigations. Here, we shall consider the problem of the change of the conformational structure of heme oxygen carriers upon their interaction with ligands. This problem is of prime importance for the elucidation of the mechanism of the electronic-conformational interaction [223] and for the estimation of the contribution of the reorganization of the medium in processes involving metalloenzymes (Sect. 11.1).

There are numerous data which indicate a substantial change in the tertiary and quaternary structures of the Hb macromolecule upon binding of O_2 and other ligands. Binding of O_2 is accompanied by a change in some properties of Hb, such as thermal stability, the ability to undergo protolysis, isotope exchange of the hydrogen atoms of the peptide groups and the reactivity of the β-93 SH groups present at the point of contact of the α- and β-subunits of Hb, etc. [345]. Upon binding of O_2 there has been observed a noticeable immobilization of the spin label added to the β-93-SH group of Hb [128]. This immobilization was caused by the emergence of the residue β-145 Tyr from the hydrophobic pocket at the site of contact between the subunits and by the incorporation of the nitroxide fragment of the label into the pocket. A more detailed investigation, which made use of the hybrid forms of hemoglobin, in which the α-and β-fragments were in different ferro and ferri states, and also a study of chemically and genetically modified preparations have made it possible to follow the concrete paths of transmission of allosteric effects from one subunit to another. This has revealed the important role of the amino acids located at the boundary between the α- and the β-subunits [346]. Here, the area of the contact between these subunits is significantly changed [346].

The most detailed data on the conformational changes in Hb have been obtained by means of X-ray diffraction analysis [327, 332]. According to the data obtained by Perutz through a comparison of human oxyhemoglobin and deoxyhemoglobin, the binding of O_2 to the heme group involves the following sequence of events. The O_2 molecule first binds to the α-subunit, whose heme pocket allows the molecule to enter without hindering the coordination sphere of the Fe atom. The displacement of the Fe atom to the centre of the porphyrin ring causes the displacement of the proximal His F8, the F-helix, and the knocking of tyrosine 145 out of the cavity betwen the F- and H-helixes and the rupture of the bond between Arg 141 and the α-NH_2 group of Val 1 and Asp 126 of the neighbouring α-chain. Here, substantial shifts take place at the contacts between the α- and β-subunits, amounting to 0.7 nm. In the β-subunits the cavity of the heme pocket extends over a distance sufficient for the entry of the O_2 molecule. Then, the Fe atom is displaced together with the helix, Tyr 145 is knocked out of its cavity and the salt bridge between His 146 and Asp 194 is ruptured. Substantial changes in the conformation of Hb have been observed upon binding of other ligands (CO, imidazole) and also, possibly, upon transition to the meta-form (a comparison of the structures of human deoxyhemoglobin and horse methemoglobin).

The Perutz scheme accounts for the cooperativity of O_2 binding as follows: the binding of O_2 to one of the α-chains induces conformational transitions first in the α-subunits and then in the β-subunits. The quaternary structure as a whole assumes the conformation of the oxy-form after the second oxygen molecule is bound. The so-called Bohr effect (an increase in the oxygen affinity of hemoglobin at acidic and alkaline pH) has been accounted for within the framework of the Perutz scheme. The protonation and deprotonation of the corresponding protein groups cause the rupture of the salt bridges in the deoxy-form of Hb and an increase in the oxygen affinity of the heme group as a result of subsequent conformational changes.

Taking into account the above mentioned data obtained by an EXAFS study of the ligand environment of the iron atom in Hb and the data indicating the retention of the cooperative properties of the cobalt derivatives of Hb, one may conclude that further investigations are required for the complete elucidation of the detailed mechanism of electronic-conformational interactions in hemoglobins. However, the bulk of the data presented in the remarkable works of Perutz, which demonstrate drastic changes in the conformation of Hb upon binding of ligands, do not cause doubt. The Perutz model is the basis for a detailed interpretation of the cooperative effects in hemoglobin.

Physical methods also reliably record a change of the electronic properties, in particular the spin state of Fe and the ligand affinity of heme groups upon binding of allosteric receptors, such as IHP, and also upon hybridization of the α- and β-chains of Hb in the oxy- and deoxy-forms [327].

Detailed data have recently been obtained on the kinetics of carboxyhemoglobin dissociation which reflects the conformational reconstruction of the macromolecule upon change in the electronic state of the heme [352]. The absorption of a light quantum results in the occurrence of the following events: the release of CO from the complex (0.35 ps) with the formation of a state in which CO is present in the heme pocket; the conformational transition $R \rightarrow T$ (1 μs); the binding of CO to the heme group in the initial R-state (6 μs); the release of CO from the pocket (10 μs).

In myoglobin and in the individual subunits of hemoglobin relatively weak, albeit distinct conformational transitions are observed upon binding of ligands. For example, a transglobular effect has been detected by means of the spin-label technique [99], which consists in a change of the local conformational state of the sperm whale Mb macromolecule in the region of the histidine groups located at a distance of more than 1.5 nm from the heme group upon binding of cyanide. The change of the conformation of the macromolecule in the region of the C^{13}-modified Met 55 and 131 in going from metmyoglobin to carboxymyoglobin has also been detected by NMR [348]. Resonance Raman investigations have demonstrated changes in the configuration of the heme of monomeric insect hemoglobins at Co ligation [354]. In monomeric hemoglobins acidic and alkaline Bohr effects are detected associated with the rupture of the salt bridges [355]. The NMR method has also detected a change in the state of protons which are spatially remote from the heme. The data on the conformation change of myoglobin will be considered in Chapter 10 in connection with the problem of the intramolecular dynamics of proteins.

In the subsequent sections of this monograph we shall return repeatedly to data on hemoglobins and myoglobins, which are a reliable basis for the discussion of the problems of structure and mechanism of action of heme enzymes and electron carriers.

6.1.2 Cytochromes

Cytochromes are very widespread at all levels of living organisms, animal, plants and microbes, where they function as one-electron carriers.

Cytochromes *a* and a_3 enter into the composition of the terminal cytochrome *c* oxidase of the mitochondria [68, 349, 350]. The prosthetic group present in them is the so-called heme a, which contains a formyl residue in the side chain; cytochrome a_1 has been detected in a number of bacteria. The physico-chemical properties of the heme groups in cytochrome *c* oxidase are given in Table 6.1; the nature and arrangement of the ligands established by EXAFS are shown in Fig. 9.3. As seen in Fig. 9.3, the heme in cytochrome *a* in the fifth and sixth positions contains histidine residues. In cytochrome a_3 the fifth position is occupied by an imidazole ligand and the sixth position by the sulphur of the cysteine. The high-spin ($S = 2$) iron of cytochrome is linked to the copper atoms via a sulphur atom. A strong exchange interaction between atoms having an exchange integral of $I \approx 200\ cm^{-1}$ leads to the anti-ferromagnetic state of a pair of atoms recorded by magnetic susceptibility, EPR and NMR measurements [350, 351]. The function of cytochrome *a* is to transfer an electron from cytochrome *c* to the other constituents of cytochrome *c* oxidase. The EPR method has shown that the O_2 molecule becomes bound to the heme of cytochrome a_3 [353]. This involves the rupture of the bond with the sulphur atom, and the adjacent copper atom assumes a paramagnetic state recorded by EPR. The macromolecular structure of the cytochrome *c* oxidase complex, its localization in the membrane and the mechanism of O_2 oxidation to water will be discussed in Section 9.1.

The cytochromes *b*, which contain non-covalently linked protoheme IX, occur widely in anaerobic and aerobic bacteria, photosynthetic microorganisms and plant and animal tissues [356].

Cytochrome b_5, which is a one-electron carrier in the microsomal hydroxylation system, is a membrane-bound protein of molecular mass 16000 which contains one heme group per macromolecule. About one-third of the protein moiety of the cytochrome (5000) is a hydrophobic portion, which is buried in the microsomal membrane. The hydrophobic portion can be split off from the hydrophilic portion, which contains a heme.

The hydrophilic portion of cytochrome b_5 has been isolated and crystallized and its structure has been studied using X-ray diffraction [357]. The heme ring has been found to be localized in the hydrophobic pocket and linked by two histidine ligands. The edge of the heme group is close to the surface of the macromolecule, but only the two propionic groups of the heme can be in direct contact with the solvent. There is a narrow groove between the solvent and the heme iron which is lined with hydrophobic groups. Spectrosocopic investigations have led to the suggestion that there is an equilibrium between the high- and low-spin forms of the iron atoms (Table 6.1).

A large number of cytochromes *b* have been detected in bacteria; for example, three types of cytochromes *b* have been isolated from mitochondria [358]. The most studied example is cytochrome b_{562} [359]. It is a protein with a molecular mass of 12000, which contains a heme group; the fifth and sixth coordination positions of the iron atom are occupied by methionine and histidine residues. The redox potential of cytochrome b_{562} is equal to 110 mV.

Cytochrome of type *b* enters into the composition of the ubiquinone-cytochrome *c* reductase of the mitochondria (complex 3) [358]. The cytochrome *b* from bovine heart has a molecular mass of 28000 and a redox potential of 340 mV. The localization of cytochrome *b* in the membrane and its possible participation in oxidative phosphorylation will be discussed in Section 9.1.

The class *c* cytochromes contain a mesoheme (Fig. 6.1) linked to the macromolecule via a thioether linkage between the vinyl groups of the heme and the sulphhydryl groups of the protein [360, 361]. Cytochromes *c* occur widely at all levels of living organisms. Comparison of their primary and molecular structures in various organisms provides valuable information on the evolution at the molecular level [360].

X-ray diffraction models have been constructed for a number of cytochromes: horse and tuna cytochromes *c*, cytochrome c_2 from *Rhodospirillum rubrum* and cytochrome c_{555} from *Paracoccus denitroficans*. The type *c* cytochromes from different sources have a similar structure.

In horse cytochrome *c* (molecular mass 12300) the heme group is linked to the protein by the sulphur of Met 80 and the nitrogen of His 18 which act as ligands. The heme is in contact with nearly 16 non-polar amino acids. the important features of the structure, which are evidently of prime significance for the functioning of cytochrome *c*, are the emergence of the heme edge to the globule surface and the presence of a cluster of positively charged lysine groups in the heme region. Both in the oxidized and the reduced form the iron of the heme group is in a low-spin state. Detailed information has recently been obtained on the coordinates of the atoms of protein groups and water molecules in an X-ray diffraction analysis of tuna cytochrome *c* [362] (Fig. 6.5). It has been found that in the oxidized form the heme group is tightly packed in the hydrophobic groove and is folded to a certain extent in the form of a saddle. The iron atom is localized in the heme plane. The water molecules are strongly bound to three amino acids (Asp 52, Tyr 67 and Try 78). The authors of the work cited have succeeded in following the fine conformational changes in the redox transitions taking place in the cytochrome. The $Fe^{2+} \rightarrow Fe^{3+}$ transition involves small but distinctly observable displacements of a large number of atoms, e.g. Tyr 67, Asp 62 and His 41 and the heme group, which amount to 0.02–0.05 nm. A displacement of the water molecule localized between Asp 52, Tyr 67 and Try 78 to the side of the heme group is also observed, which becomes more hydrophilic as a result of the transition.

The X-ray structural analysis of bacterial cytochromes *c* has revealed their dimer state. Each of the subunits of the cytochromes contains protoheme IX linked to the macromolecule in the same way in which the heme is bound in cytochromes *c* [363].

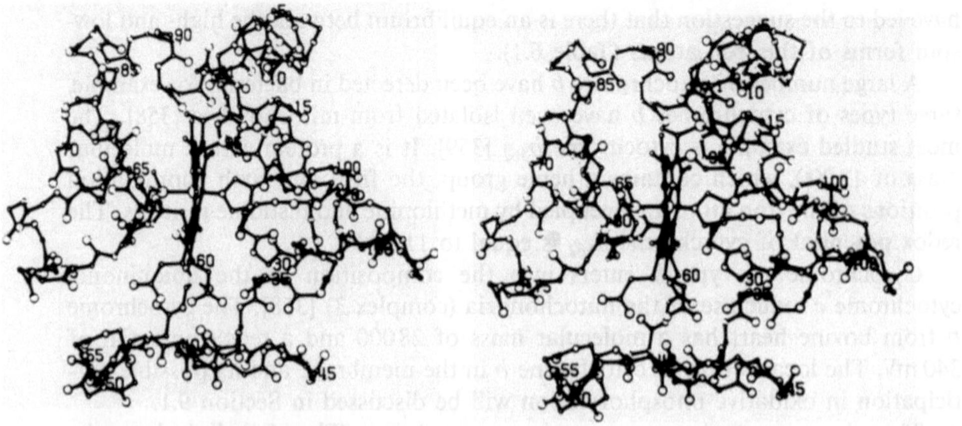

Fig. 6.5. The X-ray diffraction mode of cytochrome *c* from tuna [362]

Apart from "ordinary", high potential cytochromes *c* (Table 6.1), low potential cytochromes have been described, e.g. cytochrome c_3 isolated from *Desulfovibrio desulfuricans,* which contains four heme groups [364] and hexaheme cytochrome c_{552} [365]. There are abundant literature data, which indicate the change of the conformational state of cytochromes *c* with a change of the valency of the iron atom [361]. This involves a change of thermal stability, the rate of proteolysis, the pK of protein groups, the exchange rate, the binding of ions and of the rate of chemical modification. The effect of the conformational factors on the rate of electron transfer involving cytochromes *c* and the possible mechanisms of the transfer and the structures of the complexes of cytochromes *c* with other transport proteins will be discussed in Chapters 9 and 11.

6.1.3 Peroxidases

Peroxidases have been detected in many plant, animal and bacterial organisms [47, 49, 366–369]. The most well-studied peroxidases are horseradish peroxidase (HRP) and cytochrome *c* peroxidase (CCP). Chloroperoxidase, myeloperoxidase and lactoperoxidase are now receiving much attention. Peroxidases catalyze the oxidation of electron-donor molecules (DH_2) by hydrogen peroxide according to the following scheme:

$$DH_2 + H_2O_2 = 2H_2O + D\ .$$

Various electron donors may serve as substrates for peroxidases: phenols, amines, alcohols, acids, inorganic ions, etc.

Horseradish peroxidase contains one protoheme group per protein globule having a nearly spherical form with a radius of about 2.9 nm and a molecular mass of 40000. Apart from the amino acids, HRP contains an appreciable number of carbohydrate residues (mannose, glucosoamine). At least seven HRP iso-

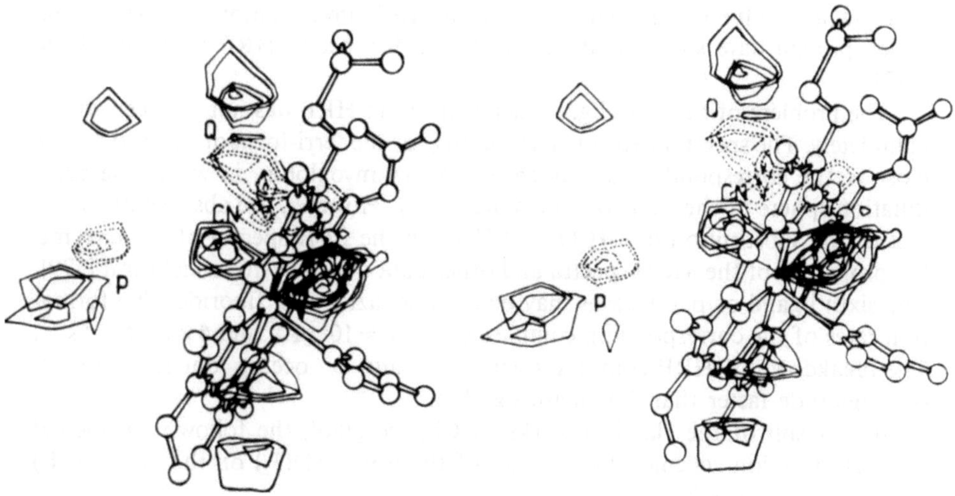

Fig. 6.6. The X-ray diffraction model of cytochrome *c* peroxidase [369]

enzymes have been detected, which differ in structure and composition and also in physico-chemical properties [368].

The X-ray diffraction analysis of cytochrome *c* peroxidase [369] and a large number of physico-chemical investigations have provided sufficiently reliable data on the basic aspects of the structure and action mechanism of peroxidases. In cytochrome *c* peroxidase from bakers' yeast the heme group becomes linked to the protein via a proximal ligand (histidine) and a large number of hydrophobic contacts (Fig. 6.6). The iron-nitrogen distance is 0.23 nm; the corresponding figure for myoglobin is 0.21 nm. The distal histidine is localized near the other side of the heme ring in the vicinity of the tryptophan and arginine residues.

There are many arguments stating that the histidine imidazole is the fifth ligand of the heme in HRP. This is evidenced by the similarity in the spectral characteristics of the heme group of the peroxidase and myoglobin (Table 6.1), the presence of a hyperfine splitting from the nitrogen nuclei in the EPR spectra of cobalt-substituted preparations [370] and in nitroxide complexes [367], the breakdown of the histidine upon irradiation of the heme [371], etc. According to the data obtained by magnetic circular dichroism (MCD) [372] and resonance Raman scattering [373], the heme ring is slightly deformed [373]. The shift of the two methyl groups of the ring, the non-equivalence of the β-CH_2 protons of the propionic-acid residues and the NMR characteristics of the HRP-CN^- complex are close to the corresponding parameters established for cytochrome *c* [374]. According to these NMR data, phenylalanines are expected to be present in the nearest environment of the heme group. An analysis of the EPR spectra of the high-spin state of HRP indicates the rhombic distortion of the axial symmetry of the iron ligands [375]. The heme group is buried in the protein globule and at neutral pH; it is inaccessible to iodide and ferricyanide ions [376]. Experiments with iron-depleted HRP have shown that these ions do not quench the luminescence of the apoenzyme. The accessibility increases upon protonation of the group with pK 4.5 [376]. The introduction of a nitroxide radical into the propionic-acid residue of the heme and further EPR investigations of the spin-labelled protein have shown that the label-iron distance in HRP is about 0.9 nm [377].

The problem of the sixth ligand-heme group in HRP has not yet been solved completely. The spectral dependencies on pH for the ferri-form of the enzyme are close to the corresponding dependencies for metmyoglobin, in which the coordination sphere of the iron contains water [378]. However, the absence of the effect of the substitution of $^{16}H_2O$ by $^{17}H_2O$ on the EPR spectra of the enzyme, the properties of the MCD spectra and other data question this conclusion [379]. The sixth ligand of peroxidases may be cyanide, azide and fluoride. The formation rates of the corresponding complexes are $k_1 \approx 10^5$, 10^7 and $5 \times 10^5\,M^{-1}\,s^{-1}$. The breakdown of HRP complexes with CN^- and F^- occurs four to six orders of magnitude faster than for metmyoglobin.

As a result of the classical works of Chance [380], the following principal chemical reaction scheme of oxidation of the donor (DH_2) on the enzyme (E) has been constructed:

$$
\begin{aligned}
&E + H_2O_2 \xrightarrow{k_1} E_I\,; \\
&E_I + DH_2 \xrightarrow{k_2} E_{II} + DH_2^{\cdot +}\,; \qquad (6.1) \\
&E_{II} + DH_2 \xrightarrow{k_3} E + DH_2^{\cdot +}\,.
\end{aligned}
$$

According to scheme (6.1), the interaction of H_2O_2 with the enzyme gives "compound I" (E_I), which has two oxidizing equivalents. The oxidation of the donor molecules leads to "compound II" (E_{II}), which oxidizes the second donor molecule or causes the first molecule to be fully oxidized. The E_I state can be obtained upon interaction of the enzyme with organic hydroperoxides and peracids and also with oxidizing agents, such as HOCl, $NaClO_2$, $KClO_4$, O_3, etc. [368].

Hydroperoxides and other ligands become bound to the enzyme in a unionized form. The rate constants of E_I formation are rather large. The rate constants for the reactions of HRP with H_2O_2, peroxybenzoic acids and with methyl and ethyl hydroperoxides are respectively: $k_1 = 1-2 \times 10^7\,M^{-1}\,s^{-1}$; $k_1 = 0.1-10 \times 10^7\,M^{-1}\,s^{-1}$; $0.13 \times 10^7\,M^{-1}\,s^{-1}$; and $0.4 \times 10^7\,M^{-1}\,s^{-1}$. The general tendency is that the value of k_1 increases with increasing hydrophobicity and decreasing nucleophilicity of the reagents [368].

There are serious arguments in favour of the direct oxidation of the heme group in E_I with the formation of iron oxide containing one oxygen atom [381–384]. The reaction of HRP and CCP (cytochrome *c* peroxidase) with H_2O_2 does not involve the binding or release of protons [378]; when 1 mol m-nitroperbenzoic acid acts on 1 mol enzyme, 1 mol of the acid is released. Analogous stoichiometry is observed upon action of ethyl hydroperoxides [382]. According to the data on the Mössbauer effect, MCD and resonance Raman scattering, the E_I and E_{II} states correspond to Fe^{4+} with a spin of $S = 1$, though the first compound contains two oxidizing equivalents and the second only one [368, 383, 384]. On the basis of this fact and the results of a number of physicochemical investigations, one may conclude that there is a cation-radical form of the porphyrin ring or protein group in E_I. It is evidently this form that takes part in the one-electron oxidation of substrates with the formation of E_{II}. The question of the exact localization of the cation-radical state has not yet been solved completely. An EPR signal with $g \approx 2.00$ has been detected in the E_I of CCP, which corresponds to one equivalent of the radical per one heme [367]. According to Mössbauer spectroscopy, this radical is at a considerable distance (>0.8 nm) from the iron atom [383]. In E_I samples of HRP the intensity of the radical signal is insignificant; when the heme is irradiated, a new intense signal appears, which is ascribed to the radical located at a distance of about 1.0 nm from the iron. The NMR spectra of E_I HRP do not show changes that are uniquely expected for the cation-radical of porphyrin. However, a spin-spin interaction between [^{14}N] and β-protons of the heme group of E_I HRP with an unpaired electron of the cation radical has, however, recently been detected with the aid of the ENDOR technique [385]. The incorporation of the [^{17}O] atom of an oxidizing agent into the composition of E_I HRP has been accomplished [386]. These authors maintained that one of the oxidizing equivalents is localized at the [^{17}O] atom and the other at the porphyrin π-ring. The latest data obtained on this subject have been reviewed elsewhere [387].

It should be noted that according to model compounds, the electronic state of the π-cation radical may be described as the $^2A_{2u}$ or $^2A_{1u}$ state (D_h symmetry) [388]. It may be presumed that the discrepancy in the treatment of the experimental data is due to the differences in the preparation of the samples, the recording of the spectra and in the nature of the object under study. A theoretical treatment based on the extended Huckel method of the molecular orbitals of the E_I state within the framework of the model of interaction of $Fe^{4+}O$ with the π-cation radical has made it possible to describe a complex of the physico-chemical properties of this state [389]. The MO (molecular orbitals) scheme of the E_I state is given in Fig. 6.3.

Of importance for the analysis of the possible mechanisms of peroxidase reactions are data on the redox potentials, whose values for HRP are as follows: −250 mV (E), 970 mV (E_I) and 950 mV (E_{II}) [390]. The E_I and E_{II} states are thus strong oxidants. The reactions of E_I with phenols and anilines proceed at high rates, depending on the structure of the substrate. For example, $k_2 = 2.3 \times 10^8\ M^{-1}\ s^{-1}$ for p-aminophenol, $k_2 = 4.2 \times 10^7\ M^{-1}\ s^{-1}$ for p-methylphenol, $k_2 = 10^4\ M^{-1}\ s^{-1}$ for tyrosine and $k_2 = 8.5 \times 10^4\ M^{-1}\ s^{-1}$ for the ferricyanide ion. For the same substrates the value of k_3 for reactions with E_{II} is three to ten times lower. A decrease in the electron-acceptor properties of the substituents of phenols and anilines leads to a decrease in the rate of oxidation. In such cases, the value of k_2 correlates well with the parameter σ in the Hammett equation [391].

The quantities k_2 and k_3 are not the elementary rate constants; they incorporate combinations of the rate constants for the steps of formation, breakdown and conversion (k_{cat}) of enzyme-substrate complexes. For example, for ferricyanide at pH 5 the value of $k_{cat} = 3 \times 10^2\ s^{-1}$, and $K_M = 4 \times 10^{-4}$ M. These values depend significantly on the presence of o-dianizidine: the value of k_{cat} increases approximately 100 times [392]. The formation of enzyme-substrate complexes has been directly recorded by NMR [393, 394]. It has been found that the substrate fragments that are subjected to oxidation (resorcinol, for example) are located perpendicular to the heme ring at a distance of 0.84–0.91 nm from the iron atom.

The possible mechanisms of the elementary steps of peroxidase reactions will be discussed in Chapter 11 with account taken of the participation of distal histidine groups. Here, we shall confine ourselves to chemical reaction schemes that are consistent with the available experimental data and theoretical conceptions (X = porphyrin or aromatic group, e.g. the tryptophan group in cytochrome *c* peroxidase):

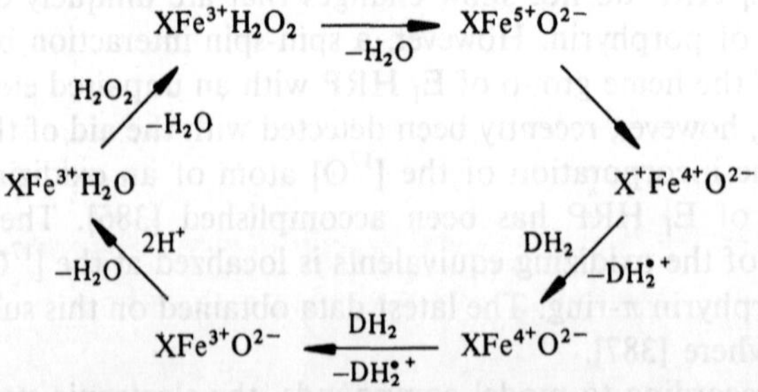

A number of works have been devoted to the problem of the similarity and difference between peroxidases (see, for example, [395–400]). The general conclusions that can be made from the investigations carried out are the following. Despite the fact that the fifth ligand in most peroxidases in the imidazole group, their physico-chemical properties differ appreciably from those of hemoglobins and myoglobins. These differences are presumably due to the presence in peroxidases of a system of hydrogen bonds in the region of distal histidine and to the high activity of the fifth ligand [395].

By its physical parameters, chloroperoxidase, which is similar in many respects to cytochrome P-450, stands out. It is therefore concluded that in these enzymes in the fifth coordination site of the iron of the heme group is occupied by thiolate rather than by histidine [396]. A point worth mentioning is that the last two enzymes are functionally similar, i.e. they are capable of involving very inert molecules in a chemical reaction. For example, chloroperoxidase catalyzes the halogenation of trans-cinnamic acid and monochlorodimedone under the action of the inert chloride ion and hydrogen peroxide. There are also differences between other peroxidases. For example, the properties of compound I (cytochrome *c* peroxidase) are similar in many respects to those of compound II (horseradish peroxidase). This is due to the fact that in the former case the free valence is localized not on the porphyrin ring, but on one of the protein groups.

6.1.4 Catalase

Enzyme catalase occurs widely in animal tissue, microorganisms and fungi [401, 402]. It catalyzes the chemical reaction of hydrogen peroxide decomposition into O_2 and H_2O. Catalase was one of the first enzymes to be isolated, purified and crystallized. The most thoroughly studied catalases are those isolated from bovine liver, the microorganism *Micrococcus lyscadeiticus,* yeasts and the fungus micro-

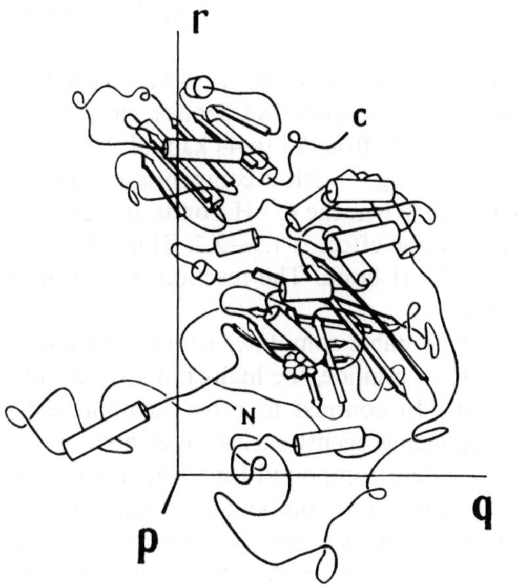

Fig. 6.7. The X-ray diffraction model of the catalase *Penicillium vitale* [406]

organism *Penicillum vitale.* The molecular mass of the enzyme is 230000–240000; the enzyme is made up of four identical subunits, each of which contains one protoheme IX group [401, 402].

The physico-chemical characteristics of the heme group in catalase and its complexes are given in Table 6.1. X-ray diffraction models have recently been constructed for catalases from two sources: from bovine liver at a resolution of 0.25 nm [403, 404] and from *P. vitale* at a resolution of 0.35 nm (Fig. 6.7) [405, 406]. The catalase molecule from bovine liver (BLC) contains four subunits. The heme group is linked to the protein via the oxygen atom of Tyr 357. The distal region contains His 34 and Asp 147. The heme group is buried 2.0 nm in the protein globule and communicates with the surface through the hydrophobic channel. The α-helical portions are similar to the E, F, Cl and H helices in myoglobins and hemoglobins.

The catalase of *P. vitale,* which has a molecular mass of 280000, has a similar structure (Fig. 6.7) [405, 406]. The main difference is that the c-terminal domain observed in this enzyme is absent in BLC. The heme groups of the subunits are 3.1, 3.5 and 4.6 nm apart from one another.

The mechanisms of the catalase reaction and the structure of the intermediate oxidation states are similar in many respects with those of peroxidases [402]. Reactions involving H_2O_2 proceed according to a scheme similar to that of reactions (6.1):

$$\begin{aligned} &E + H_2O_2 \xrightarrow{k_1} E_I + H_2O \; ; \\ &E_I + H_2O_2 \xrightarrow{k_2} E + O_2 \; . \end{aligned} \tag{6.2}$$

Compound I (E_I can also be obtained in reactions with organic hydroperoxides. The rate constants for the formation of compound I for H_2O_2, methyl and ethyl hydroperoxides are 6×10^7, 8.5×10^5 and $2 \times 10^4 \, M^{-1} \, s^{-1}$ respectively. The effective rate constant for the decomposition of H_2O_2 is $k_{eff} = 10^7\,M^{-1}\,s^{-1}$. Thus, catalase is a sort of record-holder of catalytic activity.

Compound I has two oxidizing equivalents and oxidizes effectively a number of compounds: alcohols, organic acids, hydroxylamines, hydrazones, etc. The activation energy of oxidation for alcohols ranges from 25 to 44 kJ mol^{-1} and the activation entropy is -90 to -40 kJ mol^{-1} deg^{-1} with a compensation dependence between these parameters (Chap. 10). When the C–H bonds are replaced by C–D bonds in alcohols, the kinetic isotope effect is 1.9–2.2. Thus, the rate-limiting step involves the rupture of the C–H bond. The reaction of hydrogen atom abstraction is strictly stereospecific.

Like other hemoproteins, catalase forms stable complexes with ligands, such as cyanide, fluoride and azide. In the case of cyanide, the high-spin state of iron (Fe^{3+}) is converted to the low-spin state. In contrast to other hemoproteins, catalase is not reduced even by strong reducing agenst such as dithionite.

The physico-chemical properties of catalase compound I are similar in many respects to those of the corresponding peroxidase compound. It may also be presumed that the catalytic mechanism of the catalase and peroxidase reaction are generally similar. The main difference between these reactions consists presumably in that the oxidation reactions of compounds I and II are localized in the active site without the release of the $HO_2^{\bullet}$ radical from the cage of the active site located in the bulk of the protein globule:

$$\underset{E_I}{X^+Fe^{4+}O^{2-}} \xrightarrow[-H^+,\,-H_2O]{H_2O_2} \underset{E_{II}}{XFe^{4+}O^{2-}} \xrightarrow[-O_2]{HO_2^{\bullet}} \underset{E}{XFe^{3+}O^{2-}} .$$

The mechanisms of these reactions will be discussed in more detail in Chapter 11.

A new type of catalase, the so-called pseudocatalases containing manganese, has recently been reported [405]. The enzyme from *Lactobacillus plantarum,* with a molecular weight of 172000, is made up of six identical subunits and contains about six manganese atoms. The activity of the enzyme approaches that of hemin catalases and is not suppressed by such complexones as azide and cyanide. An efficient inhibitor is hydroxylamine. In the opinion of some authors [405], during the catalytic decomposition of H_2O_2 the manganese at the active site cycles between the redox states Mn(III)↔Mn(V). According to the data of a preliminary X-ray analysis, at the active site of the enzyme there are two closely spaced Mn atoms [892].

6.1.5 Cytochrome P-450

A powerful enzyme hydroxylation system of organic substrates, i.e. hydrocarbons, organic acids, alcohols, amines, is operative in animal and plant tissues and bacteria [407–411], which catalyzes oxidation processes according to the following general scheme:

$$DH_2 + RH + O_2 \rightarrow R{-}OH + H_2O + D \ , \qquad (6.3)$$

where RH is the substrate; DH_2 is the reducing agent (NADH or NADPH). The action of this system on compounds having unsaturated bonds may result in thier expoxidation or dimethylation with methylamines.

The hydroxylation reaction is directly effected by an enzyme-hemoprotein containing protoheme IX. The reduction of the enzyme involves flavin reductases and electron carriers, such as adrenodoxin, rubredoxin and cytochrome b_5. Cytochrome P-450 is so named because its reduced complex with carbon monoxide shows maximum absorption at 450 nm. The various forms of cytochrome P-450 from liver microsomes have a molecular mass of about 49000 [409]. A preparation from *Pseudomonas putida* has approximately the same molecular mass. One of the subunits of the cytochrome P-450 of the mitochondria of the adrenal cortex has a molecular mass of 55000. There are one heme and six cysteine groups per one protein macromolecule.

The protein globule of the enzyme from the adrenal cortex consists of two fragments linked via a peptide portion which is easily hydrolyzable by the action of trypsin. The hydrophilic fragment F_1 has a molecular mas of 27000 and contains a heme and an adrenodoxin-binding site [412, 424]. The hydrophobic fragment F_2 (molecular mass 22000) binds the enzyme to the membrane. The presence of extensive hydrophobic portions has also been detected in liver cytochrome P-450 [412].

The spectral and other characteristics of cytochrome P-450 are given in Table 6.1. The reduced form of the enzyme is easily bonded to carbon monoxide, giving two bands at 450 and 555 nm. Upon isolation of cytochrome P-450 from the membrane, the band at 450 nm changes to a band at 420 nm. The EPR spectrum of the oxidized native form is characterized by three values of the g-factor: $g_z = 2.4$, $g_y = 2.2$ and $g_x = 1.9$, which unequivocally indicates the low-spin state of iron in the complex. Various preparations have been found to contain various amounts of another form, the high-spin form, with g-factors of 1.7, 3.7 and 8.0 [413].

The equilibrium constant between the high-spin and low-spin forms of the oxidized enzyme is a rather sensitive parameter and is readily influenced by external factors: temperature, pH, binding of substrates and inhibitors, the conversion of the enzyme to an inactive form by blocking of the SH group, etc. [407]. The reduction of the enzyme converts it to the high-spin form. The quantitative data on the change of the spin state of the enzyme will be discussed in Chapter 10 in connection with the problem of electronic-conformational interactions.

According to the effect the heme group has on the spectral characteristics, all substrates are classified into two groups. The first group of substrates, hydrophobic compounds, is characterized by the induction of new bands in the differential spectrum, a maximum at $\lambda_{max} = 387$ nm and a minimum at $\lambda_{min} = 422$ nm and by the "high-spin" EPR spectrum. This group of substrates is capable of binding to carboxy P-450. The substrates of the second group (aniline, metapyrone) remove carbon monoxide from the coordination sphere of iron and are characterized by $\lambda_{max} = 430$ nm and $\lambda_{min} = 395$ nm in the differential spectrum and by the low-spin state of the heme.

It has recently been found that the fifth axial ligand in the heme of cytochrome P-450 is the SH group of cysteine. This is evidenced first of all by the similarity in the optical, MCD, EPR and Mössbauer spectra of oxycytochrome P-450 and model heme compounds containing sulphur ligands (e.g. butylmercaptan) in the fifth position [415, 416]. Direct evidence in favour of the presence of an Fe–S–C linkage in the enzyme has been obtained in an investigation of cytochrome P-450 enriched with the isotopes [^{57}Fe] and [^{34}S] and by resonance Raman spectroscopy [416]. In the EPR spectra of P-450 samples from *P. putida,* in which the iron atoms have been replaced by manganese or cobalt, the parameters of the hyperfine structure correspond to the interaction of iron with sulphur atoms. The spectral properties of the preparations differ from those of the corresponding Co- and Mn-substituted myoglobins, hemoglobins and cytochrome *c* peroxidases [417], which contain a histidine group as the fifth ligand. An EXAFS study has recently provided direct evidence for the presence of a sulphur atom as an iron ligand [420]. The theoretical aspects of the effect of this ligand on the physical and catalytic properties of the enzyme have been considered elsewhere [421–423].

The question as to the sixth ligand in the active site of the enzyme has not yet been solved completely. The spectral properties of low-spin forms of the enzyme and heme linked via thiol and hydroxyl ligands, e.g. methanol, are similar [418]. The modification of tyrosine in the active site of the enzyme leads to its deactivation [419]. According to ENDOR and NMR data the Fe ligands are H_2O or OH^- [426]. The change to the high-spin form is presumably accompanied by the release of the sixth ligand.

According to the data obtained by the electron spin-echo, NMR and spin-label methods, the substrate-binding site is in the immediate vicinity of the heme group [427, 428]. As has been shown [429], the thermodynamic regularities in the binding of linear hydrocarbons of various length are in good accord with the model of simple hydrophobic binding: for example, in octadecane the enthalpy (ΔH_0) and entropy (ΔS_0) of binding are $+60$ kJ mol^{-1} and 262 eu respectively. The values of ΔH_0 and ΔS_0 increase linearly with increasing chaing length.

The localization of the substrates acetanilid, xylidine and aminopyrine in the active site of the enzyme at a distance of about 0.3 nm from the heme group has been demonstrated by elegant experiments on the effect of substrates in solutions on the NMR spectra [427, 431]. The spin-labelled substrates of the first and second types have also been shown to be localized near the heme group [427].

In the case of acetanilid the protons of both the methyl and aromatic groups are located at a distance of 0.7–0.8 nm from the iron atom of microsomal cytochrome P-450 [431]. Probably, in order to achieve direct contact between the substrate protons and the reactive oxygen group on the iron atom, it is necessary that a conformational transition be realized.

The most detailed information on the structure of cytochrome P-450 from *P. putida* has been extracted by using X-ray analysis with a resolution of 0.26 nm (Fig. 6.8) [52]. According to the X-ray structural model, the heme group is in a hydrophobic environment and is buried in the protein globule to a depth of about 0.8 nm and coordinated at the fifth position to the cysteine residue. The substrate molecule (camphor) is located in the hydrophobic pocket in the distal region in such a manner that the C-5 group can be attacked by the "activated" oxygen attached to the iron atom. The precise orientation of the substrate is favoured by

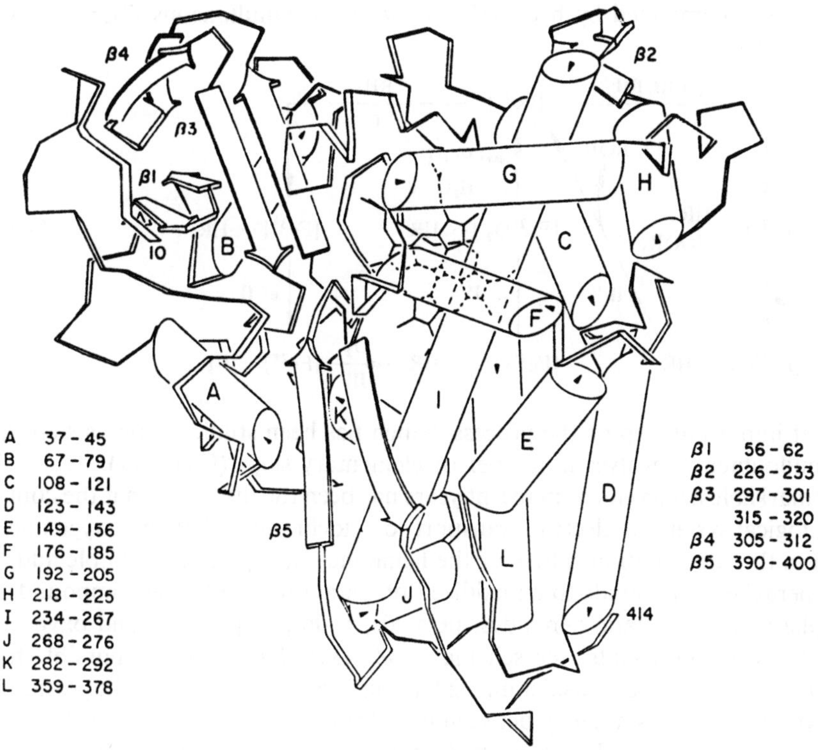

Fig. 6.8. X-ray model of cytochrome P-450 from *P. putida* [52]

the formation of a hydrogen bond between the hydroxyl group of tyrosine 86 and the carbonyl oxygen of camphor. In contrast to catalase and peroxidase, in the region of the heme group of cytochrome P-450 there are no histidine, arginine or any other polar groups. A significant similarity in the amino-acid sequence between the camphor cytochrome P-450 and an enzyme from microsomes and mitochondria in the proximal and distal regions has been reported [52]. The same authors also pointed out that there are differences in the amino-acid sequence between various preparations in the substrate-binding region.

The mechanism of substrate hydroxylation has been studied by many authors. The results of recent works have been analyzed in review arcticles [407–411, 420–432]. Here, we shall consider briefly the fundamental factual data. The structure and functions of the complete hydroxylation system and the possible elementary mechanisms of the process will be discussed in Chapters 9 and 11.

The first step of the process is the binding of the substrate with a transition of the heme electron system from the high- to the low-spin form. This is followed by the one-electron reduction of the enzyme (step II) and the binding of O_2. The first two steps proceed rapidly and take 10^{-2} to 10^{-3} s; the binding of a ligand to iron via an oxygen molecule is also accomplished rather effectively ($k_1 = 10^5$ to $10^6\ M^{-1}\ s^{-1}$ [407]). The next step (IV) is a relatively slow reduction process of the ternary complex P-450/RH/O_2 (for the enzyme from *P. putida*, $k = 4\ s^{-1}$).

The process is completed by hydroxylation with the simultaneous liberation of water (step V):

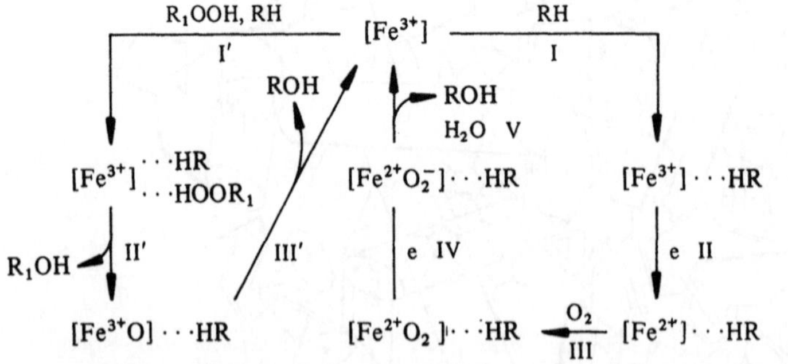

(6.4)

The most important step of the process, which has been studied little, is step V, which undoubtedly involves a number of elementary steps (Chap. 11).

A remarkable achievement in recent years has been the discovery that the complex biological system of electron transport to cytochrome P-450 and oxygen activation in the coordination sphere of the heme may be replaced by simple addition of peroxide compounds to an oxidized preparation of peroxide compounds, e.g. aliphatic and aromatic peroxides, peracids, hydrogen peroxide and oxygen-containing oxidizing agents (see scheme) [408, 433]. This makes it possible to simplify the system to a considerable extent and, what is especially important, points to a functional similarity between cytochrome P-450 and the well-studied peroxidases and catalases. It is presumed, in particular, that the above mentioned reagents oxidize the heme group of the enzyme to the $Fe^{3+}O$ state, which is analogous to compound I (scheme 6.1). The reaction of compound I with substrates leads to the appearance of hydroxylation products, e.g. according to the following scheme [408]:

$$Fe^{3+} + O_2 + 2H^+ \rightarrow Fe^{3+}O + H_2O \;;$$
$$Fe^{3+}O + HR \rightarrow Fe^{3+} + HOR \text{ ; or} \qquad (6.5)$$
$$Fe^{3+}O + HR \rightarrow Fe^{3+}OH + R^{\bullet} \;.$$

An alternative mechanism of O_2 activation and hydroxylation is the Hamilton mechanism [434], which includes the incorporation of an oxygen atom (oxenoid) into the complex across the C–H bond without preliminary cleavage of the O–O bond. Another mechanism, which has been suggested [429], involves radical "abstraction and addition" steps:

$$Fe^{2+}O_2^- RH \rightarrow Fe^{2+}O^- - OH\,R^{\bullet} \rightarrow Fe^{2+} + HOR \;. \qquad (6.6)$$

The occurrence of radical mechanisms is evidenced by the presence of considerable kinetic isotope effects, $k_H/k_D = 2-10$ [432, 435], and also by a change in the configuration of the asymmetric molecules as a result of oxygen incorporation. For example, norborane upon microsomal oxidation gives the value of $k_H/k_D = 11.5$ and epimerization [436]. However, the change in the configuration in camphor is not accompanied by an appreciable isotope effect [437]. (For explanations of this phenomenon, see Chaps. 10 and 11).

It should be noted that there is no complete correspondence between the functional properties of native and artificial systems. For example, in a number of cases there are differences in the composition of hydroxylation products [407]. When a system containing a substrate and cytochrome P-450 preparations is treated with iodosobenzene labelled with [^{18}O], the oxygen from water was incorporated into the substrate molecule (e.g. camphor or cyclohexane), which is accounted for by a rapid exchange between the oxygen atoms in the composition of iron oxide in the active site ($Fe^{3+}O$) and water [438]. However, in the complete enzyme system [^{18}O] from the labelled molecular oxygen is quantitatively incorporated into the substrate and no exchange is thus observed.

Ingenious experiments carried out recently [439] have established the participation of a carboxyl group in the activation of O_2 in the active site of cytochrome P-450 from *P. putida.* The hydroxylation of camphor by oxygen [^{18}O] was carried out in the presence of lipoic acid (R_2COOH). It turned out that as a result of one turnover of the enzyme, one [^{18}O] atom is incorporated into camphor and the other into the carboxyl fragment of lipoic acid. The authors suggested the following alternative scheme of O_2 activation:

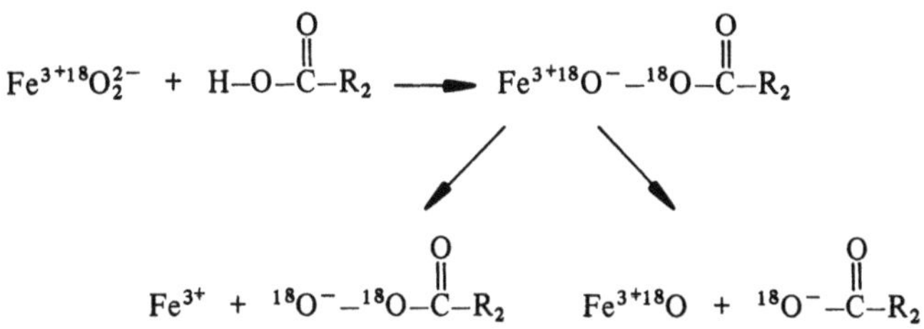

We shall return to the problem of the reduction mechanism of cytochrome P-450, O_2 activation and hydroxylation in Chapter 11. In conclusion, it should be noted that the hydroxylation process in biological and model systems may be accompanied by a number of side oxidation processes associated with the appearance of reactive particles, such as hydrogen peroxide and superoxide and hydroxyl radicals. The incorporation of such particles in the presence of impurities of transition metals and other components may substantially complicate the picture of chemical conversions in the active site of cytochrome P-450.

The similarities and differences in the functioning mechanism of a hemoprotein family may be illustrated by the following scheme. Enclosed in rectangles are the reactions specific only to a given protein that distinguish it from the other proteins of the family; alternative mechanisms are given only for cytochrome P-450 (see Scheme p. 148).

It may be presumed that the main component of the systems under consideration is a unit consisting of a heme group and a histidine, methionine, cysteine or tyrosine group in the fifth and sixth positions of the heme iron. The evolutional changes in the amino-acid composition of proteins could have contributed to the enhancement of some of the chemical functions and to the weakening of others by way of inclusion of the corresponding proximal or distal ligands or by way of the change of the extent of hydrophobicity and polarity of the surrounding amino acids. The replacement of side amino-acid residues has provided a wide range of protein surface properties that enable them, depending on their function, to be

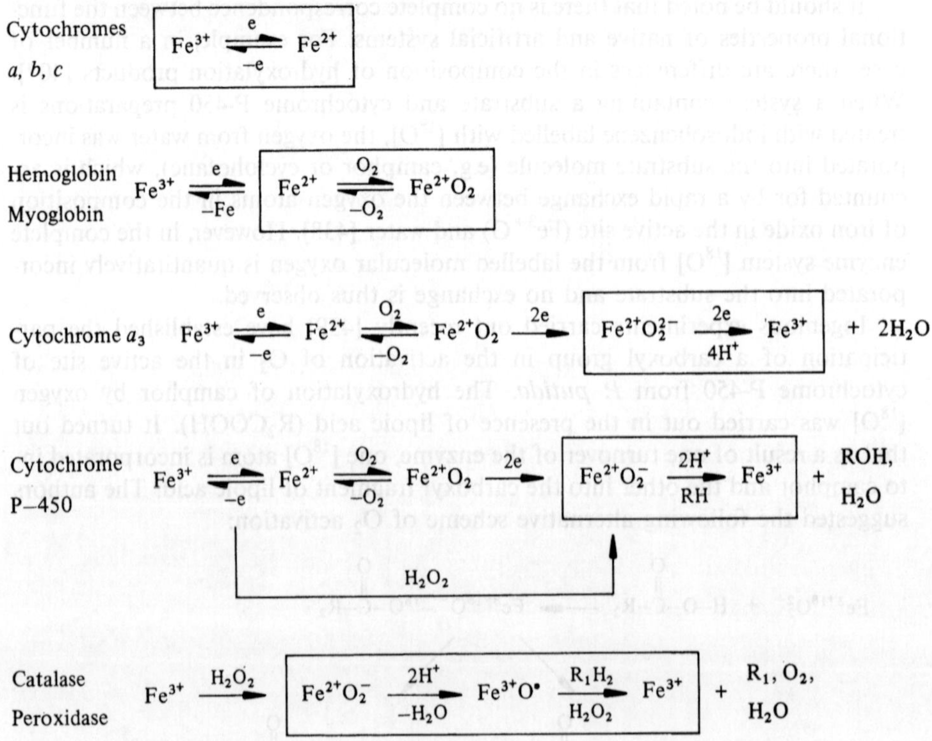

in the aqueous phase, to form stable or labile complexes with each other, to be bound tightly into the membrane or, in contrast, to move freely in it.

6.2 Non-Heme Proteins

6.2.1 Ferredoxins

Ferredoxins constitute an extensive class of proteins, i.e. electron carriers. In plant, animal and bacterial cells the reduced ferredoxins form a powerful electron beam, which is distributed depending on the concrete conditions for the occurrence of a number of reduction processes [28, 467–470]. As seen from Fig. 1.2, which shows only some of the pathways for electron transport, such processes are: the reduction of molecular nitrogen, biological oxidation and hydroxylation, formation of NADH and molecular hydrogen, fixation of CO_2, etc.

In spite of the large variety of iron-sulphur enzymes and carriers, which differ from one another by the molecular mass of elementary subunit (5000–30000), the amino-acid composition and the number of iron atoms per macromolecule (1–82), there are common features in their structure and functioning mechanism. In almost all cases, the iron atom is linked to four sulphur atoms, and in the elementary act only one electron is transferred. There are four types of elementary units in ferredoxin (Fd) (Fig. 6.9): mononuclear (rubredoxins), binuclear (Fe_2S_2

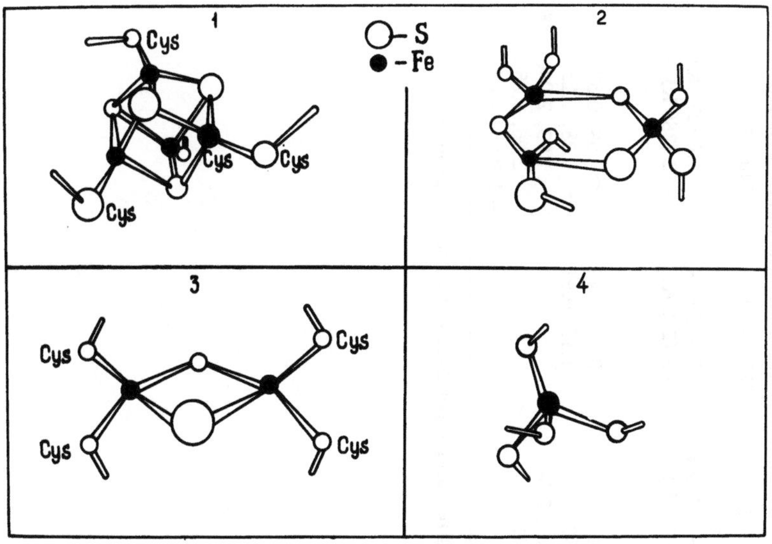

Fig. 6.9. Types of polynuclear iron-containing centres [28, 29]: *1* = 4Fe−4S; *2* = 3Fe−3S; *3* = 2Fe−2S; *4* = 1Fe

ferredoxins), trinuclear (Fe_3S_3 ferredoxins) and tetranuclear (Fe_4S_4 ferredoxins) units, which should be considered separately. The physico-chemical properties of ferredoxins are given in Table 6.2 and Fig. 6.10.

Rubredoxins. Two types of rubredoxins have been isolated. The first type isolated from *Clostridium pasteurianum* contains one iron atom per protein molecule having a molecular mass of 6000 [469−473]; the second type, which has been detected in microorganisms and which effects an electron transfer to the cytochrome P-450 system, which ω-hydroxylates alkanes, contains two iron atoms per macromolecule (molecular mass 19000). However, the functional role is played only by one of these atoms, the second being located at a remote chain portion [31].

The active site of mononuclear rubredoxin, as evidenced by X-ray structural and chemical data [28, 473], consists of an iron atom at the centre of a distorted tetrahedron, with sulphur atoms at the apexes (Fig. 6.8). It is noteworthy that in the oxidized state one of the Fe−S bonds is 0.04 nm shorter than the other three bonds (0.24 nm). As has been reported [471], such a structure is entatic, i.e. intermediate between the oxidized and the reduced forms observed in model systems. No change in the protein structure in the crystalline state has been detected upon its reduction. A set of physical properties of rubredoxin (Table 6.2) shows convincingly the presence of the high-spin of the iron atom with S = 5/2. According the the X-ray diffraction model, the active site is located practically at the surface of the enzyme and is surrounded on the protein side by hydrophobic amino-acid residues [473]. This circumstance and also the fact that the active-site structure undergoes no change upon reduction are evidently responsible for a rather high rate of electron transfer by rubredoxin. Binuclear rubredoxin from *Pseudomonas*

Table 6.2. The physical properties of ferredoxins[a] [28]

	Rd_{red}	Fd_{ox} (Fe$_2$S$_2$)	Fd_{red} (Fe$_2$S$_2$)	$HiPIP_{ox}$ Fd_{sox} (Fe$_4$S$_4$)	$HiPIP_{red}$ Fe_{ox} (Fe$_4$S$_4$)	$HiPIP_{sred}$ Fe_{red} (Fe$_4$S$_4$)
Absorption spectrum λ_{max}, nm	311 333	325 – 333 410 – 425	320, 400 470, 530	325, 375 450, 640	388 – 400	340 – 350
Magnetism μ_e (β)	5.05 (280 K)	1.2 (290 K)	2.8 (298 K)	1.9 (200 K)	0.9 (280 K)	1.6 (275 K)
NMR spectra ~CH$_2$S, ppm (280 K)	2.5	−37	−13 to −21	−23 (280 K)	−8 (270 K)	−12 (280 K)
NGR						
δ, mm s^{-1}	0.58 – 0.65	0.18 – 0.30	0.22 – 0.29	0.31	0.42, 0.43	0.59, 0.57
ΔE_Q, mm s^{-1}	3.1 – 3.2	0.60 – 0.66	0.60 – 0.80	0.80	1.13, 0.91	1.28, 1.25
EPR, g-factor	—	—	1.89, 1.94 2.04	2.12, 2.04 2.01	—	1.93, 2.04, 1.88 2.06, 1.88
Potential E_0, mV			−240 to −432	350		−280 to −490

[a] The subscripts sred, red, ox, and sox signify, respectively, the super-reduced, reduced, oxidized and super-oxidized forms of proteins.

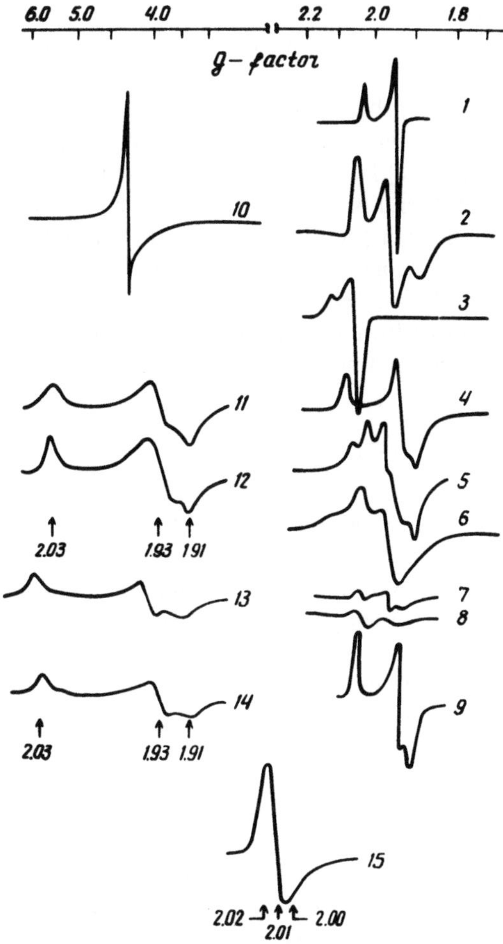

Fig. 6.10. The EPR spectra of iron- and sulphur-containing centres: *1* adrenodoxin; *2* spinach ferredoxin; *3* high-potential iron protein (HiPIP); *4* *B. stearothermophylus* ferredoxin; *5* *C. pasteurianum* ferredoxin; *6* *Chromatium* ferredoxin; *7* xanthine oxidase FeS(I); *8* xanthine oxidase FeS(II); *9* dehydrogenase succinate (reduction by dithionite); *10* rubredoxin; *11* mitochondrial complex II; *12* dehydrogenase succinate FeS(II); *13* dehydrogenase succinate FeS(II) (reduction by dithionite), pH 7; *14* dehydrogenase succinate FeS(II), pH 10; *15* dehydrogenase succinate, signal III

displays optical and EPR properties which are identical to the properties of mononuclear rubredoxin. As has been shown [473], the molecule of molecular mass 19000 can be split into two parts, one containing an iron atom and possessing all the functional properties of mononuclear rubredoxin.

Binuclear (Fe_2S_2) Ferredoxins. Plant cells, blue-green and green algae and a number of aerobic bacteria (e.g. *P. putida, Azotobacter*) and the adrenal cortex tissues have been found to contain non-heme iron-sulphur carriers containing two iron atoms and two labile sulphur atoms of (inorganic) sulphide per macromolecule [28, 467–477]. Since the most thoroughly studied preparation is one of spinach leaves, binuclear ferredoxins are called plant ferredoxins.

The molecular mass of binuclear ferredoxins usually lies in the range 10300–12600 and the number of amino-acid residues ranges from 97 to 114. An exception is *Azotobacter* ferredoxin, whose molecular mass is 21000. Two iron atoms of oxidized ferredoxin may accept only one electron. The thus obtained preparation has a potential of −240 to −400 mV (depending on the source [28]) and is thus a fairly strong reducing agent capable of reducing $NADP^+$ and NAD^+ and liberating hydrogen from water with the aid of the corresponding enzymes (reductase and hydrogenase).

The chemical properties of ferredoxin that should be noted are the ability of the labile sulphur atoms to undergo exchange with water, ^-OH and the halogens and the ability of [^{56}Fe] to exchange with [^{57}Fe] from solution [28]. The reagents for the sulphphydryl groups, such as PCMB and other mercurials, are capable of being bond by the cysteine groups of the protein and displacing the iron from the active site. The iron can be extracted from the active site by complexones of the type of o-phenanthrolines and dipyridyls. A number of interesting reactions have been described, which involve the transfer of iron and sulphides from the active site of ferredoxin to model iron-sulphur compounds [474]:

$$[Fe_2S_2(SR)_4]^{2-} + 4R_1H \rightarrow [Fe_2S_2(SR_1)_4]^{2-} + \text{Apoferredoxin} .$$

These reactions have been proposed as control reactions for the determination of the clustering nature of the enzymes.

Crystals suitable for X-ray diffraction analysis of this type of ferredoxin have been obtained [876]. It was on binuclear ferredoxins the the effectiveness of modern physico-chemical structural methods was demonstrated; the structure of the binuclear active site has been determined sufficiently reliably [28].

Investigations using the isotope [^{33}S], the replacement of sulphur by selenium, EPR and ENDOR measurements, Mössbauer parameters, magnetic susceptibility data, spin labels and X-ray photoelectron spectroscopy have all confirmed the binuclear structure of the active site of ferredoxin as predicted [476] (Fig. 6.8). Experiments devoted to the temperature dependence of the magnetic susceptibility and EPR spectra and resonance Raman scattering spectroscopy (the last two methods were used to measure the Orbach effect; see Sect. 2.7.3) have detected the presence of an anti-ferromagnetic bond between the iron atoms with $J = -180\ cm^{-1}$ for an oxidizing sample and $J = -90\ cm^{-1}$ for a reduced sample of spinach ferredoxin [479].

In the oxidized state both iron atoms are almost equivalent; however, in the reduced state these iron atoms differ in a number of properties, in particular, in the magnitude of the magnetic hyperfine splitting. The application of ENDOR and NGR methods has shown that iron forms a nearly covalent bond with sulphur and that in the reduced enzyme the unpaired electron is localized mainly at the d_{z^2} orbital [28]. The X-ray structural model of binuclear chloroplast ferredoxin has also been obtained [882].

Ferredoxins from different sources have generally a similar structure and their function is to effect one-electron transfer, but there are differences in the spectral and magnetic characteristics between preparations from different sources [28]. For example, adrenodoxin differs from spinach ferredoxin by the values of J and the symmetry of the ligand environment of iron, which is distinctly manifested in the presence of two values of the g-factor instead of three (Fig. 6.9). Recent investigations have shown that in a binuclear Rieske protein one of the Fe atoms is coordinated to the imidazole residues [478].

Tetranuclear (Fe_4S_4) Ferredoxins. There are known tetranuclear ferredoxins from the sulphate-reducing bacteria *Desulphovibrio* and nitrogen-fixing bacteria, a so-called high-potential iron protein (HiPIP) from *Chromatium* and group of ferredoxins from anaerobic and aerobic photosynthesizing bacteria, which contain eight or four iron atoms and eight or four sulphide sulphur atoms per macromolecule [462–470, 480–484].

In bacteria the ferredoxins perform a number of important functions, such as electron transfer in the phosphoroclastic reaction of pyruvate cleavage into acetyl phosphate and CO_2 to nitrogenase; the participation as a reducing agent in carboxylation during the synthesis of α-ketoglutarate and fumarate; the transfer of electrons to sulphate; electron transport in photosynthetic green and red bacteria, etc.

Some of the molecular and physico-chemical properties of tetranuclear ferredoxins are given in Tables 1.1 and 6.2. The mechanism of electron transfer with the participation of ferredoxins will be discussed in Chapter 11. The chemical properties of tetranuclear ferredoxins are similar to those of binuclear ferredoxins. The only significant difference, however, is that in tetranuclear ferredoxins there is only one cysteine group per iron atom in the active site of tetranuclear structures (Fig. 6.7). Many analogues have been synthesized, which contain four iron atoms, sulphide and sulphphydryl groups per molecule and which have a three-dimensional structure very similar to that of natural compounds. The general formula of these compounds is $[Fe_4S_4(SR)_4]^{2-}$, where $R = C_6H_5$ or $-CH_2-C_6H_5$ groups or cysteine residues of a twelve-membered peptide of the following structure: tert-butyl-gly-cis-cys-gly-gly-cys-gly-gly-cys-gly-gly-cys-gly. Exchange reactions with the analogues have been suggested as the extrusion method for the study of tetranuclear proteins [474, 480]. Analysis of the reaction products permits one to distinguish between tri- and tetranuclear ferredoxins. The analysis is made on the basis of the optical properties of the resulting iron-sulphur complexes and with the aid of the NMR method. In the latter case, use is made of fluorine-containing compounds. In oxidizing media, e.g. in air, the Fe_4S_4 clusters can be converted to Fe_2S_2 clusters, thus chemical methods have recently been proposed, which make it possible to avoid these undesirable reactions [480]. The physical approaches that enable the determination of the number of iron atoms in clusters include the study of the effect of an electric field on the EPR spectra, making it possible to analyze the symmetry of the complexes and also the shape of the EPR spectra of proteins in 80% dimethyl-sulphide [480], and resonance Raman spectra [485].

A remarkable achie vement has been the construction of X-ray diffraction models for iron-sulphur proteins: ferredoxin from *Peptococcus aerogenes* and HiPIP from *Chromatium* [481, 482]. The structure of ferredoxin is shown in Fig. 6.11. According to these data, the iron atoms are located at the corners of a somewhat distorted cube. The Fe–S distance in reduced samples of HiPIP is 0.222 and 0.232 nm and the Fe–Fe distance is 0.281 nm. In oxidized samples the bonds are shorter by about 0.007 nm. In oxidized ferredoxin the corresponding distances are close to those observed for reduced HiPIP.

As follows from the X-ray diffraction model, the active centre site of HiPIP, which is a distance of 0.045 nm from the surface, is surrounded by a shell of

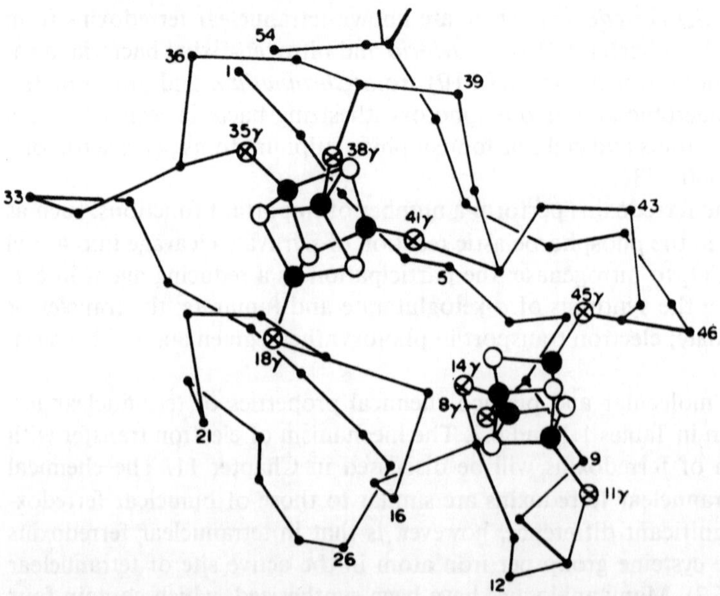

Fig. 6.11. The X-ray diffraction model of *C. pasteurianum* ferredoxin

hydrophobic residues, which include one tyrosine ring, three tryptophan rings and two phenyl rings. Both tetranuclear centres of ferredoxin are also surrounded by hydrophobic shells, which incorporate tyrosine groups, and are 1.2 nm apart from each other (the distance between the centres); they are buried to a certain extent in the protein matrix.

The surprising structural similarity of the Fe_4S_4 cluster in ferredoxin and HiPIP, with their E_0 values differing by about 700 mV, has been accounted for within the framework of the assumption of the presence of many redox forms in the tetranuclear complex [483]. According to this assumption, ferredoxin has four forms: reduced Fd_{red}, oxidizing (Fd_{ox}) and superoxidized (Fd_{sox}), the last two forms coinciding in the oxidation state with the super-reduced, reduced and oxidized forms of ($HiPIP_{sred}$, $HiPIP_{red}$ and $HiPIP_{ox}$) of the high-potential iron protein. The super-reduced form of ferredoxin (Fd_{sred}) has been obtained upon reduction of Fd_{red} by solvated electrons. The following distribution of iron atoms by their valency is presumed: 3 Fe^{3+}, 1 Fe^{2+} for Fd_{sox} and $HiPIP_{ox}$, 2 Fe^{3+}, 2 Fe^{2+} for Fd_{sox} and $HiPIP_{red}$, and 1 Fe^{3+}, 3 Fe^{2+} for Fd_{red} and $HiPIP_{sred}$ and 4 Fe^{2+} for Fd_{sred}.

A large series of physico-chemical studies on tetranuclear ferredoxins and their models, including EPR, NGR, ENDOR and X-ray photoelectron spectroscopy [28, 83], have shown that, just as in binuclear structures, all the four iron atoms are in the field of the covalent ligands of distorted tetragonal symmetry and have high-spin forms of iron in the oxidized and reduced states, which are coupled by an anti-ferromagnetic interaction. In the oxidized state there are two types of iron atoms, which differ in the parameters of the magnetic hyperfine structure; in the reduced state the extra electron is uniformly distributed over all four iron atoms.

An interesting phenomenon that arises in bacterial ferredoxins with eight iron atoms per molecule, as compared with tetranuclear ferredoxins, is a dipole-dipole and an exchange spin-spin interaction between two iron clusters [484]. This interaction manifests itself distinctly in a comparison of the EPR spectra of tetranuclear ferredoxin from *Bacillus stearothermophilus* and octanuclear ferredoxins (Fig. 6.9). The spectra of the first preparation are characterized by the line shape typical for non-interacting spins, whereas the spectra of the other ferredoxins are complicated by the interaction. Upon partial reduction, when states with one electron per eight iron atoms predominate, or upon addition of demethylsulphoxide, spectra of the first type are observed. Another important property of bicluster ferredoxins is a rapid exchange ($\tau < 10^{-5}$ s) of electrons between the clusters inside the molecule, which is indirectly recorded from NMR parameters [840].

Several types of membrane-bound iron-sulphur proteins (designated as the A and B centres) and the "Rieske" protein of molecular mass 24500 have been detected in photosynthetic systems [486]. For the first two types $E_0 = -500$ to -600 mV, the g-values are 1.86, 1.94, 2.05 and the EPR signal is relatively narrow ($\Delta H_{1/2} \approx 1.5$ mT). The third type, high-potential iron protein ($E_0 = 290$ mV) has the following values of the g-factor: 1.78, 1.89, 2.02 and $\Delta H_{1/2} \approx 4$ mT. Upon illumination reversible redox changes of bound ferredoxins are observed, which is evidence of the involvement of the active sites of these proteins in the electron-transport chain. The orientation of the g-tensors of spinach ferredoxins with respect to the membrane plane in chloroplasts has been established [487].

Trinuclear Ferredoxins. Data have lately been obtained which indicate the existence of a new type of iron-sulphur cluster, $[Fe_3S_3]$ [30, 31]. This type of cluster has been detected in ferredoxins from *A. vinelandii* (Fd_{Av}), *D. gigas* (Fd_{Dg}), in microbacteria and in bovine heart aconitase (A_{Ba}).

The ferredoxin FD_{Av} is a protein of molecular mass 14500 containing seven iron atoms and seven atoms of labile sulphur. X-ray studies at a resolution of 0.25 nm have revealed the presence of two clusters, $[Fe_4S_4]$ and $[Fe_3S_3]$, which are at a distance of about 1.0 nm from each other (Fig. 6.8). The $[Fe_4S_4]$ cluster has a potential of 320 mV and a set of physico-chemical properties which are similar to those of the cluster from HiPIP. The $[Fe_3S_3]$ cluster has E_0 -420 mV and gives an EPR signal with $g = 2.01$ in the oxidized state; the Mössbauer parameters are as follows: $\Delta_Q E = 0.63$ mm s^{-1} and $\delta = 0.27$ mm s^{-1}. Substantial differences between the spectral properties of the $[Fe_3S_3]$ cluster are observed and, in particular, the NGR and MCD spectra, and the corresponding properties of known ferredoxins. The spectral data indicate the high-spin state of the reduced forms of ferredoxin [470]. The ferredoxin Fd_{Dg} is made up of identical subunits with a molecular mass of around 6000. The tetrameric form of Fd_{Dg}(II) functions as an electron-transfer agent between cytochrome *c* and sulphite reductase. The three subunits each contain an $[Fe_4S_4]$ cluster and one subunit contains an $[Fe_3S_3]$ cluster.

Upon extrusion the new cluster produces products which are analogous to those obtained for Fe_2S_2 clusters [480]. The unexpected fact is that trinuclear and tetranuclear clusters are capable of interconversion. For instance, the oxidation of tetranuclear clusters of ferredoxin from *C. pasteurianum* by ferricyanide leads to the transformation $[Fe_4S_4] \rightarrow [Fe_3S_3]$. The $[Fe_3S_3]$ cluster of Fd_{Dg} and aconitase is converted to the tetranuclear form under the action of Fe^{2+} and dithiotritol. Experiments with iron isotopes have demonstrated that the newly introduced $[^{57}Fe]$ atom differs appreciably from the other three in the Mössbauer parameters [470].

6.2.2 Iron-Sulphur Enzymes

Hydrogenase (Hgn). This enzyme has been detected and isolated from bacterial sources, including anaerobic and aerobic microorganisms and photosynthetic bacteria [32, 34, 488, 489]. It was detected in 1974 (Chen and Mortenson) in *C. pasteurianum* and contained 4 Fe and 4 S_2^{2-} per molecular mass of 60000. The enzyme from *D. vulgaris* contains 12 iron atoms and 12 S^{2-} per molecular mass of 52000.

The enzyme catalyzes effectively the reversible reaction of H_2 abstraction under the action of reducing agents, such as ferredoxin and reduced methylviologen. The most active hydrogenase preparations mentioned above are characterized by the turnover number of $2 \times 10^5\ s^{-1}$ per $[Fe_4S_4]$ centre. Preparation containing four iron atoms were less active (by a factor of about 100).

Experiments on the extrusion of iron have demonstrated the presence of $[Fe_4S_4]$ clusters in native hydrogenase preparations. The absorption spectra of the enzyme consist of a broad maximum between 350 and 400 nm. The enzyme gives an EPR signal with g = 1.94 in the oxidized state $[4Fe-4S]^{1+}$ and with g = 2.01 in the $[4Fe4S]^{2+}$ state.

The redox potential of hydrogenase is −420 mV at pH 7. Carbon monoxide suppresses competitively the catalytic activity of the enzyme and simultaneously has an effect on the EPR signal of the preparation from *C. pasteurianum.*

It has been shown [490] that hydrogenase catalyzes the isotope exchange between D_2 and H_2O, which is an indication of hydride formation as a product of the interaction of the oxidized form of the enzyme with molecular hydrogen. On this basis, the authors presume a hydride mechanism of H_2 evolution in the reaction with water protons.

Data have recently been obtained on new species of hydrogenases. Anomalous properties of hydrogenases from *D. vulgaris* (Hildenbourgh) containing 12−14 Fe atoms per two subunits (46000 and 135000) have been reported [34]. The authors presumed the existence of Fe_6S_6 clusters in the enzyme. Graf and Thauer were the first to detect nickel in hydrogenase preparations from *Methanobacterium thermoautotrophicum* [491]. An Ni-containing hydrogenase was isolated from other species of bacteria. A preparation from *D. salexigens* (M_r = 98000) contained selenium in addition to Ni [69]. The hydrogenase from *D. gigas* contained 1 nickel atom, 11 iron atoms and 11−12 sulphide per two subunits of molecular weight 63000 and 26000 daltons [70]. The iron atoms were found to be distributed among three clusters, 2 $[Fe_4S_4]$ and $[Fe_3S_x]$. The latter cluster in the oxidized state gave an isotropic EPR signal with g = 2.02 (at T < 30 K). The Ni-containing enzyme complex is characterized by three signals: a signal A with the g-factors

2.31, 2.23 and 2.02 (at 120 K); a signal B with g = 2.33, 2.16 and 2.0; and a signal C with g = 2.19, 2.16 and 2.02. The change in the redox states of the Ni atom and the clusters during the reaction with H_2 may be repesented by the following scheme:

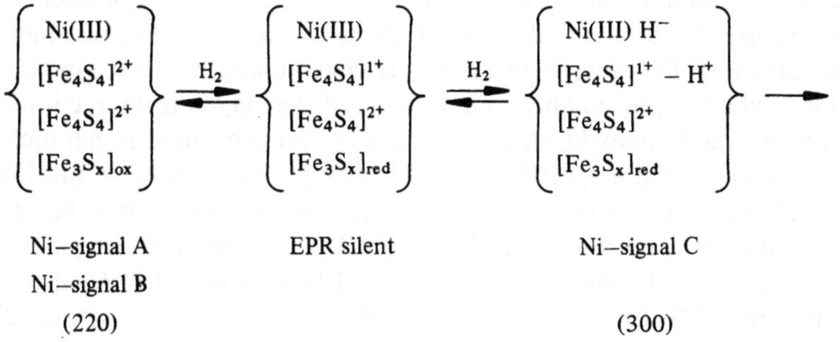

Enclosed in parentheses are the values of E_0^0 (in mV) for the corresponding states of Ni.

According to EXAFS data [492], the most probable ligands of the nickel atom are S (four atoms) and N or O (two atoms), in which case R_{Ni-S} = 0.220 − 0.228 nm and $R_{N-(N, O)}$ = 0.192 − 0.193 nm. An estimate of the Ni-[Fe_4S_4] cluster distance made on the basis of the effect of the cluster on the rate of spin relaxation of Ni led to the value 1.2 nm [493].

According to their functional properties, two types of hydrogenases are distinguished; a dihydrogen-evolving hydrogenase and a dihydrogen-uptaking hydrogenase, the former containing only Fe and the second Fe and Ni.

Taking into account the number of oxidation states in the tetranuclear complex non-Ni-containing hydrogenase, it is possible to suggest a scheme of the hydrogenase reaction involving one-electron and two-electron reducing agents, cytochrome c_3 and NADH respectively. The scheme includes the formation of four clusters forms of hydrogenase (Fe_4S_4): the oxidized form (Hgn_{ox}), the reduced form (Hgn_{red}), the super-reduced form (Hgn_{sred}) and the hydride form ($Hgn_{sred} - H$):

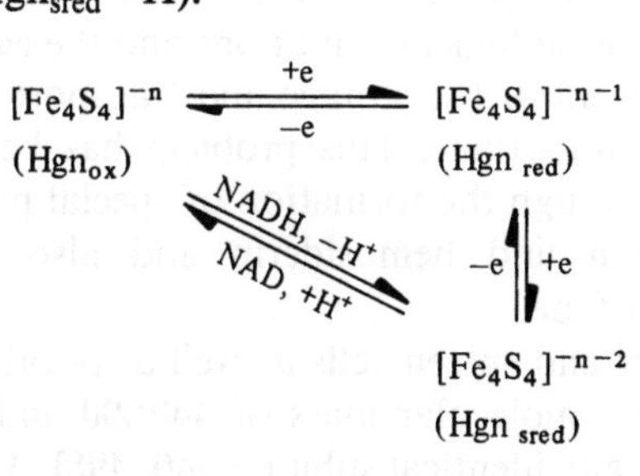

The high catalytic activity of three-centre hydrogenases and also some kinetic data provide evidence for the possible interaction of Fe_4S_4 clusters during the functioning of hydrogenases.

Other Enzymes. Apart from hydrogenase, several enzymes are also known containing non-heme FeS clusters in the active site (Table 1.1). Adenylyl sulphate reductase contains Fe_nS_n clusters and FAD [504]. The bovine heart aconitase contains three iron atoms per molecular mass of 83000 in the form of an Fe_3S_3 cluster [495]. Fumarate reductase from the microorganisms *Vibrio succinogenes* has one Fe_4S_4 cluster and one FAD molecule per subunit of molecular mass 79000, the other subunit containing an Fe_2S_2 cluster [496]. Experiments at helium temperatures have revealted the differences in the saturation parameters of the EPR spectra of the clusters in the whole enzyme and in the composition of the subunits, which is presumably an indication of the magnetic interaction between the centres. The ferredoxin-nitrite reductase of spinach contains, per molecular mass of about 61000, a specific prosthetic group called siroheme [an iron-tetrahydroporphyrin of the isobacteriochlorin type having eight carboxylate side chains as substituents) and an Fe_4S_4 cluster [497]. The non-heme enzymes involved in the oxidative phosphorylation chain will be considered in Chapter 9.

Among the special features of enzymes containing Fe_3S_3 clusters, we may distinguish the following: the ability to be converted to Fe_4S_4 clusters; the possibility of the transition from a triangular to a linear configuration with a change of the spin state (S) from 1/2 to 5/2; the ferromagnetic interaction between a pair of atoms, each of which in turn interacts anti-ferromagnetically with the third iron atom [470]; the ability of iron atoms to add water and substrates and thus to participate in a chemical reaction as a Lewis acid [494]. The last-named property has been revealed by the ENDOR method with the use of the isotope [^{17}O].

A special position among iron-containing enzymes is occupied by sulphite reductase. The active site of the enzyme contains both an Fe_4S_4 cluster and siroheme, which have a common sulphur ligand. There is a strong anti-ferromagnetic linkage between the centres [470, 499].

6.2.3 Transport and Storage of Non-Heme Model Proteins

The necessity of the presence of a considerable amount of iron for the building of the active sites of hemoproteins and non-heme proteins, the insolubility of inorganic compounds of trivalent iron under physiological conditions and the ease of oxidation of divalent iron make the problem of the storage and transport of trivalent iron in biological systems vitally important. This problem has been solved in nature in a very remarkable way through the formation of special proteins for the storage of inorganic ferritin and hemosiderin and also of iron(III)-binding proteins of the transferrin type.

Ferritin, which has been detected in liver and spleen cells as well as in other tissues and blood serum, is a protein with a molecular mass of 440000 and a diameter of 12 nm and consists of 24 chemically identical subunits [40, 498]. The protein is literally stuffed with ferrihydroxyphosphate, the number of molecules of which amount to 4500 per macromolecule. Inorganic iron fills up the protein core, having a diameter of about 7.5 nm, and may be present in it in the form of a single crystal and in the polycrystalline state. Iron ions are delivered to the protein through six channels between the subunits. The X-ray diffraction model of the protein has been obtained for the protein at a resolution of 0.28 nm [500]. The cylindrical subunits have a size of 5.5 and 2.7 nm, the channels with an outer diameter of 1.0 nm become narrover down to 0.7 nm and extent to 1.6 nm in the region of the core.

The iron-transport protein transferrin from blood serum has a molecular mass of 77000 and contains two centres which bind trivalent iron(III) very strongly ($K \sim 10^{-30}$ M) [498]. A number of data indicate that the iron atoms are separated by a considerable distance. The transport of iron is also accomplished by uteroferrin which contains two closely spaced iron atoms, presumably linked by an oxygen bridge [502].

The plasma of some purple bacteria contains in oxygen-transport protein, hemerythrin [42, 501]. The molecular mass of this protein ranges from 66000 to 11000 in various preparations, but in all cases each subunit of molecular mass 11500 to 13500 contains two iron atoms, which bind one oxygen molecule. In the oxidized state both iron atoms are trivalent and are anti-ferromagnetically linked, which is convincingly evidenced by the EPR and NGR spectra and optical measurements. In oxyhemerythrin the same methods and also the data obtained by resonance Raman spectroscopy suggest the structure $Fe^{3+}\,O_2^-\,Fe^{2+}$, in which the oxygen has a partial peroxide nature and the spins of the iron atoms are linked by an anti-ferromagnetic interaction [42]. According to the X-ray diffraction model of the protein, the two iron atoms are linked to each other via an oxygen atom and two carboxyl bridges. One of the bridges is cleaved upon binding of O_2 to one of the iron atoms. The other ligands for iron are histidine groups (Fig. 6.12).

The first model iron-sulphur protein was synthesized on the basis of serum albumin. The procedure used includes the reduction of the $-S-S$ bonds in albumin in the presence of urea and followed by the reconstitution of the protein macromolecule by addition of $FeCl_3$ and Na_2S in an aerobic medium. The resulting preparation contained eight iron atoms and sulphide per macromolecule; the molecule has a nearly spherical shape [503] and some of its physical properties are close to those of natural ferredoxins.

The multi-nuclear structure of the model protein has been proved by five independent methods [503]: (1) by the absence of a magnetic HFS in the NGR spectra at $T = 4.2$ K; (2) by the presence of a strong exchange interaction between the spin labels that replace the iron atoms; (3) by the sharply pronounced cooperativity in the study of the kinetics of binding of nitroxide radicals by the sulphhydryl groups of the protein, which form complexes with iron; (4) by the value of the coefficient in the HILL equation (3.7) in which it has been found that at least three SH groups are in the cooperatively interacting centre; (5) by the dependence of the degree of the inductive-resonance quenching, by the active site, of the luminescent labels distributed statistically over the macromolecule on the number of labels in the molecule.

Let us summarize the main results of studies on iron-containing proteins. Iron-containing proteins play the most important biological role as oxidases, electron and oxygen carriers, and iron-transport and -storage proteins. In spite of the large variety of structures and functions, these proteins are made up of a relatively small number of elementary units. In hemoproteins the heme group is linked in the fifth position to nitrogen, sulphur or oxygen atoms. In non-heme proteins

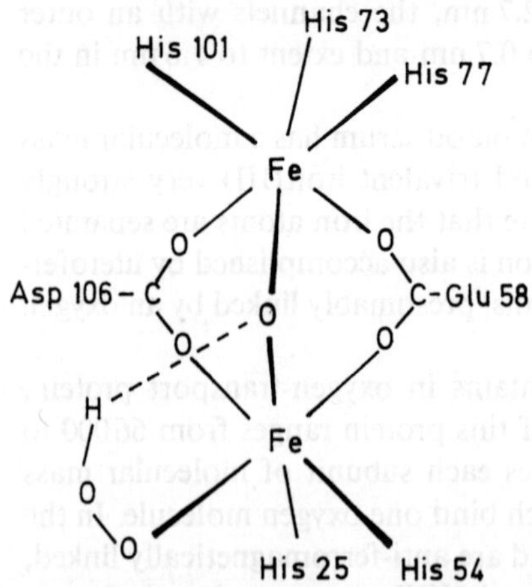

Fig. 6.12. The supposed model of the binuclear iron-containing centre of oxyhemerythrin [42]

there are mono-, bi-, tri- and tetranuclear structures, in which the iron atoms are present in the distorted tetrahedral environment of sulphur atoms and are antiferromagnetically linked.

Each unit may accept or donate only one electron in one step. In polynuclear complexes there are many oxidation states. Apart from the "ordinary" oxidized and reduced forms, under certain conditions super-oxidized and super-reduced forms can be obtained. This property, coupled with the possible impact of the local environment of the protein matrix, is a powerful means of controlling the redox potential and kinetic properties of ferredoxins. The replacement of amino-acid residues of the protein moiety during evolution has given rise to a large variety of protein structures required for the occurrence of biochemical reactions.

Further investigations of model proteins may be associated with the study of their catalytic properties. The presence of the protein matrix makes convenient the location, near an iron-sulphur cluster, of other metal-containing reaction centres, which will enable the modelling of polynuclear assemblies of the third type (Chap. 1). The procedure used to obtain model proteins may prove useful for producing clusters of other transition metals, e.g. molybdenum and manganese.

The similarities and differences in the elementary mechanisms of chemical reactions involving iron-sulphur proteins will be discussed in Chapter 11.

CHAPTER 7

Iron- and Molybdenum-Containing Oxidoreductases

Eight molybdenum-containing enzymes have been studied so far. Two of these enzymes are associated with nitrogen metabolism. Nitrogenase reduces molecular nitrogen to ammonia; nitrate reductase converts nitrate to nitrites. The other enzymes, xanthine oxidase, aldehyde oxidase, xanthine dehydrogenase, formate dehydrogenase and sulphate reductase, catalyze the corresponding redox processes. The active-site structures and the mechanism of action of these enzymes have many common features, this being an indication of a close evolutional and functional relationship between the molybdenum-containing enzymes.

In this chapter we shall be concerned with the basic principles of the structure and mechanism of action of this interesting class of enzymes. Special emphasis will be placed on the most thoroughly studied enzymes, the central enzyme of biological N_2 fixation, nitrogenase and xanthine oxidase.

7.1 Xanthine Oxidase

7.1.1 General

The most widespread source of xanthine oxidase is cow's milk. A rather large amount of xanthine oxidase is present in animal liver. Appreciable amounts of xanthine oxidase have been detected in the liver and kidneys of birds, in insects and bacteria [60, 504–511, 516]. The main function of the enzyme is to oxidize hypoxanthine to xanthine and then xanthine to uric acid. Apart from this "classical" function, a number of authors ascribe the role of a generator of superoxide radicals O_2^- to xanthine oxidase [508]. The xanthine oxidase reactions may be described by the following general equation:

$$D-H+A+H_2O \rightarrow D-OH+AH_2 \ . \tag{7.1}$$

The role of the donor DH may be played by purines, pyrimidines, pteridines or aldehydes, and the acceptors may be O_2, tetrazolium salts, methylene blue and the derivatives of indophenol. For example, for the main biological substrate, xanthine, and oxygen as an acceptor, the following reaction occurs:

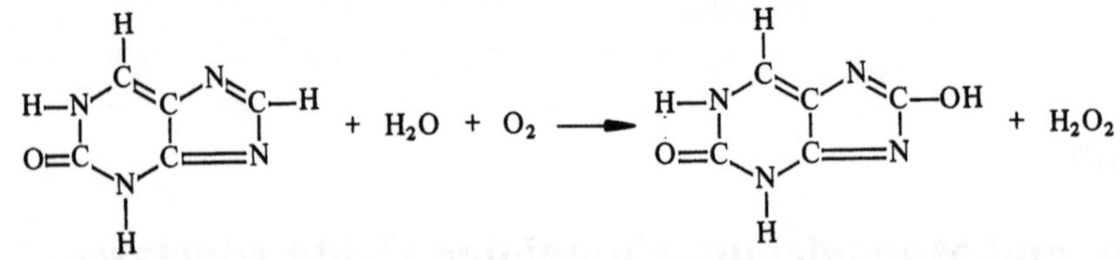

Aldehydes are oxidized in an analogous way; an example is acetaldehyde. It has long been shown that the oxygen incorporated into the aromatic ring comes from water [511]. The role of molecular oxygen is to oxidize the enzyme reduced by the substrate. A molecule of milk xanthine oxidase contains two Mo atoms, eight atoms of non-heme iron, eight atoms of inorganic sulphur and two molecules of flavin adenine dinucleotide (FAD) per molecular mass of around 300000. The enzyme consists of two identical chains folded into three blocks: C, E and F [512]. These blocks are packed into a single structure due mainly to the interaction between the side chains. The enzyme fragment E–C, obtained by partial protolysis of the enzyme, contains Mo and Fe centres and has a xanthine-dehydrogenase activity, i.e. the ability to catalyze the oxidation of xanthine in the presence of artificial acceptors but not O_2. The flavin component is presumably localized on E or F fragments. It has been suggested [513] that such a subunit structure was formed as a result of the gene combination and that each subunit carries a certain prosthetic group.

In the oxidized state molybdenum has a valency of +6; upon reduction by substrates a change to the +5 state is observed: the EPR parameters vary somewhat, depending on the conditions (Fig. 7.1). From a set of the EPR parameters (g-factor = 1.973–1.977, weak hyperfine splitting) it follows that molybdenum is in a rhombic environment [504, 514]. Studies on the xanthine oxidase of rat liver have shown that molybdenum in the enzyme may be replaced in vivo by

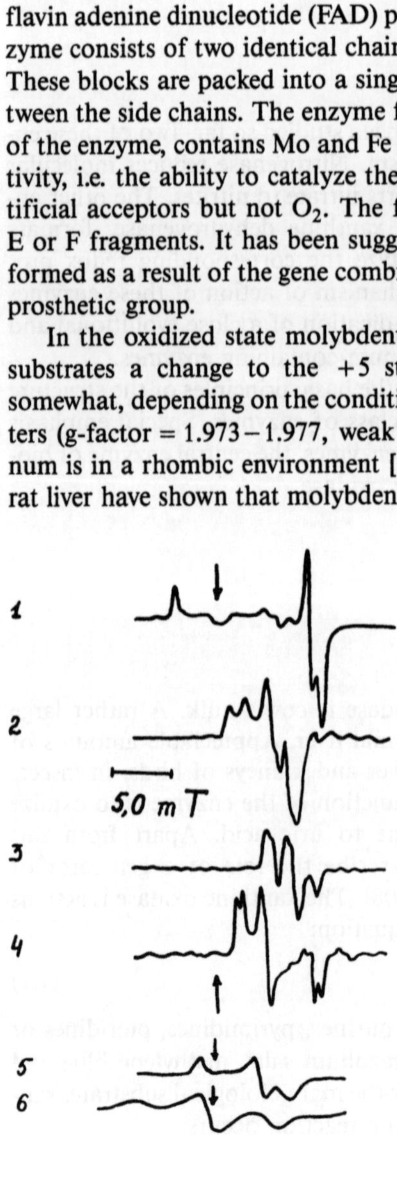

Fig. 7.1. The EPR spectra of the paramagnetic centres of xanthine oxidase [505]: *1–4* signals from Mo^{5+} (*1* "very fast", *2* "fast", *3* "slow", *4* "inhibited"); *5* und *6* centres of FeS(I) and FeS(II). The *arrows* show the position of the diphenylpicrylhydrazyl lines

tungsten, the properties of the FAD and Fe system remaining unchanged. The modified enzyme losses the ability to oxidize xanthine, but it can oxidize NADH in the presence of ferricyanide.

It was discovered in 1970 [515] that a number of molybdenum enzymes, including xanthine oxidase from cow's milk, interstine xanthine oxidase, aldehyde oxidase from rabbit liver, chicken xanthine dehydrogenase, sulphite oxidase from bovine liver and nitrate reductase from *Neurospora,* restore the activity of nitrate reductase from the nit-1 mutant after being subjected to acid treatment. It is presumed [515, 517] that the acid treatment of each of the enzymes mentioned above produces a Mo-containing cofactor MoCo, which may serve as the missing block for the mutant reductase.

The Mo fragment from xanthine oxidase gives an EPR Mo^{5+} signal and does not contain Fe. There are data indicating that MoCo includes a molybdenum fragment of low molecular mass, and possibly, a high molecular mass peptide [517]. The MoCo fraction has been reported to contain a fluorescing organic component, presumed to belong to the pteridine class [517, 518]. The ligands of Mo are two sulphur atoms of pteridine and two sulphur atoms of the protein [510].

EXAFS studies have recently provided data pointing to a ligand environment of molybdenum in xanthine oxidase [519]. In the native enzyme the Mo atom is surrounded by four sulphur atoms with S–Mo distances being equal to 0.225, 0.25 and 0.29 nm and by one oxygen atom with a Mo–O distance of 0.175 nm. One of the sulphur atoms is very labile and is easily replaceable by oxygen. The presence of oxygen in the coordination sphere of molybdenum has been detected by EPR from the magnitude of the hyperfine splitting in the Mo^{5+} EPR spectrum after the incorporation of the isotope [^{17}O] [520]. One of the ligand sulphur atoms has an enhanced chemical activity in the reaction with cyanide.

The optical and EPR characteristics of xanthine oxidase leave no doubt that the iron in this enzyme belongs to the class of non-heme iron-sulphur complexes. This is evidenced by the absorption spectrum of a flavin-free form of the enzyme with maxima at 315, 420 and 467 nm, the circular dichroism spectrum and the EPR spectra in the reduced state with a g-factor of 1.95–2.11 [58, 60, 505, 524]. According to the set of parameters, the cluster in xanthine oxidase may be classified as belonging to the [Fe_2S_2] type (Chap. 6). The kinetic and EPR data indicate the there are two types of centres, FeS(I) with $g = 1.95$ and FeS(II) with $g = 2.11$, which differ also in the temperature dependence of the signal intensity and relaxation time.

A FAD group can be reversibly removed by the action of salts, e.g. potassium iodide or potassium chloride [505]. The enzyme obtained in this way can oxidize xanthine in the presence of artificial acceptors but not O_2. Besides, the enzyme loses the ability to oxidize NADH.

The EPR method has provided extensive information on the structure of metal-containing portions and the mechanism of electron transfer in the system of active sites of milk xanthine oxidase and on the interaction between the centres. It has been established that there are several types of EPR spectra attributed to Mo^{5+}, depending on the substrate, the reducing agent, reaction time, pH and the conditions under which the spectrum is recorded [60, 504, 505]. The corresponding EPR spectra are given in Fig. 7.1. Four types of signals have been detected:

"very rapid" (I), "rapid" (II), "slow" (III) and "inhibited" (IV). In the following we shall designate these signals by the corresponding Roman numerals.

Experiments involving substrates containing D instead of H as the atom being split off have demonstrated the presence of a hyperfine interaction between the electron spin localized on Mo^{5+} and a proton transferred from the substrate. This interaction manifests itself in an isotropic splitting of the EPR lines. The proton in the region of Mo is capable of exchanging with the protons of the medium. The interaction with protons is also manifested in the spectra of electron-electron double resonance (ELDOR) spectra [521].

The following scheme has been proposed to illustrate the origin of various signals from Mo^{5+} [60]:

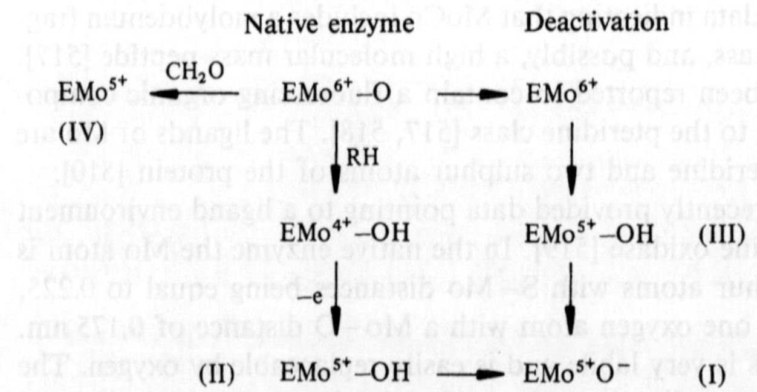

7.1.2 Kinetics and Mechanism of the Xanthine Oxidase Reaction

The xanthine oxidase reaction is described from a formal kinetic standpoint by a scheme involving two substrates: a reducing agent (S) and an oxidizing agent (O_2). The general scheme may be divided into three main parts [511]: (1) the substrate oxidation centre; (2) the electron-transport chain; (3) the centre of O_2 reduction to H_2O_2 or $HO_2^{\bullet}$.

The rates of electron transport between the centres have been found to be so fast ($k = 200\ s^{-1}$) and the equilibrium sets in so rapidly that the concentration of each of the forms is determined by the equilibrium thermodynamic potential of the system of centres. The rate of electron transfer to O_2 is also fast ($k = 200\ s^{-1}$). With saturating concentrations of the substrates S and O_2, the rate of the process is entirely determined by the oxidation of S ($k = 24\ s^{-1}$), i.e. presumably by the cleavage of the C–H bond.

The joint use of potentiometric titration in the presence of dithionite and dye mediators and the quantitative estimation of the reduced components by EPR spectroscopy has provided the values of the half-reduction potentials for the centres of xanthine oxidase (in mV) [522]: $Fe_{(I)}$ (–343), $Fe_{(II)}$ (–303), $FADH/FADH_2$ (–236), Mo^{6+}/Mo^{5+} (–355) for the centre characterized by the "rapid" EPR signal and Mo^{6+}/Mo^{5+} (–440) and Mo^{5+}/Mo^{4+} (–480) for the centre, whose behaviour is described by the "slow" signal.

Kinetic and thermodynamic data are in good agreement with the following electron-transfer sequence:

$$DH \rightarrow Mo \rightarrow Fe_2S_2(I) \rightarrow Fe_2S_2(II) \rightarrow Fl \rightarrow O_2 \ .$$

There are a number of serious arguments in support of the view that the oxidation of the substrate by enzymes involves a two-electron hydride transfer of hydrogen to the molybdenum complex: (1) molybdenum, which is the first of the redox centres, accepts electrons from the substrate; (2) an isotope effect is observed with respect to K_M; (3) the hydrogen is found to be in the immediate vicinity ($\approx$0.3 nm) of Mo, this being evidenced by additional splitting of the EPR line from Mo^{5+}; (4) the free-radical state of the substrate cannot be recorded; (5) the enzyme is reduced by typical hydride reagents, such as sodium borohydride; (6) the radical mechanism is energetically unfavourable (Chap. 11).

The following mechanism of hydride oxidation of xanthine on the enzyme seems to be the most probable (only the fragment entering into the reaction is shown):

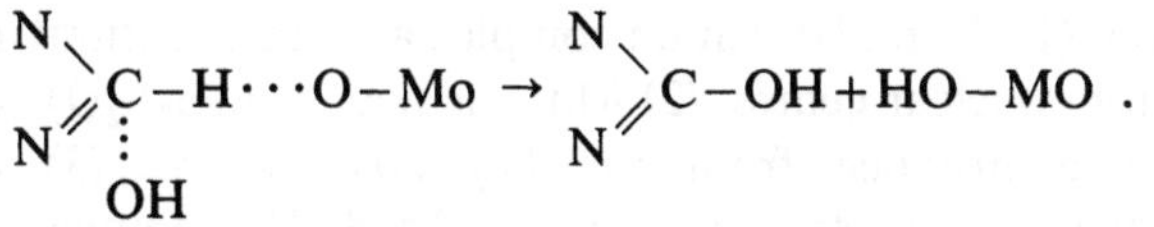

The hydride is transferred to the oxygen atom, this being accompanied by the attack of the nucleophile (the hydroxyl ion or the inner protein group) on the nitrogen atom of the substrate. It has been presumed [511] that the persulphide group may act as the nucleophile. Recent works, have, however, shown that the active sulphur detectable by the reaction with cyanide is one of the Mo ligands [519, 520]. The reducing end of the system can operate either by a one-electron or two-electron mechanism. In the first case, the reaction product is the superoxide radical and in the second, hydrogen peroxide. The results of a number of investigations [508, 511] indicate that the first pathway is only a small fraction of the second if the rapid stage of electron transfer from the reduced enzyme to O_2^- takes place.

To account for these regularities, a scheme has been advanced, which assumes the retention of the superoxide O_2^- in the coordination sphere of FADH in order that the second electron has a chance to reduce it to H_2O_2. An alternative explanation, which seems to be the more preferable, is the presence of the binuclear $FADH_2$, centre, in which a more effective two-electron reduction of O_2 may take place than in the mononuclear centre, with rapid disproportion of two O_2 molecules in the cell. It is obvious that under physiological conditions the system must avoid the involvement of the reagent O_2^-, which is "dangerous" to the cell. The release of O_2^- into the medium is probably a "flaw" in the action of the enzyme. The binuclear centre at the terminal portion is an excellent means of reducing the "flaw". The close location of FAD is evidenced by the complete inhibition of the enzyme upon blocking of one flavin molecule by acetamide per enzyme macromolecule [523].

A series of remarkable radiospectroscopic and biochemical studies [506, 521, 524–526], including the analysis of the effect of the state of certain centres of xanthine oxidase on the spectral properties of other centres, have clearly demonstrated the occurrence of a spin-spin interaction between the centres. With ac-

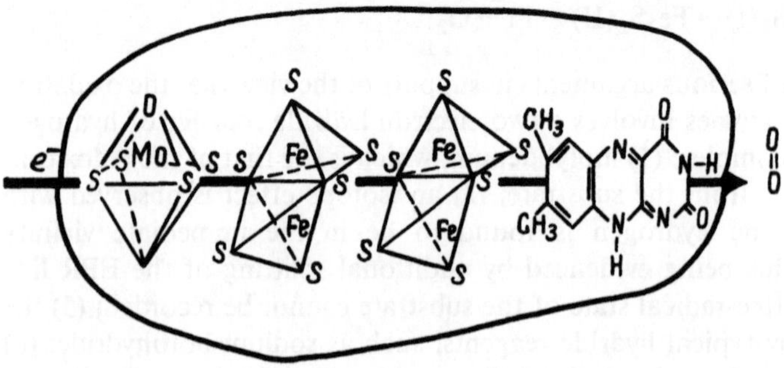

Fig. 7.2. The model of the active site of xanthine oxidase. One of two identical halves of the site is shown

count taken of the results of these works, it is possible to summarize the basic data on the active site of XO: (1) the molybdenum complex acts, in structural and functional respects, as a mononuclear centre; (2) Mo^{5+} is near the Fe_2S_2(I) centre (~1.5 nm) and is at a distance from the Fe_2S_2(II) centre; (3) the paramagnetic centre (FAD) and not Mo^{5+} is near the Fe_2S_2(II) complex; (4) upon binding of one of the two flavin groups the activity of the enzyme molecule is completely switched off; (5) the enzyme oxidizes the substrate by the hydride two-electron mechanism; (6) the enzyme can reduce oxygen by a one- or two-electron mechanism, the second path being preferred under native conditions. All these facts are explained by the model of two converging electron-transport chains, one of which is shown in Fig. 7.2.

The mechanism of such a reaction is described by the scheme [2→1→1→1] for the reaction leading to the superoxide and by the scheme:

$$\begin{bmatrix} 2\to1\to1 \\ 2\to1\to1 \end{bmatrix} 2$$

for the generation of hydrogen peroxide.

7.2 Other Molybdenum-Containing Oxidoreductases

Aldehyde Oxidase. This enzyme catalyzes the reaction

$$CH_3C{\overset{O}{\underset{H}{\lessgtr}}} + O_2 + H_2O \rightarrow CH_3C{\overset{O}{\underset{OH}{\lessgtr}}} + H_2O_2 \ .$$

Aldehyde oxidase occurs in animal, plant and microbial tissues as well as in insects [58, 60, 528]. The molecular mass of the enzyme isolated from hen liver and the fruit fly is about 300000. The enzyme macromolecule is made up of two identical subunits. The prosthetic groups are the same as in xanthine oxidase (2Mo, $2Fe_2S_2$ and 2FAD). The substrates may be various purines, which are oxidized to 8-hydroxypurines, aldehydes and N-methylnicotinamide. Cyanide and methanol are the inhibitors of the enzyme.

The kinetic behaviour of the EPR signal from Mo^{5+} is generally similar to that observed for the xanthine oxidase signal [58]. Treatment with methanol gives rise to a signal similar to the Mo^{5+} (III) signal with a larger splitting on the proton than for the corresponding signal for XO. There is every ground for believing that the macromolecular structure, the active-site structure and the mechanism of catalyzed reactions in aldehyde oxidase are similar to those for xanthine oxidase.

Xanthine Dehydrogenase. This enzyme, which catalyzes the transfer of hydrogen from xanthine to NAD^+ and other acceptors, has been detected in bacteria, animal livers and insects [60, 507, 529].

The substrate (donor) utilized by the enzyme is xanthine and its derivatives, which are oxidized by it to 8-hydroxyl derivatives. The molecular mass of the enzyme isolated from *Drosophila melanogosa* and hen liver is 280000. The enzyme does not contain small subunits and, like xanthine oxidase, is most probably made up of two, large, indistinguishable subunits and contains two FAD groups, two Mo atoms and two iron atoms [528].

Xanthine dehydrogenase is inhibited by cyanide, arsenite and methanol and, by analogy with xanthine oxidase, it is likely to contain an active sulphide group. However, different acceptors respond differently to the blocking of this group [529]. The first group of acceptors is not sensitive to blocking, whereas the second is. The molybdenum atom participates in the oxidation of xanthine, while the flavin transfers electrons to NAD^+. Indeed, the removal of FAD prevents the oxidation of NADH but not of xanthine or methyl nicotinamide in the presence of other acceptors. Methylene blue and ferricyanide are presumed by the authors to remove electrons from the iron-containing cluster. The rate of the process with $k = 10\,s^{-1}$ is evidently limited by the detachment of the hydride from the substrate. The optical and EPR parameters of the molybdenum portion of the enzyme and their variation under different conditions is very similar to those observed for xanthine oxidase.

It has been reported [505, 514] that xanthine dehydrogenase from *Veilloniella alcalescens* has one type of the FeS signal of the axial type. The signal from Mo^{5+} recorded at helium temperature undergoes changes analogous to those observed in XO. This is apparently an indication that the Mo atom is close to FeS in the enzyme. The two enzymes have also similar genetic properties and a similar electron-transport chain structure:

$$DH_2 \rightarrow [Mo \rightarrow Fe_nS_n \rightarrow FAD] \rightarrow NAD^+ \ .$$

The reaction is described by the scheme [2→1(2)→2]: the molybdenum oxidizes the substrate by the hydride mechanism, two electrons are transferred separately via Fe_nS_n to FAD, whose reduced form transfers the hydride to NAD. The presence of one chain is probably the main difference between xanthine dehydrogenase and xanthine oxidase, in which the chain portions with FAD provide jointly a two-electron reduction of O_2.

Nitrate Reductase. The assimilation of nitrogen by way of the reduction of nitrate starts with the nitrate reductase reaction [58, 507, 517, 527]:

$$DH_2 + HNO_3 \rightarrow D + HNO_2 \; .$$

This reaction occurs in bacteria, moulds, higher plants and algae. The donor is NADH in algae and plants and NADPH in moulds. In bacteria and blue-green algae the ferredoxins act as the donors.

The prosthetic groups of the enzyme are FAD groups, cytochrome b_{557} and Mo. The electron transfer from NADH and other donors to nitrate is effected in the following sequence:

$$NADH \rightarrow [FAD \rightarrow Cyt\ b_{557} \rightarrow Mo] \rightarrow NO_3^- \; .$$

Functionally, the enzyme is made up of two portions, which can act independently: (1) the so-called diaphorase portion, which catalyzes the oxidation of NADH or methyl viologen by acceptors, e.g. ferricyanide; and (2) terminal nitrate reductase, which transfers an electron from the external donor to the nitrate. System I is blocked by reagents for SH groups and system II is competitively inhibited by metal-chelating compounds, such as azide, phosphate carbamyl and cyanate; cyanate inhibits the reduced state of the enzyme in a non-competitive manner.

Tungsten can replace Mo in portion II upon biosynthesis; the diaphorase function remains practically unchanged in this process when the ability to reduce nitrate is lost. The Mo atom is incorporated into the enzyme via the molybdenum cofactor, which is analogous to the xanthine dehydrogenase and sulphite oxidase factors. The cofactor is localized preferentially on the protein subunit with a molecular mass of about 3000 [517].

Nitrate reductase is an enzyme complex, which has a molecular mass from 160000 to 600000, depending on the source [58, 530–532]. The isolation of the enzyme from *E. coli* has been reported [58]; its monomer has 12 atoms of non-heme iron, 12 labile sulphur atoms, 24 cysteine residues and 1 Mo atom. The monomer consists of a large subunit (150000) and a small subunit (55000) in the ratio of 1 : 1.

The kinetics of the nitrate reductase reaction is described by the Michaelis equation with the following values of K_M: 0.19 μM for NO_3^-, 4 μM for methyl viologen, and 11 μM for NADH [58]. Upon oxidation of reduced pyridine nucleotides a stereospecificity was observed with respect to hydrogen when nitrate reductases from different sources were used. The enzyme is competitively inhibited by derivatives of NAD with the substituent in the pyridine ring [533].

The energy relief of the nitrate reductase reaction is characterized by the thermodynamic ease of occurrence of the overall reaction, which is energetically favourable by about 0.9 eV, and also of the intermediate steps. The reaction involving a two-electron reducing agent may be described by the scheme:

$$[2 \rightarrow 1 \rightarrow 1 \rightarrow 2] \; .$$

Sulphite Oxidase. This enzyme contains molybdenum and a heme group of the b_5 type per molecule (molecular mass 115000) [58, 60, 505, 534]. A number of factors indicate that the structure and some of the functional properties of this enzyme are similar to those of xanthine oxidase and other molybdenum-containing enzymes.

For example, the Mo^{5+} EPR signal appears upon reduction of sulphite and disappears upon further action of dithionite. The magnitude of the HFS splitting is close to that of the splitting in XO. Addition of ferricyanide eliminates the signal. Obviously, the molybdenum in the enzyme can assume the states $Mo^{6+} \leftrightarrow Mo^{5+} \leftrightarrow Mo^{4+}$. The kinetic behaviour of the signal from Mo^{5+} is to a certain extent analogous to the bahaviour of the "rapid" and "slow" signals in xanthine oxidase. The enzyme is found to contain a rapidly exchangeable proton, which gives a splitting of 1.0–1.2 mT in the EPR spectrum and, possibly, an active sulphide group, since the cyanide and arsenite inhibit the enzyme irreversibly. It is presumed that sulphite interacts with the oxygen attached to molybdenum; this oxygen abstracts the hydride from the substrate. The dioxygen evidently interacts with the heme group.

A new molybdenum-containing protein containing a non-heme iron cluster of the Fe_2S_2 type has been described in the literature [527].

The problems of the biological activity of molybdenum compounds and the physiological role of molybdenum-containing enzymes, their structures and genetic specificity have been described elsewhere [58, 60, 517].

7.3 Nitrogenase

7.3.1 General

The central enzyme of biological fixation of molecular nitrogen catalyzes three coupled reactions [24, 56, 535–537]: (1) the reduction of N_2 to NH_3 under the action of natural electron donors (ferredoxins, flavodoxin) or artificial electron donors (dithionite, reduced methyl viologen) in the presence of ATP; (2) the hydrolysis of ATP to ADP and inorganic phosphate in the presence of electron donors; (3) the ATP-dependent evolution of H_2 under the action of hydrogen donors. The active form of the enzyme nitrogenase is formed upon combined action of two components: an iron-containing protein (FeP) and an iron-molybdenum protein (FeMoP).

Since the publication of the pioneer works of Bulen and Mortenson and coworkers [17, 18], who reported the isolation of the first partially purified preparations of FeP and FeMoP, the efforts of many biochemists have been concentrated on the preparation of the individual components of nitrogenase and on the study of their structure and mechanism of action. The achievements in this field have been reviewed in a number of papers [24, 56, 308, 535–540].

These achievements were paralleled by the developments in the field of chemical fixation of N_2 to ammonia, hydrazine and organic nitrogen-containing compounds. The currently available chemical models of nitrogenase [305, 538–543] make it possible to effect the reduction of atmospheric nitrogen to hydrazine and to ammonia in protic media under mild conditions.

The components of nitrogenase have been isolated and purified from a number of microorganisms, both anaerobic and aerobic: *Azotobacter vinelandii, Clostridium pasteurianum, Klebsiella pneumonia,* etc. The components of bacterial nitrogenases from the root nodules of legumes have been isolated and

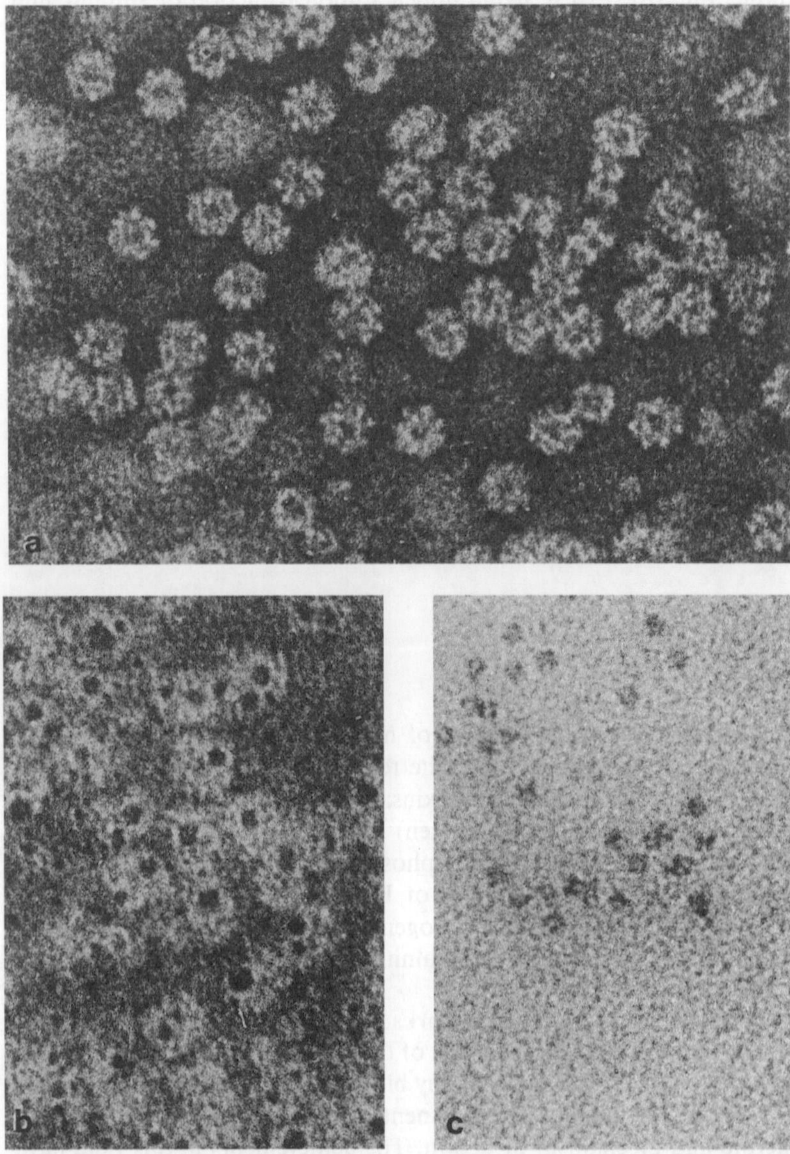

Fig. 7.3 a – e. Electron micrographs of nitrogenase [135, 136, 139]. Magnification ×700000. **a** Native nitrogenase. Negative contrasting by 2% phosphotungstic acid (PTA). **b** The nitrogenase macromolecule modified by 100 equivalents of p-chloromercuribenzoate (PCMB), contrasting by 1% Na_2MoO_4. **c** Same as **b**, without contrasting. **d** Nitrogenase macromolecules modified successively by two equivalents of an electron-scattering mercarbidolabel and 100 equivalents of PCMB, contrasting by 1% Na_2MoO_4. **e** Same as **d**, without contrasting (preparation without separation into components)

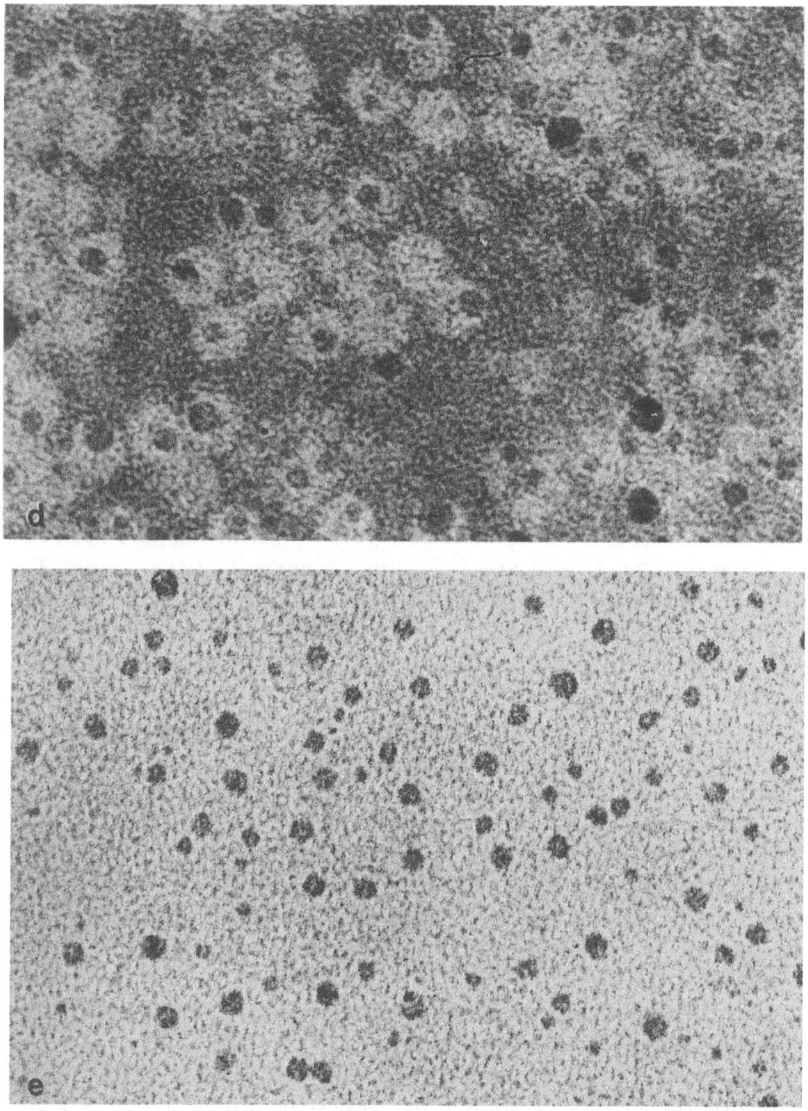

Fig. 7.3 d, e

their properties described [544]. Depending on the source and methods of measurement, the molecular masses, the subunit composition and the content of metal atoms cited by different authors differ somewhat.

The molecular mass of FeP is 55000–65000. This component is made up of two subunits and contains four atoms of non-heme iron, four atoms of sulphide sulphur and 4–18 cysteine groups. The FeMoP component has a molecular mass of 200000–230000 and consists of four subunits of two species with similar molecular masses. There are 30–32 atoms of non-heme iron, 24–26 S^- and 1–2 molybdenum atoms per four subunits. The presence of two molybdenum atoms per four subunits is typical of thoroughly purified preparations with a minimally distorted structure. Methods have been worked out for the isolation of the nitrogenase complex [545, 546] and of FeP and FeMoP [17, 18, 56, 537, 545]. It has been presumed [547] on the basis of kinetic data that FeP is eliminated from the complex during the reaction. The native complex contains a third component, a non-heme protein, Shetna II.

The redox potential of FeP from *C. pasteurianum* is $E_{0(pH\ 7.5)} = -250$ to -290 mV and after Mg ATP is added it is equal to -400 to -490 mV; for the FeMoP preparation from *K. pneumonia* the redox potential is -60 and -260 mV. For the FeMoP preparation from *A. vinelandii* oxidized by thionine two values of E_0 (pH 8) have been observed: -300 and -460 mV, to each of which six electrons correspond [548, 549]. The Bulen complex is characterized by values of E_0 equal to -300 mV (four electrons) and -465 mV (12 electrons) [549].

In nitrogenase preparations isolated from various sources in the presence of dithionite EPR spectra are observed with the following values of the g-factor: 4.30–4.32, 3.65–3.70, 2–2.015 [56, 550]. This combination of g-factors is not characteristic of the non-heme iron, therefore special experiments have been carried out with the isotopes [^{95}Mo] and [^{57}Fe] for assignment of the signals. The incorporation of the isotope [^{57}Fe] leads to a broadening of the component with $g = 2.01$ by about 0.6 mT and the introduction of [^{95}Mo] has practically no effect on the EPR spectra of the preparations from *C. pasteurianum.* An analogous broadening has been observed upon incorporation of [^{57}Fe] for *A. vinelandii* FeMoP. The replacement of iron atoms in FeP and FeMoP from *A. vinelandii* by spin nitroxide radicals, which are the derivatives of PCMB (parachlormercury benzoate), by means of a specific mercuration reaction, has led to spin-labelled preparations, whose EPR spectra are similar at room temperature and a temperature of 77 K to the spectrum of spin-labelled pea ferredoxin and indicate a strong exchange interaction [308, 551]. All these factors indicate that FeP and FeMoP belong to the class of non-heme iron-sulphur proteins. The Fe protein contains one or two clusters of the Fe_4S_4 type [550, 552]. The FeMo protein contains clusters of a new type [550].

The enzyme nitrogenase has been studied by electron microscopy and X-ray analysis [553, 554] (Fig. 7.3).

The quaternary structure of nitrogenase proteins has been studied by means of electron microscope methods [553] and by X-ray structural analysis [554]. According to data [553], the micrographs of an iron-molybdenum-containing protein display a series of particles in the form of parallelograms, rectangles, triangles and crosses. The analysis of these projections and also of planar crystals of FeMo proteins allowed the authors to propose a model of the quaternary structure of the protein, in which two α-subunits are displaced with respect to two β-subunits along the second-order axis in such a manner that the protein molecule is on the whole a distorted tetrahedron. Such a model agrees well with X-ray data [554] and with neutron small-angle scattering data [555]. The electron micrographs of the preparations containing a mixture of FeMo and Fe proteins in the ratio 1 : 1 have revealed two additional extreme of electron density in the central portion, which are ascribed by the authors to molecules of an iron-containing protein.

7.3.2 Kinetics and Mechanism of Nitrogenase Reactions

The results of the first kinetic studies of the nitrogenase reactions have been summarized in the form of the Hardy-Night scheme [556]. According to this scheme, the electrons from the donors are transferred to the electron-activating centre, on which the coupling of electron transfer takes place with the cleavage of the high-energy phosphate bond of ATP. The "activated" electrons are then transferred to the N_2-fixing centre, where the activation and reduction of N_2 occurs. The activity of the enzyme depends on the reaction conditions, the ratio of the components, pH, temperature, etc. The activity under optimal conditions is about 2 mol N_2 mol^{-1} enzyme s^{-1}.

The Arrhenius graphs for the reduction of N_2, evolution of H_2 and the cleavage of ATP show a kink at 20 °C. The values of the activation energy above and below the turning point are 60 and 190 kJ/mol^{-1} respectively. The calculations of the entropy change at the same portions of the curve are 40 and 320 eu [557]. The value of $K_{M(N_2)}$ depends only slightly on temperature [535, 557].

Apart from N_2, nitrogenase reduces substrates that contain triple bonds (HCN, N_2O, HN_3, etc., and also cyclopropene) and are also inhibitors of nitrogen reduction [56, 556, 559]. The kinetics of nitrogenase reactions and inhibition is rather complicated and depends on the ratio of the components of the nitrogenase complex, the concentrations of the substrates being reduced and ATP, temperature, pH and the reaction time. The size of the substrates and inhibitors must not exceed 0.5 – 0.6 nm. The relations between the substrates and inhibitors of nitrogenase are illustrated in Fig. 7.4.

In nitrogenase reactions are accompanied by an isotope exchange between the hydrogen atoms of water and molecular hydrogen [558 – 562]. The exchange is small in the absence of N_2 but it is strongly accelerated in an atmosphere of N_2. Other substrates do not favour the formation of HD from D_2 and H_2O. It has

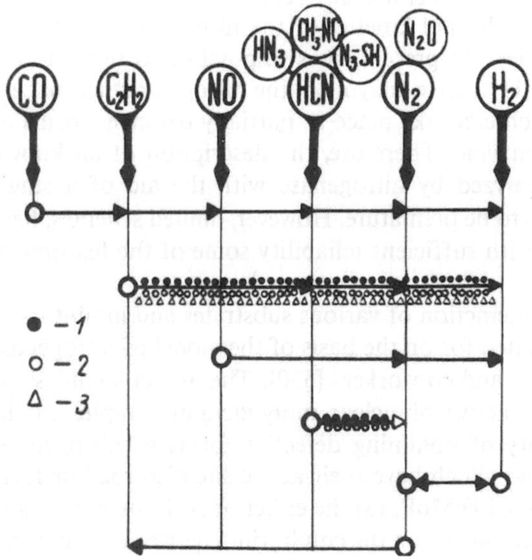

'ig. 7.4. Mutual inhibition of arious substrates of nitrogenase eactions [24]. The *beginning* of he *arrow* corresponds to the inibitor and the *end* of it to the reacion substrate. *1* Reactions that de›end on the ATP concentrations; *2* eactions that depend on the FeP/ ЛoFeP ratio; *3* reactions that de›end on the presence of H_2

been shown convincingly [561] that the formation of one HD molecule requires the consumption of one electron. The reaction of HD formation, which is dependent on N_2, is irreversible. Experiments with radioactive T_2 have shown that tritium is not incorporated into water during the reaction.

Apart from a few special cases, the total number of reducing electrons that pass through the active sites of the enzyme does not depend on the substrate used. Thus, the reduction of the substrate is not the rate-limiting step of the process. However, the distribution of electrons between the substrates depends appreciably on the turnover number, which is determined by the ratio of the components and the other conditions indicated above. The general tendency is that the greater the total turnover number, the larger the fraction of electrons transferred for the reduction of N_2 [563]. It has been experimentally found that H_2, N_2H_4 and NH_3 are produced during the lag period (100, 200 and 400 ms respectively) after the start of ATP hydrolysis [537, 564]. In the limit the ratio of the products is $2NH_3 : H_2$ [560]. It has been found experimentally that the formation of H_2 and NH_3 is preceded by a lag period of 100 and 400 ms respectively, after the onset of ATP hydrolysis (a preparation of *K. pneumonia,* 230°, pH 7.4). The rapid cessation of the nitrogenase reaction in an acidic and an alkaline medium made it possible to separate hydrozine (N_2H_4); the lag period of its formation is about 200 ms [562]. The lag periods for the isolation of HD and reduction of C_2H_2 are equal to 320 and 100 ms [568] respectively. The works just cited are of fundamental importance since the authors demonstrate unambigously the necessity of a step-by-step accumulation of electrons in active centre for N_2 to be reduced to hydrazine and then to ammonia.

Among the other kinetic features of nitrogenase reactions the following should be distinguished [542, 561, 568]. The reactions of NH_3 and HD evolution ar identically inhibited by carbon monoxide; the ratio NH_3/HD increases with increasing ratio of the Fe fractions to FeMO proteins, in which case D_2 becomes a weaker inhibitor of the N_2 reduction; the values of the Michaelis constants $K_{M(N_2)}$ for the formation of HD and NH_3 are different.

The interpretation of the complicated kinetics of the nitrogenase reaction is associated with certain difficulties. At present, it is impossible to separate the reactions that occur in a completely native enzyme in the nitrogen reduction centre from the side processes, which can take place at partially deficient forms or at other cluster centres of the enzyme. Therefore, the description of all known chemical processes that are catalyzed by nitrogenase with the aid of a single generalized kinetic scheme seems to be premature. However, limited schemes, have been proposed, which describe with sufficient reliability some of the features of nitrogenase reactions.

The specific features of the interaction of various substrates and inhibitors of nitrogenase may be partly accounted for on the basis of the model of nitrogenase action proposed by Likhtenshtein and co-workers [530]. The model includes: (1) the formation of a catalytically active binuclear molybdenum complex in the native enzyme; (2) the possibility of obtaining defective forms of nitrogenase, which have lost one Mo atom and which have their active site disturbed or form complexes with an excess amount of FeMoP; (3) the effect of ATP, substrates and concentrations of the protein components on the equilibrium between these forms and on the state of the binuclear centre inside the complex. Thus, the forms with one molybdenum atom per macromolecule are capable of reducing H^+ or other two-electron substrates, including hydrazine, and effecting a slow deuterium exchange. The whole form can also reduce molecular nitrogen by a four-electron mechanism and catalyze a rapid deuterium exchange in the presence of N_2. The ability to reduce N_2 is lost when the structure of the binuclear centre is locally disturbed.

In an analysis of the above indicated kinetic features of the inhibition of nitrogenase reactions, it is reasonable to take into account the possibility of the existence of several oxidation states of the site for the binding and reduction of the substrates. If, for example, it is presumed that the elementary acts of binding and reduction of molecular nitrogen requires the four-electron (E_1) state of the enzyme and that the inhibitors and other substrates can bind to the oxidized or the half-reduced forms (E_0), then the process of interest may be described by the following scheme [24]:

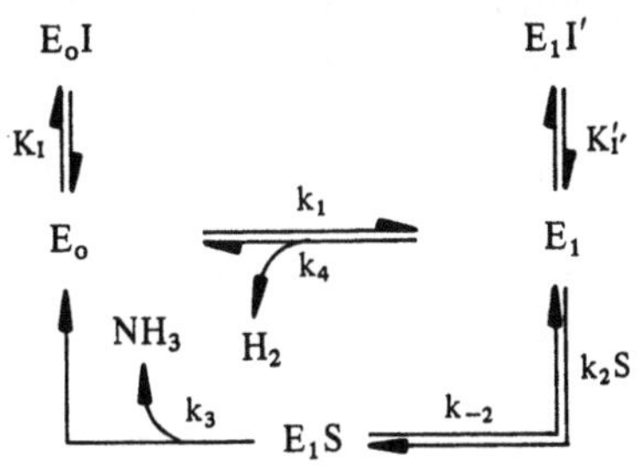

where E_0I and E_1I' are nitrogenase complexes with different inhibitors or substrates, the reduction of which requires one or two electrons; E_1S is a nitrogenase complex with N_2 or substrates, which can be reduced only by four electrons.

Analysis of the scheme given above allows one to account for a number of kinetic features of nitrogenase reactions: (1) provided that $[I]/K_I \gg [I']/K'_I$, the inhibitor exerts a retarding effect on the maximum rate of the process and is thus formally a non-competitive inhibitor. In this case, the inhibitor may be reasonably termed the electron "interceptor" or "interrupter"; (2) at $[I']/K'_I \gg [I]/K_I$, an expression is obtained characteristic of an ordinary competitive inhibitor; (3) in the absence of an inhibitor, if the rate constant for the evolution of H_2 (k_4) is greater than k_1, but lower than k_3, the following expression is obtained:

$$V = k_1 [E_0] [S]/(K_{app(N_2)} + [S]) ,$$

where $K_{app(N_2)} = K_{M(N_2)} k_4/k_3$. Since $k_3 > k_4$, it follows that $K_{app(N_2)} < K_{M(N_2)}$, i.e. the Michaelis constant measured experimentally for N_2 is lower or is a true Michaelis constant. The physical significance of this relation is that in the system considered, as the substrate concentration increases, the maximum rates are attained as a result of the complete capture of electrons being transferred in the rate-limiting step and not as a result of the saturation of the active sites by the substrate molecules.

An experimental testing has been made [546] for the binding of a non-competitive inhibitor of the reduction of N_2, azidodithiocarbonate (ADTC), by nitrogenase from *A. vinelandii* to the oxidized form of the enzyme by the photoaffinity labelling technique. The results obtained indicate the binding of the inhibitor to the oxidized form of the enzyme, which must lead to a formally non-competitive inhibition in spite of the addition of the substrate and inhibitor at the same spatial centre.

Jensen and Burris [560] have suggested the following scheme for discribing the relationship between the reduction of N_2 and the exchange with the isolated HD:

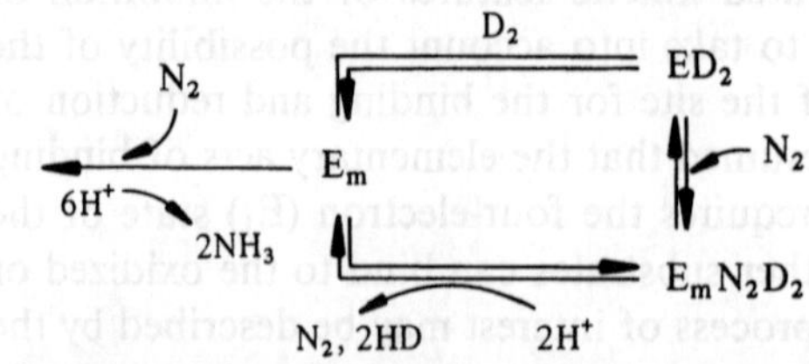

According to this scheme, the state E is responsible for the reduction of N_2, while the state $N_2E_mD_2$ is responsible for the evolution of HD.

The problem of the electron-transfer sequence within the active-site system of nitrogenase is closely related to the problem of the functional importance of its individual components.

Syrtsova and co-workers [551] used the inhibitor method based on the specific action of reagents on the various functional portions of the active site of the enzyme. In particular, FeP and FeMoP were treated separately with a CO-specific inhibitor of the nitrogen-binding site or with the PCMB inhibitor of the ATPase site. Then, a complementary component was added to the inhibited preparation. The experiments have shown that CO suppresses the reduction of N_2 only if the FeMoP component was preliminarily incubated. PCMB was used as an inhibitor if the FeP component of nitrogenase was pretreated. The authors presume that the nitrogen-binding site of the complex-forming CO resides at the FeMoP component and the SH groups of the ATPase site at the FeP component. A series of remarkable works [549, 550, 564, 565] carried out by means of EPR and Mössbauer spectroscopy have shown that the hydrolysis of ATP is accompanied by an electron transfer from the Fe component to the FeMo component, which has been reduced by dithionite. This leads to the formation of the super-reduced form of the active site. The variation of the spectra parameters of the iron-containing enzyme clusters is shown in Fig. 7.5, which was plotted on the basis of the data obtained in the works cited. It is essential that substrate reduction involves only the super-reduced forms of the enzyme. For one electron to be transferred, two ATP molecules are required. The addition of ADP inhibits the process.

According to the available kinetic data [56, 549], the characteristic time of electron transfer from FeP to FeMoP in the presence of ATP is close to 10 ms under conditions of preliminary incubation of the components. These values are much lower than the total reaction time of electron transfer from the reducing agent to the substrate (about 100 ms per electron). From this it follows that the transfer of one electron from FeP to FeMoP does not limit the process as a whole. On the other hand, we have already cited the kinetic data indicating that the direct

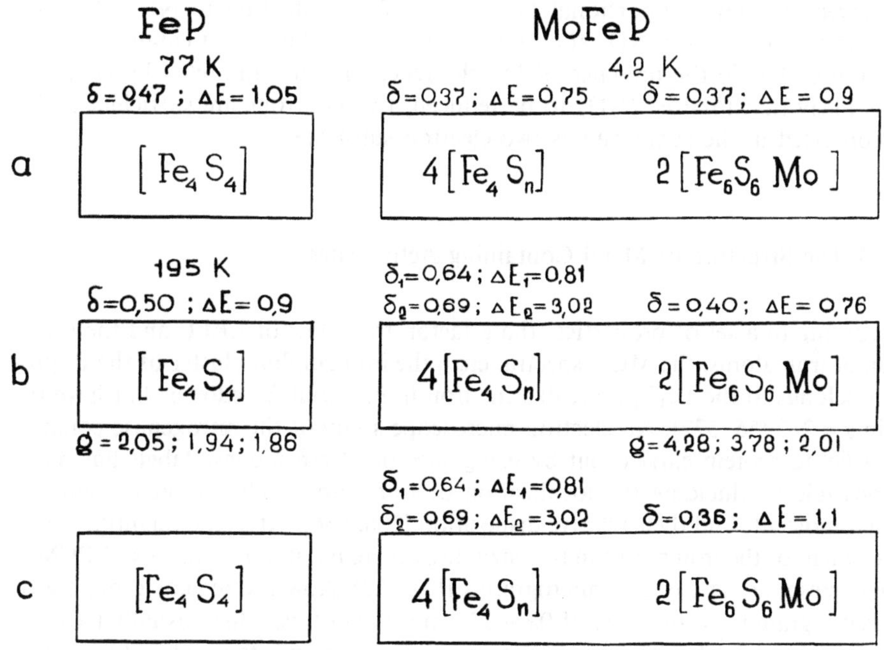

Fig. 7.5 a – c. The values of the parameters of the EPR and Mössbauer spectra of various forms of nitrogenase: **a** native form; **b** reduction by dithionite; **c** reduction in the presence of Mg^{2+} ATP. The Mössbauer parameters are given in mm s^{-1} [549, 550, 565]

oxidation of reduced nitrogenase by protons and substrates is not the rate-limiting step. This controversy can be removed if it is assumed that one super-reduced centre recorded by the disappearance of the EPR signal of FeMoP is insufficient for the substrates to be reduced. The overall reaction is presumably limited by the transfer of a second electron, which is followed by the rapid steps of substrate reduction and hydrolysis of the enzyme-product complexes.

Some other presumptions have been advanced concerning the nature of the limiting step. According to some authors [564], the slowest step of the process is the dissociation of the nitrogenase complex with the oxidized state of the Fe protein. However, there are data that cast doubt on this conclusion. In particular, it has been shown [569] that the nitrogenase complex in vivo is localized in the membrane and is at the same time characterized by activity, which exceeds its activity in solution [570]. Under these conditions it is difficult to expect that the dissociation rate of the complex in the membrane will increase as compared with the solution. The activity of the enzyme is retained upon incorporation into liposomes as well.

The elementary mechanisms of N_2 reduction will be discussed in Chapter 11. As has been shown [24, 538, 566, 567] from the thermodynamic and kinetic viewpoints, the most probable mechanism is the so-called four-electron mechanism involving the formation of a hydrazine derivative. The appearance of the intermediate hydrazine derivative has been demonstrated experimentally [564]: upon "quenching" of the nitrogenase reaction by an acid the evolution of hydrazine was observed at a rate proportional to the rate of N_2 reduction to NH_3. The free hydrazine is reduced to NH_3 at a much slower rate and does not catalyze the formation of HD in the presence of D_2; the reaction is not inhibited by H_2 [561]. It is likely that for free N_2H_4 to be reduced, two electrons are sufficient and it is converted in the same way as two-electron substrates.

7.3.3 The Structure of Metal-Containing Active Sites

According to a set of properties (the g-factor, the effect of LEFE and the extrusion of iron atoms, the MCD spectra, etc.), the iron-sulphur cluster of the Fe protein belongs to the Fe_4S_4 type, like the iron in bacterial ferredoxins but have the spin $s = 3/2$ [56, 572]. An electron microscope study of the nitrogenase complex of a FeMo protein carried out by using specific electron-dense labels has made it possible to elucidate the localization of iron-sulphur clusters at the enzyme macromolecule and their relative disposition [308, 536]. After the quantitative replacement of the iron atoms in the enzyme by mercury atoms by means of PCMB, the corresponding electron micrographs (Fig. 7.3) show the regular grouping of mercury granules with a size of 0.6–0.7 nm into clusters, which usually have the shape of quadrangles. The size of the granules corresponds to an assembly of four to eight mercury atoms. Two important conclusions can be made on the basis of this result: (1) the nitrogenase contains clusters of non-heme iron of the Fe_4S_4 type; (2) the "clusters", in turn, are grouped into assemblies. Experiments on the separation of modified nitrogenase into its components with the aid fo DEAE cellulose have shown that the major part of such an assembly is formed by FeMoP. The presence of such clusters in FeMoP has been demonstrated by EPR and NGR [56, 549, 550]. The results of these experiments are presented in Fig. 7.5.

A nuclear gamma resonance study of the FeMo protein has revealed four types of iron-sulphur clusters in preparations reduced by dithionite (the content of iron atoms per protein macromolecule is enclosed in parentheses): M4 (4) M5 (12–14), M6 (12–14), S (2). A magnetic interaction has been observed between the M4 and M5 types. It has been presumed that the M4- and M5-type signals reflect the presence of four P-clusters, each of which contains three M5 atoms and one M4 atom. The conclusion as to the presence of Fe_4S_4 clusters in the FeMoP protein has been made on the basis of extrusion data [573]. It has recently been found that the P-clusters may be in four oxidation states, from [$4Fe^{3+}$] to [$4Fe^{2+}$] [549]. By the set of the EPR and NGR parameters the P-clusters differ substantially from the known Fe_4S_4 and Fe_2S_2 clusters and evidently contain not only sulphur ligands. The variation of the spectral parameters of nitrogenase during the oxidation-reduction transformations of nitrogenase is shown in Fig. 7.5.

It has been demonstrated in the works cited above that there is a distinct correlation between the presence of a hyperfine magnetic structure in the NGR spectrum for the M6 cluster and the EPR signal with $g_y = 4.6$ and $g_x = 3.3$. These two effects have been observed in samples reduced by dithionite; they disappear upon electron transfer from the Fe protein to the FeMo protein, which is coupled with the hydrolysis of ATP.

A considerable advance in the understanding of the structure of the molybdenum-containing site of nitrogenase has been achieved due to the extraction and characterization of the FeMo cofactor of nitrogenase (FeMo-Co) [550, 565, 574–581]. The presence of the cofactor was first demonstrated in the works of Shah and Brill [574], who worked out a procedure for the isolation of FeMoCo form the FeMo protein by acidification of the preparation with citric acid followed by extraction with N-methylformamide. The FeMoCo preparation activated the cofactor-deficient nitrogenase isolated from the mutant strain UW-45 and has the ability to reduce C_2H_2 to C_2H_4 under the action of $NaBH_4$. Several procedures have been developed for the isolation of FeMoCo, including a procedure in which no organic acid is used [581].

It has been found that regardless of the source of the preparation, the EPR signal from the reduced form of the FeMo protein is due to the iron-containing cluster localized at FeMoCo. The EPR spectrum of MoFeCo in N-methylformamide was observed in the same range of g-factors ($g_y = 4.6$; $g_x = 3.3$) as the spectrum of MoFeP ($g_y = 4.28$; $g_x = 3.78$). Orme-Johnson and co-workers [573] noted that in FeMoCo the spectral lines are much more broadened. As a result of the addition of thiol compounds to FeMoCo, its EPR characteristics approach those of FeMoP. A detailed comparison of the Mössbauer spectra of whole FeMoP and FeMoCo has revealed a close similarity between the spectra of an M-cluster containing about six to eight iron atoms per macromolecule and those of MoFeCo [573, 593]. According to Mössbauer and EPR spectra parameters the M-cluster differs significantly from the well-known Fe_2S_2 and Fe_4S_4 clusters.

The isolation of a new cluster has been reported [575]; it was obtained upon treatment of FeMoP with dimethly ethyl ketone. The new cluster contains 6 Fe per Mo and gives an EPR signal of the HiPIP type ($g = 2.05$ and 2.01) and is inactive with defective nitrogenase. When the preparation was passed into N-methylformamide, the "ordinary" FeMoCo signal was recovered.

Of special interest are the data obtained by a study of the state of the molybdenum atom in the enzyme by means of EXAFS [576, 577] (Sect. 2.4). On the basis of the EXAFS spectra parameters (the position of the spectral line, the hyperfine structure and the line shape), it has been established that the molybdenum atoms in FeMoP and FeMoCo are surrounded by four sulphur atoms which are separated by a distance of 0.236 nm. Another sulphur atom is 0.25 nm from the Mo atom. The second coordination sphere incorporates several iron atoms 0.272 nm from the Mo atom. An unexpected fact is that the EXAFS parameters are practically independent of the oxidation state of FeMoCo. The three forms of the enzyme, one oxidized by dithione, the second reduced by dithionite and the third super-reduced in the presence of ATP, gave similar spectra. Presumably, as a result of the localization of electrons in the FeMoCo cluster, addition of an extra electron makes an insignificant contribution to the local state of the Mo atom. According to theoretical considerations [552, 554] and experimental data on model systems, the most probable state of the molybdenum is Mo^{3+}. According to EXAFS data for the Fe atom, the distances from this atom are (R in nm, the number of the corresponding ligands is given in parentheses) $R_{Fe-Fe} = 0.263$ (2.6), $R_{Fe-Mo} = 0.284$ (0.3) and $R_{Fe-0} = 0.187$ (2).

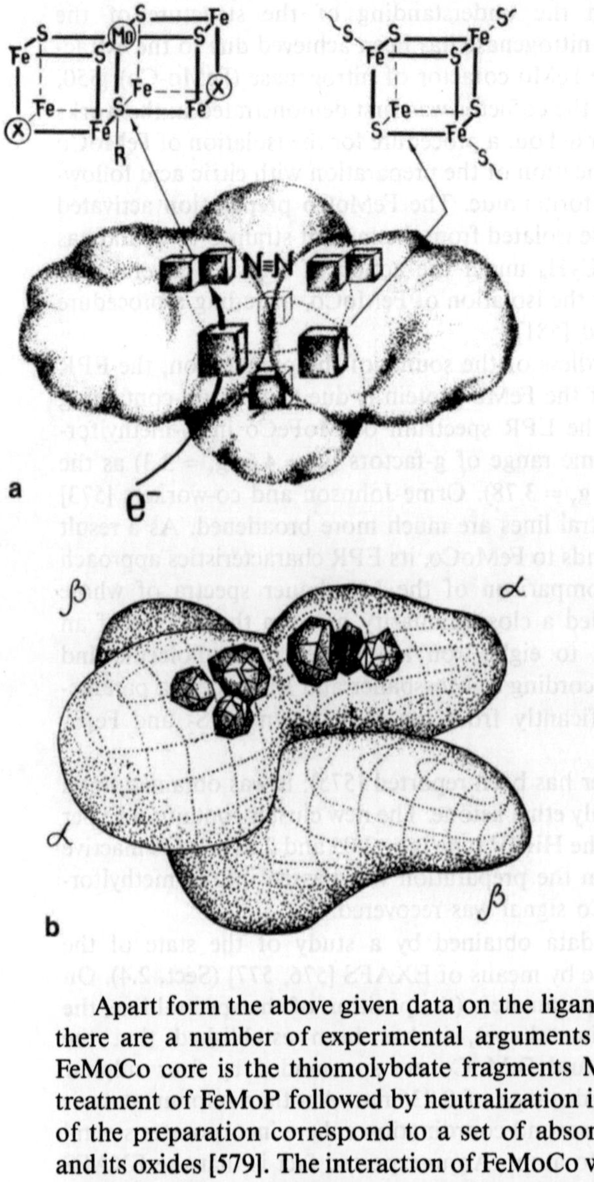

Fig. 7.6 a, b. Model of the FeMo-containing component of nitrogenase: **a** according to data of electron microscopy, spectral and physical-label methods (for explanations, see the text); **b** according to data of X-ray structural analysis [554]

Apart form the above given data on the ligand environment of molybdenum, there are a number of experimental arguments in favour of the view that the FeMoCo core is the thiomolybdate fragments MoS_4^{2-} or $MoOS_3^{2-}$. Upon acid treatment of FeMoP followed by neutralization in air, the spectral characteristics of the preparation correspond to a set of absorption spectra of thiomolybdate and its oxides [579]. The interaction of FeMoCo with fluorescein dimercuriacetate under anaerobic conditions leads to fluorescence quenching, which corresponds to the quenching of solutions of a mixture of thiomolybdates and an inorganic sulphide [580]. The EPR and Mössbauer parameters are similar (but not identical) to those of thiomolybdate-iron complexes [581–582]. The structure that corresponds best to the set of physico-chemical parameters is the FeMoCo structure shown in Fig. 7.6. It should be emphasized that the available data do not permit one to draw a final conclusion on the FeMoCo structure and therefore the model of the structure is tentative.

The relative disposition of Fe_nS_n clusters in nitrogenase and FeMoCo has been studied using the method of electron-scattering labels [585]. The introduction of PCMB into cofactor-deficient nitrogenase leads to the replacement of the iron atoms by mercury. The electron micrographs of modified preparations shows mercury clusters, whose size is appreciably smaller than that of the clusters of native nitrogenase. However, when FeMoCo is preliminarily added to defective nitrogenase, the cluster develops to the "normal" state, probably due to the mercury-modified cofactor. There is evidence that the cofactor molecules are in direct contact with another iron-containing FeMo-protein cluster: the electron transfer from the M-cluster to the P^+-cluster occurs at a high velocity [564].

The problems of the ligand environment of the iron atoms in FeMoCo, both in the isolated state and in the composition of the enzyme, and of the two molybdenum atoms have not yet been solved. It is only known that the molecules of the solvent N-methylformamide are tightly bound to FeMoCo and that according to EXAFS data, the second coordination sphere of Mo does not contain the other molybdenum molecule [576–578]. The available data on the EPR characteristics of FeMoCo and the native enzyme indicate that there is no spin-spin interaction between M-clusters sufficient for a qualitative change in the EPR spectrum upon incorporation of FeMoCo into nitrogenase. However, the above mentioned differences in the spectra (the broadening of the spectral line and the shift of the values of the g-factor) could result from a weak exchange interaction between the sites with $I = 10^7 - 10^8\ s^{-1}$. According to the experimental calibration dependence (Fig. 11.3), this interaction corresponds to the length of the non-conducting portion that separates the spins, $r = 0.6 - 0.7$ nm. Since the spin density at the Mo atom is negligible according to EPR data, the results available at present do not contradict the model, according to which two FeMoCo molecules are very close to each other, i.e. at a distance that permits the formation of a complex with the nitrogen molecule residing between the two molybdenum atoms.

All the clusters in the FeMo protein have been found to be closely spaced by X-ray structural analysis (Fig. 7.6) [554]. As can be seen from the X-ray model of the MoFe protein, all the clusters in nitrogenase are localized in a single part of the macromolecule. The regions of highest electron density, ascribed to the MoFe cofactor, are separated from each other by a small gap (0.6–0.8 nm). This result confirms the data obtained by the method of electron-dense labels [585]. According to the X-ray model, the FeMo cofactor molecules reside near the globule surface, this is confirmed by experiments utilizing the quenching of triplet probes by molecules of the MoFe protein in solution [586].

The problem of the binding nature of the cofactor to apoenzyme has been examined in a number of works. The use of spin and electron-dense labels [585] has shown that in the FeMo protein deficient in the cofactor there are closely spaced sulphphydryl groups. These groups are not accessible to modification in a whole native protein. However, the ENDOR technique has revealed no interaction between the unpaired electrons on the iron atoms of the cofactor and on $[^{35}S]$ of the sulphphydryl groups [587]. The preliminary blocking of the carbonyl and imidazole groups in the deficient FeMo protein does not hinder the incorporation of the FeMo cofactor, which is controlled by the triplet-probe technique [586]. The addition of MoS_4 and a Fe cluster impedes the incorporation of the radioactive FeMo cofactor with the formation of an active enzyme [575]. Thus, the potential ligands of the Fe-Mo cofactor in nitrogenase are sulphphydryl groups;

the participation of other groupings, e.g. peptide groupings, is not ruled out. This is evidenced by the tight binding of N-methylformamide in solutions of the cofactor.

7.3.4 The ATPase Site

The main outcome of the action of ATP is that the cleavage of the high-energy pyrophosphate bond is accompanied by an electron transfer from FeP to FeMoP with the reduction potential of the system being increased due to the formation of the low oxidation states of Fe and Mo (the super-reduced states) unusual under normal conditions. The consumption of ATP for the transfer of one electron (m) as estimated by different authors varies (2–8 ATP) and depends on the experimental conditions (pH, temperature, the ratio of the components, the nature and concentration of the substrate) [56]. The minimal value of $m = 2$ (pH 7, $t = 20\,°C$) [56, 537]. The reaction proceeds in the presence of Mg^{2+} ions, which can be replaced by Mn^{2+}, Fe^{2+} and Co^{2+} ions.

The formation of ternary ATP-Mn^{2+} nitrogenase and ADP-Mn^{2+} nitrogenase complexes has been demonstrated [583] by the variation of the rate of the proton magnetic relaxation of water protons as compared with the corresponding rates of the relaxation of the individual components. The effect of the ternary nitrogenase complex on the relaxation of the protons of water differs from the effect of the corresponding ATP-Mn^{2+}–FeP complex, which is an indication that the complexes have different microstructures.

The ATPase site of nitrogenase from *A. vinelandii* has been studied using an inhibitor analysis [584, 592]. It has been found that nucleotides (GTP, UTP, CTP, ADP) and their derivatives, containing diphosphate and triphosphate groups, are competitive inhibitors of nitrogenase, just like inorganic triphosphates and diphosphates. The study of the effect of the nature of the metal ion has led to the conclusion that the size of the ion is an essential, though not a decisive factor. The major role is played by the electronic structure, which determines the acid-base properties of the ion and its ability to participate in specific complex formation.

The most important factor in the formulation of the mechanism of the ATPase reaction is the localization of the ATPase site relative to the electron-transport chain. The solution to this problem has been sought for experimentally by using several independent methods. Experiments have been carried out to elucidate the relative arrangement of the free SH groups of the ATPase site and of the iron-containing "cluster" by modification of these sites [136]. The free SH groups were labelled with electron-scattering labels (see Sect. 3.3) and the iron-complexing SH groups were labelled with PCMB. In this case, the electron micrographs demonstrated the formation of enlarged mercury clusters localized asymmetrically at the nitrogenase macromolecule (Fig. 7.3). The structure and size (3.0–4.0 nm) of such associated "clusters" differ substantially from those of the paired granules observed upon modification of only two free SH groups by electron-scattering labels (Fig. 7.3) and from those of the tetragonal cluster detected upon modification of the iron-containing portion by PCMB (Fig. 7.3).

This result is an indication that the ATPase site of the enzyme containing free SH groups on FeP is close to the iron-containing clusters on FeMoP.

The phenomenon of inductive resonance energy transfer has been used for the estimation of the distance between the fluorescing chromophore of the labels bound by mercury and the alkylating groups [134, 536] to the SH residues of the ATPase site of nitrogenase and the non-heme iron-containing clusters of nitrogenase (Chap. 3). The incorporation of the labels into the active site was accompanied by an abrupt quenching of fluorescence; when the iron was removed, the initial fluorescence was regained. The use of the corresponding theory (Sect. 3.2.2) has enabled the determination of the distance between the chromophore centres, which was found to be equal to about 1.6 nm. Taking into account the finite size of these groups, it is possible to assume that the ATPase portion is rather close to the neighbouring non-heme group.

An analogous conclusion has been made by some authors [107, 108], who analyzed the spin-spin interaction between the nitroxide radical introduced via the parachloromercuribenzoate residue into the ATPase site and the paramagnetic centre of the enzyme. A nitroxide radical has also been used [536] as a redox reagent which was introduced into the ATPase site and bound to the triphosphate residue. An EPR study has shown that the reduction rate of this radical by nitrogenase is much higher than in the case of a free nitroxide radical. Addition of ATP sharply decreased the reduction rate of the triphosphate derivative. Since the nitroxide fragment of the label was actually at the site of the adenosine residue, this experiment has shown that an electron is transferred to a site in the immediate vicinity of the adenosine fragment of ATP.

The luminescent properties of the ethene derivatives of ATP have been used for the study of their localization on a nitrogenase macromolecule [584]. This compound is a competitive inhibitor of nitrogenase and, hence, it enters into the composition of the ATPase active site of the enzyme. It has been found that after the 1,6-ethene-ATP is covalently linked to the nitrogenase complex, it becomes bound to the two components. The luminescing fragment on FeMoP was strongly quenched by the chromophores of the protein (by more than 50 times) and, hence, was in the direct vicinity of the iron-containing cluster.

There are arguments in favour of the view that the nitrogen-binding site is remote from the ATPase site. For instance, no dipole-dipole broadening of the protons of acrylonitrile (which is an inhibitor of nitrogenase and evidently interacts with the nitrogen-binding site) could be detected [588]. By introducing paramagnetic species into the ATPase site, i.e. a spin label on the SH group or Mn^{2+} in the composition of ternary complexes (Mn^{2+}-ATP nitrogenase or Mn^{2+}-ADP nitrogenase), it was possible to estimate, on the basis of the theory of the spin-spin interactions, the distance between the paramagnetic centre and acrylonitrile in the active site, which was found to be larger than 0.6 nm in the case of Mn^{2+} and larger than 1.0 nm in the case of the nitroxide radical. Introduction of Mn^{2+} into a ternary nitrogenase complex had only an insignificant effect on the line width of the EPR signal from the iron-containing FeMoCo cluster. Estimates show that the distance between these paramagnetic centres is $r > 1.2$ nm, which evidently rules out a direct interaction between the centres.

The supposed model of the active-site system of nitrogenase with account taken of the data considered above is shown in Fig. 7.6. Figure 7.6 shows the close spacing between the two molybdenum atoms. However, the model with spatially separate FeMoCo cannot be dispensed with.

The phenomenon of isotope exchange between the oxygen of water and phosphate oxygen has been detected [589, 590] under the conditions of the nitrogenase reaction. The exchange rate was proportional to the rate of the nitrogenase reaction. The two processes were equally inhibited by reagents for the histidine group and for the carboxyl residue activated by phosphorylation (paranitrothiophenol). The exchange of radioactive phosphorus between ATP and phosphate in the presence of functioning nitrogenase has been observed [593].

The breakdown of the structure of the iron-containing site of nitrogenase under the influence of light or upon treatment with pentachlorophenol has been shown to cause the separation of the ATP-hydrolysis and electron-transfer centres [588]. As has been found [591], the super-reduced state of FeMoCo can also be obtained in the absence of the ATPase reaction under the action of a strong reducing agent, i.e. a hydrated electron. The reaction of FeMoCo with solvated electrons at 77 K leads to a decrease in the intensity of the EPR signal of the cofactor. Heating to room temperature followed by cooling to 77 K led to the recovery of the signal from the cofactor. Based on these results, on the known facts of the effect of the protonation of transition metals on their electron-accepting properties and also on the data of proton-transfer systems in the active site of enzymes via the histidine group (the charge relay system), a scheme of the coupled reaction as shown in Fig. 7.7 was proposed [588–590]. According to this scheme, the role of ATP boils down to the phosphorylation of the active-site carboxyl group with a simultaneous proton transfer from the carboxyl to the adjacent imidazole residue of histidine. The protonation of the latter, in turn, causes as forced transfer of the second proton of histidine to the iron-sulphur cluster. An increase in the positive charge on the cluster contributes to the transfer of an extra electron and to the formation of the super-equilibrium form of the enzyme after removal of the proton.

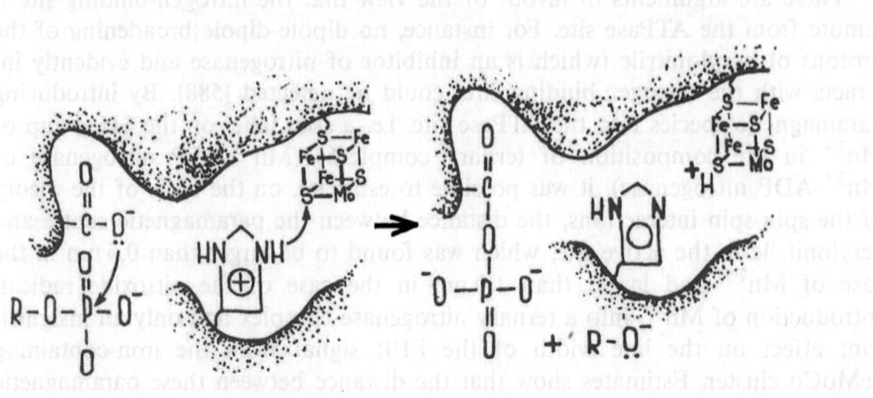

Fig. 7.7. The expected coupling scheme of ATP hydrolysis with electron transfer involving the formation of the super-reduced state of Fe_nS_n [588–591]

From the scheme shown in Fig. 7.7 it follows, in particular, that the rate of ATP-dependent electron transfer may be influenced by proton acceptor compounds in the hydrophobic portion of the active centre. It has been found [594] that pentachlorophenol and other H^+ acceptors, the derivatives of phenols, considerably reduce the rate of electron transfer from the Fe protein to the MoFe protein without affecting the rate of ATP hydrolysis. The time of phosphate and the proton release in the non-steady-state phase of the reaction is close to the rate of an elementary electron transfer (~40 ms at 10 °C, preparations from *A. vinelandii*) [595].

It should be pointed out that the problem of the existence of a phosphoryl intermediate in nitrogenase cannot be regarded as having been solved. Apart form the above given arguments in favour of the phosphorylation step of the carboxyl group under the action of ATP, the point that attracts attention is the similarity between the ATPase properties of nitrogenase and those of other ATPases, including Ca^{2+} ATPase [590]. In the case of the latter enzyme, the formation of a covalent intermediate has been proved by experiment.

According to the available data, the ATPase centre is formed at the interface between the Fe and FeMo proteins [56, 536]. Here, the ATP-binding portion is presumably localized on the Fe protein, which is evidenced by the data on the blocking of the SH groups by the inhibitor of the ATPase centre of PCMB and also by the closeness of the fragment of the amino-acid sequence of the Fe protein to the fragments responsible for the binding of ATP in the other ATPases [596]. The plausible localization of the catalytic groups of the ATPase centre is near the β-subunit of the FeMo protein. These subunits contain elements of the amino-acid sequence analogous to the corresponding ATPases [596].

Thus, the analysis of the available data at present allows one to presume the following scheme of nitrogenase reactions (Figs. 7.6 and 7.7). The electrons from one-electron reducing agents (ferredoxin, flavodoxin or model reducing agents) are transferred to the iron-containing component of the enzyme. In the ATPase site of the enzyme, which is located at the boundary between the components and is adjacent to the assembly of iron-containing FeMoP clusters, the coupling of the hydrolysis of the high-energy phosphate bond of ATP and electron transfer from FeP to the FeMoP clusters with the formation of super-reduced states takes place. The electrons are transferred along the chain of iron atoms and reach the FeMoCo cluster and are then transferred to the molybdenum atoms, which are evidently reduced to a valency of 3+. The decisive reduction step of molecular nitrogen complexed at the binuclear site occurs as a synchronous four-electron attack with electron transfer either from the two molybdenum atoms of the FeMoCo clusters or from the molybdenum and iron atoms within the same cluster takes place (for more details, see Chap. 11).

The system of polynuclear iron clusters and molybdenum-containing sites of nitrogenase is a mechanism-switching system which changes the one-electron to the four-electron mechanism; the action of this system may be described by the following scheme, which refers to the formation process of the hydrazine derivative [597, 598]:

$$\left[\begin{array}{l} 1\,(2) \rightarrow 1\,(2) \searrow \\ \qquad\qquad\qquad\quad 4 \\ 1\,(2) \rightarrow 1\,(2) \nearrow \end{array}\right] \text{ or } [1\,(4) \rightarrow 1\,(4) \rightarrow 4] \ .$$

The material considered in this chapter is convincing evidence of the structural and functional similarity between molybdenum-containing enzymes. The molybdenum centre of the enzyme enters into the composition of a special polypeptide block, which is common to all enzymes of this class. The function of molybdenum is to effect two-electron reactions of oxidation and reduction, in which the valence of molybdenum oscillates between three states: $Mo^{6+} \leftrightarrow Mo^{5+} \leftrightarrow Mo^{4+}$. The two-electron character of the process is dictated by thermodynamic necessity. The process is followed by one-electron transfer steps, which are completed with one- or two-electron reduction of O_2 (under physiological conditions, the more "harmless" two-electron reduction takes place preferentially) or two-electron reduction of NAD^+. Thus, in molybdenum-containing enzymes the mechanism-switching operation occurs twice as a rule: at the beginning and at the end of the chain. In nitrogenase the one-electron mechanism is switched over to the four-electron one.

A scheme illustrating the similarity between molybdenum-containing enzymes is given below: xanthine oxidase (I), nitrate reductase (II) and sulphate oxidase (III); the arrows indicate the direction of electron transfer. The nitrogenase reaction (IV) evidently involves two molybdenum centres of a more complex structure:

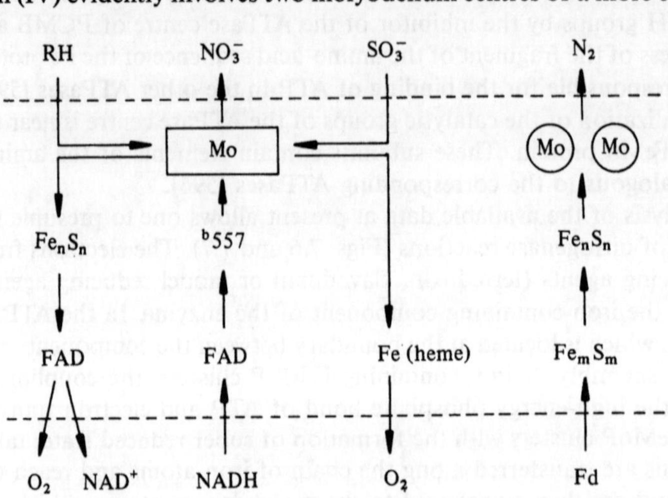

CHAPTER 8

Copper-Containing Enzymes

The pioneering works of Keilin published at the end of the 1930s [2] gave impetus to the study of copper-containing proteins and the interest in these proteins has become traditional in biochemistry. This class of proteins was found to include the proteins that were well known at the time: laccase, ascorbate oxidase and cytochrome *c* oxidase, the terminal member of biological oxidation [62, 63, 599–603]. Cytochrome *c* oxidase is known to play a key role in the reduction of oxygen to water, a process, the annual yield of which is about one hundred billion tons. It was found later that copper-containing enzymes catalyze a large number of redox reactions, which, while having a lesser tonnage yield than the cytochrome oxidase reaction, nevertheless play an important biological role. These include the reactions of oxidation of a large number of organic substrates (amines, phenols, etc.), divalent iron, the protective reactions of neutralization of the peroxide oxygen radical (the superoxide), etc.

Copper-containing enzymes may be classified into seven types, which differ in copper content, the structure of the copper-protein complex and the chemical mechanism of reactions: (1) cytochrome *c* oxidase which contains two copper atoms and two heme groups and reduces O_2 to water; (2) blue enzymes (ceruloplasmin, ascorbate oxidase, laccase) which contain four to eight copper atoms and catalyze the reduction of dioxygen to water; (3) monooxidases (dopamine oxidase, tyrosinase) which contain two to four copper atoms per macromolecule and catalyze the hydroxylation of substrates to give H_2O as the final product; (4) oxidases which catalyze the oxidation of amines and galactose and the reduction of O_2 to hydrogen peroxide; they contain one to three copper atoms; (5) superoxide dismutases which catalyze the disproportion of the superoxide radical to O_2 and H_2O_2 and contain one to two copper atoms per macromolecule; (6) oxygen-carrying proteins (hemocyanin); (7) electron carriers (they contain one to two copper atoms per protein molecule).

The material concerning cytochrome *c* oxidase will be discussed in Chapter 9 in connection with the structure and mechanism of action of electron-transport chains. Here, we shall restrict our discussion to enzymes containing only copper, with the major emphasis on proteins of the second and third types, whose active sites are true multi-nuclear centres. The problems of the energetics and detailed mechanism of action of these enzymes will be treated in Chapter 11 from the standpoint of the genral conceptions of redox processes.

8.1 Blue Copper-Containing Enzymes

8.1.1 Ceruloplasmin

This blue copper-containing protein was detected in the mid-1940s by Holmberg in mammalian blood plasma [602]. It had long been thought that the sole function of ceruloplasmin, which carriers almost 95% of the copper in the organism, was to store and transport copper required for the building of the active sites of other enzymes. It was, however, found at a later time that ceruloplasmin performs other important functions as well as the oxidation of divalent iron, phenols, amines and other compounds [62, 603]. The best substrate for the enzyme is the Fe^{2+} ion:

$$Fe^{2+} + \tfrac{1}{2}O_2 + 2H^+ \rightarrow Fe^{3+} + H_2O \quad (8.1)$$

The molecular mass of ceruloplasmin has been found to range from 125000 to 160000 according to different authors. However, recent works based on X-ray diffraction analysis and a thorough study of the measurement of sedimentaion characteristics have shown that the most reliable value is 132000 and that the protein globule is formed by a single polypeptide chain [604]. There are about 8% carbohydrates in the enzyme. The number of copper atoms per molecule is six to eight according to data from different sources.

A remarkable result of a large number of spectral-optical, radio-spectral and kinetic studies is the proof of the inhomogeneity of copper in ceruloplasmin and in other blue copper enzymes and the characterization of the structural and functional features of various copper species [62, 599–603].

The type-1 Cu, the content of which is about two atoms per macromolecule, is noted for physical parameters which are unusual for model complexes. The EPR spectrum of this type of copper is characterized by relatively low values of the g-factor and HFS constants on the copper nucleus (Fig. 8.1). Another specific feature of this species is the anomalously high values of the extinction coefficient: $\varepsilon = 4400\ M^{-1}\ cm^{-1}$ at $\lambda = 600$ nm. The blue colour disappears upon reduction of the enzyme. An analysis of these specific features has led to the conclusion that the type-1 Cu has a four- or five-ligand configuration with a distorted axial symmetry which is unusual for copper (Fig. 8.2).

The presence of sulphur in the coordination sphere of copper is evidenced by a number of indirect data [605]: (1) the position of the bands in the resonance Raman scattering spectrum; (2) the participation of cysteine in the coordination of copper for polynuclear blue enzymes, azurin and stellacyanin; (3) the unusually high extinction coefficient, which is accounted for by the transfer of an electron from sulphur to the d-orbital of the metal; (4) data on the photo-oxidation of the enzyme, (5) X-ray data on azyrin [606].

Upon treatment of the enzyme with cyanide, Cu^{2+} is not removed, but in the presence of diethyl dithiocarbamate it has been possible to remove it reversibly [607]. By analogy with electron carriers, such as azurins and plastocyanins, which have similar EPR and optical properties (see below), it may be expected that the type-1 site is in a hydrophobic environment. The EPR and spectral methods have recorded the reduction of Cu^{2+} by the action of substrates. Two values of the

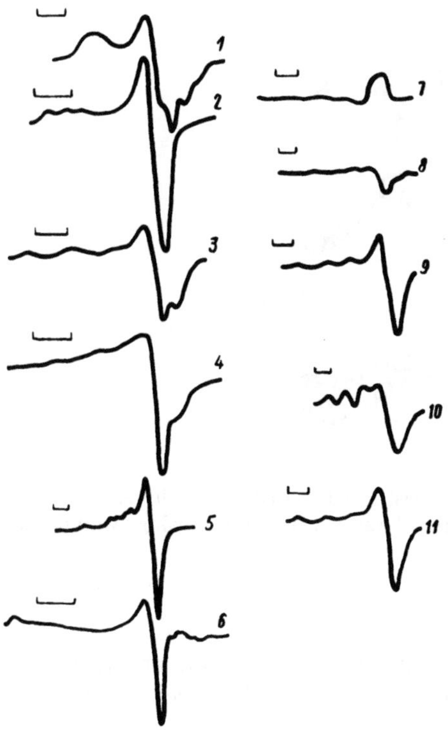

Fig. 8.1. The EPR spectra of copper-containing proteins [24]. *1* stellacyanin; *2* azurin; *3* laccase; *4* ceruloplasmin; *5* ascorbate oxidase; *6* cytochrome oxidase; *7* galactose oxidase; *8* dopamine oxidase; *9* hemocyanin; *10* erythrocuperein; *11* Cu^{2+} insulin. The scale corresponds to 10 mT

redox potential have been detected at pH 5.5: $E_0 = 490$ and 580 mV [608]. The second type, Cu^{2+}(II), manifests itself in the EPR spectrum in the form of a signal characteristic of octahedral copper complexes (Fig. 8.1). Since the reagents for histidine groups lead to the distortion of the type-2 site, it has been presumed that imidazole ligands are present in the coordination sphere of copper [609]. Binding of histidine by specific reagents has no effect on the characteristics of Cu^{2+}(I), which is an indication that type-1 and type-2 sites are far apart. This is also evidenced by the analysis of the EPR spectra of both type-1 and type-2 sites. The absence of a broadening in the HFS lines of the ESR spectra of polynuclear Cu enzymes as compared with mononuclear Cu enzymes (azurins, plastocyanins), in which case it exceeds 2.0–3.0 mT, allows one to estimate the lower limit of the distance ($r_{I,II}$) between the centres by using the calibration dependence for copper complexes and the corrections for a change from a vitrified solution to a pair of interacting centres or by using formula (3.6). The thus estimated value of $r_{I,II}$ is $\geqslant 1.0$ nm. As has been shown for the related enzyme laccase [62], binding of F^- to Cu^{2+}(II) does not affect the redox properties of Cu^{2+}(I). Examples have been described for the effect of azide, cyanide and H_2O_2 on the EPR and optical properties of this copper species [62]. The relaxation NMR method has demonstrated the accessibility of the type-2 centre to water molecules [610].

The third copper type, Cu^{2+}(III) is characterized by a number of properties, which indicate a strong interaction between the copper atoms: (1) this copper type is non-detectable by EPR; (2) it does not make a contribution to paramagnetic susceptibility; (3) it is titrated by reducing agents as a two-electron acceptor [62].

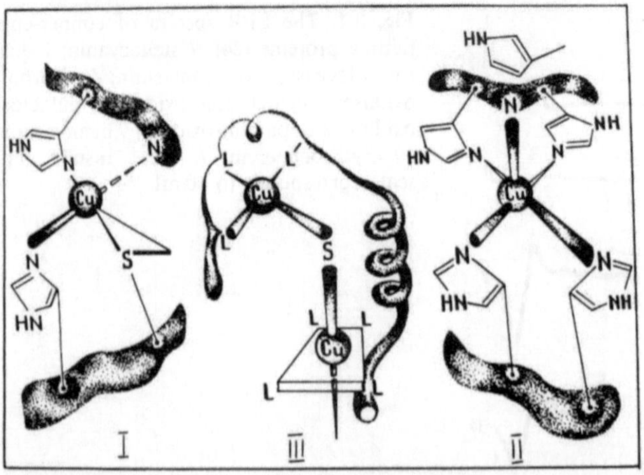

Fig. 8.2. The model of the active site of laccase. Constructed in accordance with X-ray diffraction [67, 606, 635] and EXAFS data [68]. Structure *I* was constructed by analogy with plastocyanin

Physically, the presence of this copper type is revealed upon absorption at $\lambda = 330$ nm, acception of two electrons and also in the EPR spectra of nitrosyl complexes [611].

The kinetics of substrate oxidation by oxygen in the presence of ceruloplasmin is rather complicated. The most active reaction of Fe^{2+} oxidation is accompanied by the activation of the enzyme by the substrate. The most important features of the kinetics and mechanism of action of ceruloplasmin are as follows [62]: (1) the best substrate is Fe^{2+}, which has the lowest value of K_M ($5\times10^{-7}\,M^{-1}$) and the highest value of k_{max} ($1\,s^{-1}$). The corresponding values for other substrates, amines, phenols, catechols, aminophenols, etc., are: $K_M = 10^{-5}$ to $10^{-3}\,M^{-1}$ and $k_{max} = 10^{-2}$ to $10^{-1}\,s^{-1}$; (2) the formation step of a complex with Fe^{2+} takes place rapidly at $k_1 = 1.2\times10^6\,M^{-1}\,s^{-1}$; (3) the oxidation step of substrates occurs by a one-electron mechanism, this being evidenced by the appearance of free substrate radicals; (4) high rates are also observed in re-oxidation of the reduced enzyme ($k = 5.7\times10^5\,M^{-1}\,s^{-1}$); (5) the rate-limiting step of the process is an intramolecular electron transfer from Cu^{1+}(I) to $2Cu^{2+}$(III) at $k_{max} = 1\,s^{-1}$, which is evidenced, in particular, by the slow recovery rate of the blue colour of Cu^{2+}(I) upon re-oxidation of this centre by oxygen and also by the slow appearance of the absorption band from Cu^{2+}(III); (6) oxygen reduction is found to involve an intermediate, which is similar in spectral properties to the H_2O_2-laccase complex.

All these factors and also the energy and kinetic considerations described in Chapter 11 allow one to construct the following electron-transfer sequence:

$$Fe^{2+} \rightarrow \left[Cu^{2+}(I) \rightarrow \begin{matrix} Cu^{2+}(III) \\ Cu^{2+}(III) \end{matrix} \rightarrow Cu^{2+}(II) \right] .$$

The chemical steps of O_2 reduction in blue enzymes are considered in the next section; the possible elementary mechanisms of the reduction will be discussed in Chapter 11. The expected structure of the tetranuclear unit of the active site of ceruloplasmin, constructed by analogy with the X-ray diffraction models of Cu^{2+}(I) plastocyanin and Cu^{2+}(II) superoxide dismutase and with the experimental data described below, is shown in Fig. 8.2. The model of the binuclear $2Cu^{2+}$(III) unit has been constructed by analogy with the unit of hemocyanin studied by EXAFS (see below).

8.1.2 Laccase and Ascorbate Oxidase

The physical and chemical properties of these two enzymes are similar to those of ceruloplasmin. However, they are more adapted to performing the oxidase function with respect to organic substrates and, as a consequence, their catalytic activity is one or two orders of magnitude higher. Laccase is widespread in plants and microorganisms. The functions and some of the properties of this enzyme were known in the last century [1]. The main function of laccase is evidently to catalyze the oxidation of biological phenols and amines. The functions of ascorbate oxidase, an enzyme discovered by Szent-Györgyi and detected in many plants [612], have not yet been established clearly. The two enzymes catalyze the oxidation of substrates according to the following scheme:

$$2DH_2 + O_2 \rightarrow 2D + 2H_2O \ . \qquad (8.2)$$

The substrate specificity of laccase is close to that of ceruloplasmin.

The two copper-containing enzymes have been isolated and purified [62, 601]. The thoroughly purified ascorbate oxidase has been isolated from a number of plants, including cucumber, and laccase has been isolated from plants and moulds. The molecular mass of ascorbate oxidase is 133000 and the copper content is eight to ten atoms per molecule; the plant laccase has a molecular mass of 110000–140000 and contains four to six copper atoms. The laccase molecule from moulds is almost half the size with four copper atoms. Like ceruloplasmin, the two enzymes contain a large amount of carbohydrates (10–45%).

The physical and chemical properties of copper in the active sites of ascorbate oxidase and laccase are surprisingly similar to those of ceruloplasmin. Three copper types have been detected in both enzymes [62, 608, 613–616], the species distribution being as follows: laccase, $1Cu^{2+}$(I), $1Cu^{2+}$(II) and $2Cu^{2+}$(III); ascorbate oxidase, $3Cu^{2+}$(I), $1Cu^{2+}$(II) and $4Cu^{2+}$(III). The optical and EPR characteristics of various copper atom species, just as their functional relations, are very similar in both enzymes. The redox potentials for type-1 Cu and type-3 Cu at pH 5.5 are $E_1^0 = 394$ mV and $E_3^0 = 434$ mV respectively, for plant laccase and 785 and 780 mV for a preparation from moulds. A noteworthy feature is the appreciable effect of the F^- anion added to Cu^{2+}(II) on the value of E_3^0. For instance, addition of 10^{-2} M NaF lowers E_3^0 by 45 mV for plant laccase and by 210 mV for mould laccase. This points to a considerable binding of the type-2 and type-3 centres. The F^- and CN^- ions and H_2O_2 have an effect on the EPR parameters of Cu^{2+}(II). It is interesting that upon deprotonation of a group with pK 7.45 an endogenic reduction of Cu^{2+}(I) is observed. It is presumed that the proton, while being bound to the Cu^{2+}(II) centre, prevents the electron transfer from Cu^{1+}(I) to a two-electron acceptor [62]. There are data on the incorporation of H_2O_2 and H_2O into the coordination sphere of Cu(II) [614]. Here, N_3^- interacts simultaneously with the centres of types 1 and 2 [616].

The spin-echo technique has shown the presence of histidine as ligands in the type-1 Cu and type-2 Cu in laccase preparations from *Rhus vernicifera* and pig ceruloplasmin [615]. One of the ligands of the Cu^{2+}(II) centre is water (the data on the HFS of the EPR spectra in the presence of H_2 [^{17}O]) which can be replaced by F^- and other ligands. The presence of sulphur in the coordination sphere of Cu^{2+}(I) is confirmed by a number of data given in the previous section. The CD and MCD data indicate the non-equivalence of the ligand environment of Cu^{2+}(I) [617]. By analogy with the results of X-ray diffraction studies of another blue protein, azurin (see below), it may be presumed that the copper in the Cu^{2+}(I) centre of laccase and related enzymes is bound by two histidine groups, one sulphhydryl group and one methionine group in a distorted tetrahedral configuration. The model of the tetranuclear unit of laccase is shown in Fig. 8.2.

However, as has recently been shown [893], the EXAFS data on laccase, devoid of the Cu(II) centre, point to the absence of a sulphur atom as a ligand bridge.

The kinetics and mechanism of the catalytic action of laccase have been studied by many authors [62, 618]. The basic experimental data may be formulated as follows: (1) the oxidation of the substrate by a strong oxidizing agent, the Cu^{2+}(I) centre, is effected at high rates ($k = 10^4 - 10^5\ M^{-1}\ s^{-1}$); the one-electron nature of this step is proved by the appearance of radicals and by experiments involving the direct electrochemical reduction of the enzyme. (2) The electron transfer from Cu^{2+}(I) to $2Cu^{2+}$(III) under anaerobic conditions proceeds slowly ($k_{max} = 1\ s^{-1}$) and increases by two orders of magnitude in the presence of dioxygen. (3) The O_2 molecule becomes bound to the reduced form to the enzyme, presumably to $2Cu^{2+}$(III). The introduction of F^- into the Cu^{2+}(II) centre increases the rate of binding but blocks the reduction of dioxygen. (4) Experiments on the incorporation of [$^{18}O_2$] have shown that the first water molecule as the reaction product is liberated rapidly and the second slowly [619]. The reaction of

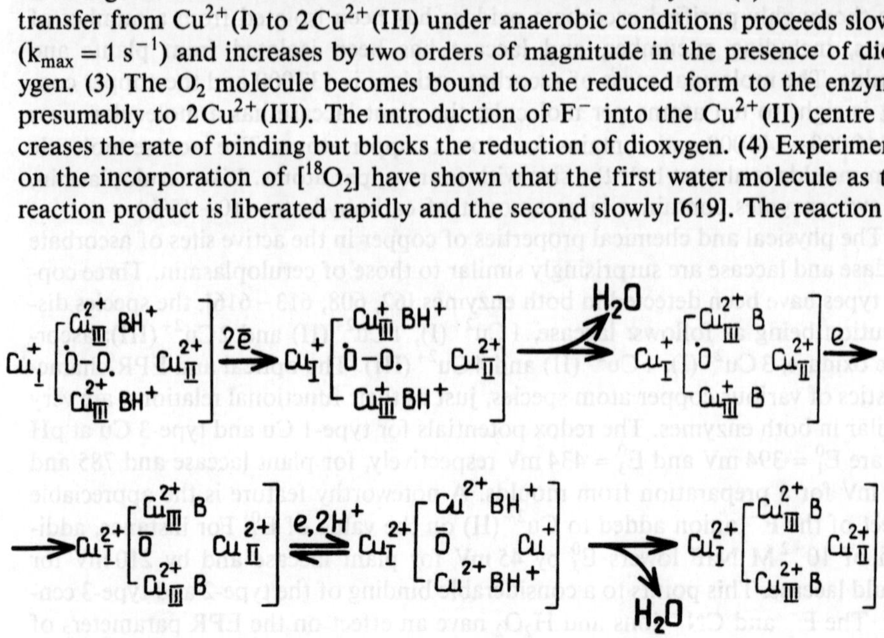

Fig. 8.3. Chemical conversions involved in the reduction of O_2 at the active site of laccase [618]

the enzyme with [$^{17}O_2$] reduced by three electrons gives an EPR spectrum, which is ascribed to the transition radical [$^{17}O_2$] state. (5) The reduction of dioxygen and of the transition state is preceded by the protonation of two ionic groups with pK 5.5 [618]. Similar properties have been detected in ascorbate oxidase [621].

The above data and other have been summarized in the form of a scheme, the main part of which is shown in Fig. 8.3. The electron mechanism of the various steps of the process is described by the following scheme (see Sect. 1.3):

$$1(2) \rightarrow 2 \rightarrow 1(2) \ .$$

8.2 Copper-Containing Monooxygenases and Oxidases

Monooxygenases catalyze the hydroxylation of substrates in the presence of oxygen and reducing agents (DH_2) according to the following general scheme [62, 63, 620]:

$$XH + DH_2 + O_2 \rightarrow X-OH + D + H_2O \ . \tag{8.3}$$

The product X−OH may undergo further conversions. At present, two enzymes of this class have been studied: dopamine β-oxygenase and phenol o-oxygenase.

Dopamine β-Monooxygenase. This enzyme, which was isolated from bovine adrenal glands, catalyzes reaction (8.3), in which the substrate X−OH is 3,4-dihydroxy-phenylethylamine (dopamine) and the reaction product is norepinephrine. Phenylethylamine, tyramine or epirine may also serve as the substrate. The best reducing agent is ascorbic acid, which can be replaced by catechol but not by pyridine nucleotides [63]. An oxygen atom is incorporated into the product from O_2 and not from water, as is the case with xanthine oxidase.

The molecular mass of the enzyme is 290000; the macromolecule consists of four identical subunits, each of which contains two copper atoms. The EPR and optical characteristics of the enzyme have been described in the literature [620]. According to EPR data, ascorbate reduces copper and the reduced enzyme reacts with dopamine and oxygen and, as a result, the following products are obtained: hydroxylated dopamine or other substrates and water. The univalent copper, which appears upon reduction, may be complexed by cyanide or carbon monoxide. The data of the kinetics of the reactions have been described in the literature [620]. According to the opinion of the authors of this work, the reaction proceeds by the ping-pong mechanism.

The rate-limiting step of the reaction is the conversion of the ternary complex, the mechanism of which is not yet clear. It may be presumed that the hydroxylation involving copper occurs by a mechanism similar to that of the reaction on cytochrome P-450 (see Chaps. 6 and 11).

Phenol-o-Monooxygenase (Tyrosinase). The enzyme is widespread in bacteria, plants, insects, animals and mushrooms. The biological role of the enzyme is thought to be the participation in the biosynthesis of lignins, bioflavonoids, tannin, adrenaline and other compounds; the starting materials for these compounds are phenols [63, 620−623]. The enzyme catalyzes the following reaction:

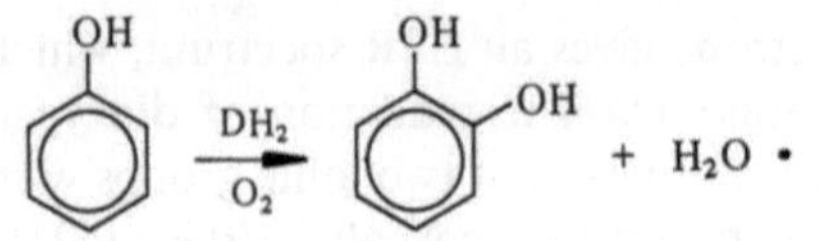

The reducing agent DH_2 may be the product biphenol. The turnover number for various substrates is $15-400\ s^{-1}$ and the value of $K_M = 0.2-5$ mM.

The molecular mass of the enzyme from mushrooms is 126000; each enzyme macromolecule contains one Cu^{2+} atom and one Cu^{1+} atom. The type-1 copper is EPR-detectable and gives a signal with $g = 2.0$ and with a hyperfine splitting of 18.0 mT [63].

In bacteria the forms of the enzyme with molecular masses of 29000 and 41000 have been detected. In all cases, on the basis of EPR data, it is presumed that one copper atom is in the Cu^{2+} state and the other in the Cu^{1+} state [620]. Of certain interest for the understanding of the mechanism of action of tyrosinase is the work of Lerch [620], who has shown that the interaction of the enzyme with hydrogen peroxide in the ratio of 1 : 1 involves both copper atoms. The spectral characteristics of the complex are similar to those of dinuclear oxyhemocyanin (see below). Thus, for tyrosinase $\varepsilon_{345} = 9550\ M^{-1}\ cm^{-1}$ and $\varepsilon_{600} = 645\ M^{-1}\ cm^{-1}$ and for oxyhemocyanin $\varepsilon_{347} = 8900\ M^{-1}\ cm^{-1}$ and $\varepsilon_{570} = 500\ M^{-1}\ cm^{-1}$. In both cases, the product is diamagnetic. In the case of tyrosinase the complex with H_2O_2 can be reduced by specific substrates, such as 4-tert-butyl-catechol and p-cresol (and not by ascorbate), presumably to water. The small distance between the two copper atoms containing NO is distinctly shown by the EPR method [622]. These results indicate that oxygen is reduced most probably on the binuclear complex first to a peroxide compound and then to water. The active-site model of the enzyme constructed by analogy with the hemocyanin model is shown in Fig. 8.4. The substrate binding

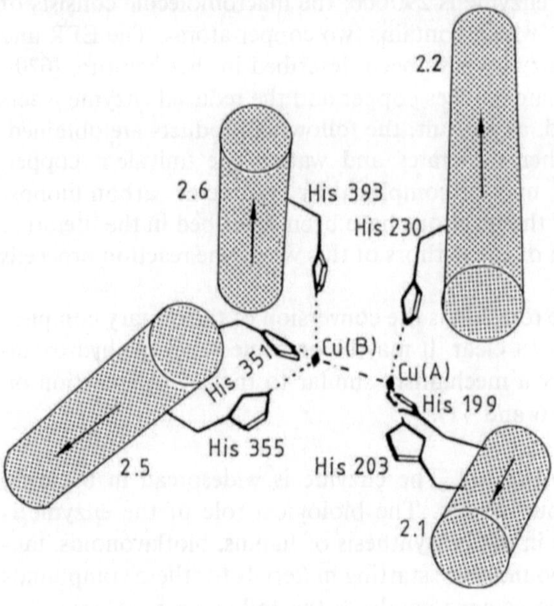

Fig. 8.4. Schematic representation of the binuclear centre in arthropodan hemocyanin [873]

leads to a change in the configuration of the ligands of Cu (II) and to the activation of O_2 [623].

Monoamine Oxidase. The enzyme catalyzes the reaction [62]:

$$R-CH_2-NH_2+O_2 \rightarrow R-C\begin{smallmatrix}\nearrow O \\ \searrow H\end{smallmatrix} +H_2O_2+NH_3 \ . \tag{8.4}$$

Monoamine oxidase from *Aspergillus* contains three copper atoms per molecule (molecular mass about 250000). But these atoms do not interact with each other. The enzyme has a pink colour with a broad absorption band between 420 and 490 nm; the band disappears upon reduction of the substrate and appears again upon re-oxidation by oxygen. The EPR spectrum of monoamine oxidase is similar to that of atoms of the Cu^{2+} (II) species and is characterized by the following parameters: $g_{\perp} = 2.07$, $g_{\parallel} = 2.31$ and $A_{II} = 16.2$ mT [62].

Galactose Oxidase. This enzyme catalyzes the oxidation of the CH_2-OH group of galactose to an aldehyde group [624]. The enzyme has a molecular mass of 42000 and contains one copper atom per molecule. The EPR parameters of copper are: $g_z = 2.058$, $g_x = 2.048$, $A_x = 28.8$ mT, $A_y = 30.1$ mT and $A_z = 17.65$ mT. The authors of the work cited suggest the pseudosquare-planar symmetry of the ligand environment.

Other copper-containing enzymes have also been described in the literature: hexose oxidase, diamine oxidases (which contain pyridoxal phosphate), nitrite reductase, urate oxidase and quarcetin-2,3-dioxygenase [62]. Data have recently been reported on a new type of Cu_4S_4 cluster in yeast Cu-thionin, in which the copper atoms are at the corners of a distorted cube, just as in the case with Fe in Fe_4S_4 clusters [625].

8.3 Electron and Dioxygen Carriers

Superoxide Dismutase. Erythrocuprein, hepatocuprein and cerebrocuprein, which were well known to biochemists in the 1930s, where then later identified as superoxide dismutases [626, 627] capable of catalyzing the following reaction:

$$2O_2^{\cdot -} \rightarrow O_2+H_2O_2 \ . \tag{8.5}$$

This reaction, which is energetically, highly favourable, nevertheless proceeds rather slowly at neutral and especially at alkaline pH because of a number of kinetic factors. The superoxide radical in biological systems is generated as a side product in a number of processes, e.g. in reactions catalyzed by xanthine oxidase and rubredoxin. The interaction of oxygen with heme proteins is accompanied by a "burst" into the medium of the superoxide radical, which is extremely dangerous to the intactness of a biological system and the processes that occur in it. It is not therefore surprising that superoxide dismutases are very widespread as protective systems which catalyze the disproportion of the superoxide ion.

Most of the enzymes under consideration are characterized by a molecular mass of about 32000 and by the presence of two copper atoms per macromolecule and one copper atom per subunit, the subunits being almost identical [62]. The optical spectra contain a broad band at $\lambda_{max} = 680$ nm, which becomes faded upon reduction. The EPR spectra are characterized by three g-factors: $g_z = 2.257$, $g_y = 2.103$ and $g_x = 2.025$ and $A_{II} = 13.2$ mT. The potential of the copper atom in the enzyme is $E_0 = 42$ mV; the reduction of copper is accompanied by the absorption of the proton. Hydrogen peroxide oxidizes one of the histidines, which are presumably bound to the copper. The presence of histidine as a ligand has been demonstrated by the spin-echo technique [628].

According to the X-ray diffraction model of a preparation from erythrocytes [67], the copper atom is bound to five histidine ligands, one of which is linked to a Zn atom. Upon replacement of Zn^{2+} by Cu^{2+} the exchange integral I is found to be equal to 26 cm^{-1} [629]. The catalytic role of copper is probably to smooth out the energy relief of the reaction according to the following scheme:

$$\begin{aligned} &Cu^{2+} + O_2^- \rightleftharpoons Cu^{1+} + O_2 \ ; \\ &Cu^{1+} + O_2^- \rightleftharpoons Cu^{2+} + H_2O_2 \ . \end{aligned} \tag{8.6}$$

The two steps proceed at high rates ($k \approx 2 \times 10^9$ M^{-1} s^{-1}). Each of the two copper atoms acts independently in the elementary step, but the reduction of one centre on the molecule immediately switches off the second centre [630]. The physiological role of this process is probably associated with the danger of producing high concentrations of H_2O_2 in the vicinity of the reduced state of copper; thus the reaction:

$$Cu^{1+} + H_2O_2 \rightarrow {}^-OH + {}^{\bullet}OH + Cu^{2+}$$

may lead to the generation of $^{\bullet}OH$ radicals which break down the system.

Oxygen Carriers. The hemolymphs of various molluscs and arthropods contain oxygen-carrying copper non-heme proteins, hemocyanins [62, 631]. The molecular mass of the smallest functional unit is 50000 for molluscs and 75000 for arthropods. Each unit contains two copper atoms and is made up of two protein subunits. In the absence of oxygen both copper atoms are in the univalent state; the EPR and magnetic susceptibility studies demonstrate the absence of paramagnetic susceptibility. The absorption spectrum of oxyhemocyanin shows two bands, one of which with $\lambda_{max} = 340$ nm is ascribed to a charge-transfer complex and the other at $\lambda_{max} = 440$ nm, which is less intense, is evidently due to the d-d transition in the ligand field of low symmetry. The incorporation of O_2 does not lead to the appearance of paramagnetism.

The results described above and also some of the properties of the model complexes have led to the suggestion [632] of a bridge model for the oxyhemocyanin complex in which the O_2 molecule is located between the two copper atoms. The transitions between the various forms of hemocyanin are shown in the following scheme:

$$
\begin{array}{ccc}
[Cu^{1+}O_2\ (\uparrow\downarrow)\ Cu^{1+}] & \rightleftharpoons & [Cu^{2+}\downarrow\ O_2^{-\,-}\ \ Cu^{2+}\uparrow] \\
\text{(II)} & & \text{(III)} \\
+O_2 \Big\uparrow\Big\downarrow -O_2 & & -2H^+ \Big\uparrow\Big\downarrow +H^+ \\
[Cu^{1+}\quad Cu^{1+}] & & [Cu^{2+}\quad Cu^{2+}] \\
\text{(I)} & & \text{(IV)}\ H_2O_2
\end{array}
$$

The transitions between the forms I, II and IV have been recorded experimentally. In particular, hydrogen peroxide can recover diamagnetic oxyhemocyanin from paramagnetic methemocyanin (IV). It is interesting that the conversion I→IV actually corresponds to the reduction of oxygen in the corresponding copper-containing oxygenases. But in the case of oxygen carriers the reaction is stopped at the structure which is intermediate between forms II and III and in oxygenases the II→IV transition is thermodynamically and kinetically favourable.

An important work has recently been published [633] in which the EXAFS method was used for the study of the active-site structure of hemocyanin from *Basycon canaliculatum.* The Cu-Cu distance is 0.339 nm in the deoxy-form and 0.368 nm in the oxy-form. The distance to the three closest ligand atoms is 0.195 nm, which corresponds to the nitrogen atom of histidine. The small distance between the copper atoms is also evidenced by the high value of the exchange integral ($I \sim 538\ cm^{-1}$) [634]. The position of the absorption band in the resonance Raman spectrum of oxyhemocyanin ($745\ cm^{-1}$) points to the presence of a peroxide O–O bond. The two copper atoms, the oxygen atom and the unknown ligand X (most likely, S or O) lie in a single plane. The other atom occupies the apex of a square pyramid [633]. All these data and X-ray results permit one to construct a model of the binuclear active site shown in Fig. 8.4 [632]. There are experimental arguments in favour of the view that the active sites of dopamine oxidase and tyrosinase have the same structure [623]. It may be presumed that the binuclear 2 Cu (III) unit in blue enzymes has a similar structure.

Electron-Transfer Proteins. Plants and microorganisms contain mononuclear and binuclear copper-containing electron-transfer proteins, i.e. plastocyanins and mononuclear azurins [63, 606, 635]. These electron-transfer proteins have a rather high redox potential (+200 to +400 mV). The optical spectra of azurins and plastocyanins show an intense band with a maximum at 600 nm, which is attributed to a transition with a charge transfer. The specific features of the EPR spectrum point to the Cu^{2+} (I) type of copper with a low symmetry of the ligand environment (Fig. 8.2). According to nuclear magnetic resonance data, the copper atom in azurin is localized in a hydrophobic portion and is not accessible to water molecules. The data on fluorescence quenching indicate the presence of tryptophan in the immediate vicinity of the copper atom.

Plastocyanin is a copper-containing protein of chloroplasts, where it transfers electrons between cytochrome *f* and pigment P-700. The spinach plastocyanin, purified by fractionation with ammonium sulphate and by DEAE chromatography, has a molecular mass of 21 000 and contains two copper atoms per molecule [63]. Plastocyanin has a blue colour in the oxidized state and is colourless in the reduced state and does not undergo autoxidation in the reduced state. The absorption spectrum of oxidized plastocyanin shows peak at 460, 597 and 770 nm with extinction coefficients of 1.2×10^3, 9.8×10^3 and 3.3×10^3 M^{-1} cm^{-1} respectively. The blue colour of the enzyme is due to a charge transfer from sulphur to the d-orbitals of the copper atom. The isoelectric point of plastocyanin lies below pH 4; its redox potential at pH 7 is $E_0 = 37$ mV; below pH 5.4 the redox potential increases with an interval of 0.06 mV per pH unit. The EPR signal from the enzyme is characterized by $g_{\parallel} = 2.226$, $g_{\perp} = 2.053$ and $A_{\perp} = 0.0063$ cm^{-1}.

The X-ray diffraction model of the protein plastocyanin indicates the presence of two histidine groups and sulphur atoms of sulphhydryl and methionine groups in the coordination sphere of the copper atom with a distorted tetrahedral geometry (Fig. 8.2) [635]. An analogous conclusion is made on the basis of an X-ray diffraction analysis of azurin [606]. The fact that azurins and plastocyanins containing, respectively, one and two copper atoms per molecule have very similar physical and chemical properties, allows one to conclude that in binuclear systems each centre operates independently.

Though there are seven classes of copper-containing proteins and electron carriers, it is still possible to distinguish some common features in their structure and in the mechanism of action of their individual elements. These features are illustrated by the following scheme:

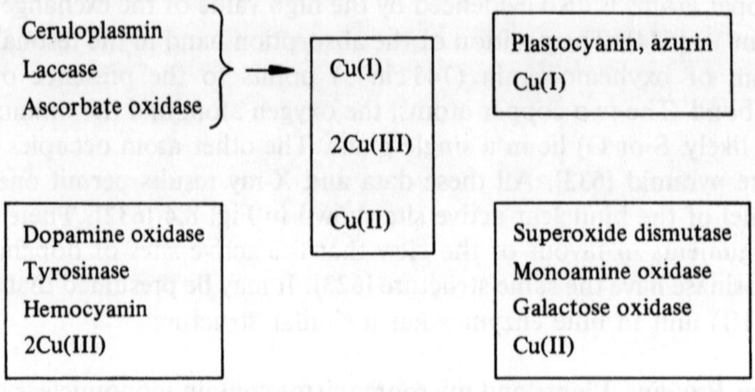

The Cu(I) type "specializes" in one-electron transfer from and to substrates; the special function of 2Cu(III) is to complex and reduce O_2; the Cu(II) type is a one-electron carrier. The two copper atoms in cytochrome *c* oxidase are close to the Cu(I) type (Sect. 8.1).

CHAPTER 9

Electron-Transport and Energy-Transformation Systems

Metalloproteins and electron-transfer proteins constitute an indispensable part of electron-transport systems. Non-heme clusters, heme and copper-containing "blocks" are the components of the electron-transfer chain. The extensive portions of these chains may be regarded as polynuclear systems of the third type made up of fragments of the first and second types (Chap. 1).

The enormous biological importance of electron-transport chains in the respiratory and photosynthetic systems and their coupling with the synthesis of ATP attracts the attention of a large number of specialists. Thousands of papers are published annually which are devoted to the various physiological, biochemical and structural aspects of this problem. In this section we shall only briefly consider the basic physico-chemical principles of the structure and mechanism of action of the following most important systems: (1) the mitochondrial respiratory chain; (2) microsomal oxidation; (3) biological oxidation of methane; (4) photoseparation of charges in photosynthesis; and (5) photodecomposition of H_2O.

9.1 The Mitochondrial Respiratory Chain

The mitochondrial electron-transport chain consists of four complexes, the functional relationship between them and the energy profile of the electron-transport process are shown in Fig. 9.1. The arrangement of the complexes in the membrane is illustrated in Fig. 9.2. As can be seen, the chemical outcome of the action of the chain is the oxidation of NADH and succinate (complexes I and II), the reduction of O_2 to water (complex IV) and the synthesis of ATP from ADP and phosphate (complexes I, II, III) [36, 636–645].

According to present-day conceptions [637, 644], the arrangement of the systems of complex I–III, II–III and IV is associated with their ability to transport electrons and to effect proton transfer across the membrane. Such a transfer is accompanied by the appearance of a difference in pH and electric potentials on both sides of the membrane, which may be the driving force for the synthesis of ATP. A detailed treatment of the coupling mechanism is not the aim of the present monograph; we shall only briefly consider this problem in Chapter 11 in connection with the mechanism of energy utilization of ATP hydrolysis for the reduction of molecular nitrogen on enzyme nitrogenase and other processes.

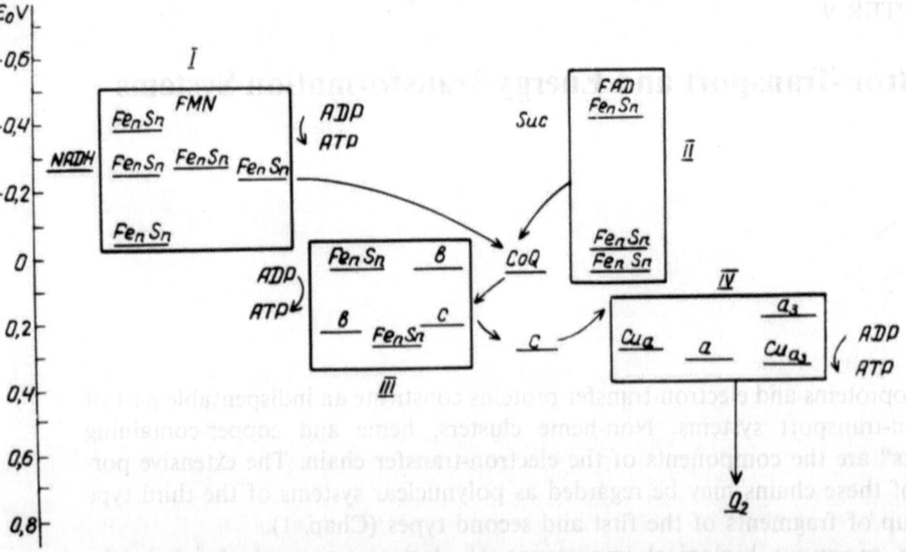

Fig. 9.1. The energy profile upon oxidation phosphorylation in mitochondria

NADH Dehydrogenase (Complex I). The oxidative part of the electron-transport chain which effects the oxidation of NADH and the reduction of ubiquinone (coenzyme Q) is a rather elaborate protein complex containing FMN (flavin mononucleotide) and non-heme iron in the ratio of 1 : 16–18 [639, 642–645].

The process of NADH oxidation occurs at a high rate and the turnover number at saturation concentrations of substrates is $8 \times 10^5\ s^{-1}$, the rate-limiting step being the transfer of electrons from the substrate [642]. Intramolecular electronic transitions and the re-oxidation of the enzyme occur even more rapidly. Ferricyanide may also serve as an electron acceptor together with CoQ. A direct attack on the substrate is effected by the flavin fragment of FMN, which can transfer hydride to other nucleotides without the participation of non-heme iron. The role

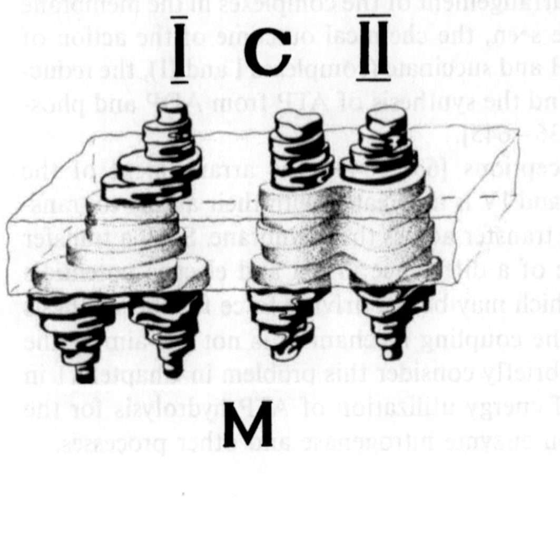

Fig. 9.2. Schematic model of the arrangement of protein complexes of the oxidative phosphorylation system in the mitochondrial membrane: *I* ubiquinone-cytochrome *c* reductase complex [657]; *II* cytochrome *c* oxidase [659]; *M* matrix; *C* cytoplasm

of the five types on non-heme iron is to transport electrons and to donate an electron to the acceptor.

Four types of EPR signals of hon-heme iron have been detected [646], differing in the position of the g-factor and in line shape: type I, $g_I = 2.022$, 1.938 and 1.92; type II, $g_{II} = 2.054$; type III, $g_{III} = 2.101$, 1.886 and 1.864; type IV, $g_{IV} = 2.103$ and 1.861. These signals differ also in their saturation ability. For example, the shortest relaxation time is displayed by the signal with g_I, which indicates the presence of a paramagnetic relaxation oscillator near the centre. The signals with g_{II}, g_{III} and g_{IV} are broadened at a temperature of 25 K. The values of E_0 for different types of non-heme iron are given in Fig. 9.1. Addition of ATP has an effect on the properties of the EPR signals with g_I and g_{II}. It has been reported [647] that complex I contains Fe_2S_2 and $3[Fe_4S_4]$ clusters.

Under certain conditions it is possible to obtain a low molecular fragment of NADH dehydrogenase which partly retains the functions and structure of the enzyme. For instance, the fragment has the ability to oxidize NADH and to reduce cytochrome *c* and indophenol and shows the four types of EPR signals. The subunit composition of the complex is very elaborate and has not yet been fully clarified. According to the latest data, there are three subunits (57000, 9000 and 24000) [648]. It is known that the flavin-containing portion of NADH dehydrogenase is located on the inner side of the mitochondrial membrane or the matrix side (M-side is the matrix side of the membrane) [650]. Experiments devoted to the study of the effect of paramagnetic Ni^{2+} ions on the EPR parameters of FeS proteins have led to the conclusion that the active sites of these proteins are located at a depth of about 1.0 nm on both sides of the membrane [649]. According to recent electron microscope studies [650], complex I from bovine heart mitochondria is a tetramer consisting of four asymmetric subunits. The length of the complex across the membrane is 8.2 nm and its dimensions are $15.2 \times 15.2\ nm^2$. The model of the complex which takes into account the data quoted above is shown in Fig. 9.2.

As a whole the polynuclear active-site system effects the switchover of the mechanism from the two-electron NADH-FMN portion to the one-electron Fe_nS_n portion (shown in the scheme at the end of this section).

Succinate Dehydrogenase (Complex II) and Coenzyme Q-Cytochrome c Reductase (Complex III). The first enzyme in mitochondria catalyzes the oxidation of succinate to fumarate and the reduction of the quinoid form of coenzyme Q; the second transfers electrons from coenzyme Q to cytochrome *c*, effecting the ATP synthesis on its pathway [639, 640].

Succinate dehydrogenase is a protein complex with a molecular mass of about 175000 daltons. One macromolecule of the enzyme contains flavin, non-heme iron and sulphide in the ratio of 1:8:8 and can be broken down into its constituent subunits. One of the subunits (molecular mass 70000) contains flavin, 4 Fe and 4 S and the other (molecular mass 27000) is made up of 2 Fe and 2 S [640]. A preparation from photosynthetic bacteria has a similar composition. Succinate is oxidized with a turnover number of $5 \times 10^3\ s^{-1}$. In an isolated system the electron acceptors are O_2 and ferricyanide. In the reduced form the enzyme gives two types of the EPR signal of the non-heme iron of Fe_2S_2 (I and II), these two types being characterized by the following values of the g-factor: $g_z = 2.03$, $g_y = 1.93$ and $g_x = 1.91$; the third type of EPR signal is also observed: $g_z = 2.03$, $g_y = 2.02$ and $g_x = 2.00$ [647]. The type II centre is reduced only by dithionite. The EPR signal from the semi-quinone radical of flavin is seen only after the reduction of the non-heme iron [647]. The values of E_0 for the components are given in Fig. 9.1. The dependencies of the EPR spectra I and II on temperature and the presence of forbidden signals in low fields in native and reconstituted systems have revealed an appreciable spin-spin interaction between these centres, which are separated by a distance of about 1.0 nm [653]. According to the literature data [647, 652], succinate dehydrogenase contains [4Fe·4S], [3Fe·xS] and [2Fe·2S] clusters, the first and the third being very close to each other.

Taking into account the one-electron nature of non-heme iron, the ability of FAD to undergo a two-electron reduction, the difficulty of abstracting a hydrogen atom or one electron from succinate and also the redox potentials of the components of complex II, which are given in Fig. 9.1, it is possible to visualize that the scheme of the switchover mechanism is succinate dehydrogenase, presented at the end of this section.

Coenzyme Q-cytochrome *c* reductase contains at least six elements [639, 642, 652–655]: three types of non-heme iron complexes, two type *b* cytochromes and type c_1 cytochrome.

One monomer of complex III has a molecular mass of 275000 and is made up of eight subunits. Three subunits carry redox centres: *b* (30000), c_1 (31000) and an iron-sulphur protein of the Fe_2S_2 type with g = 1.89 (25000). The other five subunits have a molecular mass ranging from 50000 to 80000. Cytochrome c_1 contains a myoglobin like heme and consists of two subunits, a hydrophilic section (23000) and a hydrophobic section (8000). Experiments involving chemical modification have demonstrated the localization of cytochrome c_1 on the M-side of the membrane, where it interacts with ferrocytochrome *c* with a complex-formation constant $K = 10^{-8}$ M. Proteins without active sites are localized on the cytoplasmic (C-side) side of the membrane. According to electron microscope studies of two-dimensional membrane crystals, the molecules of complex III form dimers. Each monomeric complex is located across the membrane in such a way that it projects by 3 nm above the membrane plane, the remainder being buried by 5 nm in the membrane, the embedded portions containing a Fe_nS_n protein, and by 7.0 nm to the side of the cytoplasm (Fig. 9.2). The electron transfer in complex III occurs by the one-electron mechanism.

Cytochrome c Oxidase (Complex IV). Of greatest interest from the viewpoint of the problem of the effect of the interaction between metal complexes on the reactivity of the complex is cytochrome *c* oxidase, which catalyzes the reduction of oxygen to water under the action of the one-electron carrier cytochrome *c* [43, 639, 642, 644]. The functional unit of the complex is a polynuclear system, which is composed in the oxidized state of one copper atom of type Cu_a(I) (see Chap. 7), low-spin cytochrome *a*, high-spin cytochrome a_3 and a Cu_{a_3} atom, which is linked anit-ferromagnetically to the heme iron of cytochrome a_3.

The monomer of cytochrome *c* oxidase with a molecular mass of 140000 consists at least of seven subunits with a molecular mass of 36000 to 5400. There are arguments in favour of the view that each paramagnetic centre is located on a separate subunit. Cytochrome *a* is located on the C-side of the membrane, where it oxidizes cytochrome *c*, while cytochrome a_3 is on the M-side of the membrane. Details on the localization of various subunits are given elsewhere [43, 642]. Electron microscope studies have revealed the following structure of the monomeric form of the cytochrome *c* oxidase complex [659]. The monomers, 11-nm-long, have a Y-shape. The two domains with a size of about 5 nm (the distance between the domains being about 4 nm) are partly embedded in the matrix. The large domain that forms the base of the monomer projects partly into the cytoplasmic portion. The monomers in the membrane evidently form assemblies. This is evidenced by electron microscope data on labelling with antibodies having conjugated ferritin molecules as electron-scattering labels [659] and by the large values of the times of rotational motion of the complexes [660]. The EPR and optical characteristics observed for the oxidized enzyme indicate the presence in the unit of one Cu_a(I) atom, whose EPR spectrum is noted for the unusually low value of the HFS constant [62]. At the same time, in the system the EPR signal of the low-spin heme of cytochrome *a* ($S = \frac{1}{2}$) is observed. The low-spin nature of heme a is confirmed by the magnetic circular dichroism spectra at the 428-nm band. The absence of an appreciable broadening in the EPR spectra of copper and iron (exceeding 1.0 mT) allows one to presume that these centres are separated by a distance of at least 0.8–1.0 nm.

There are serious arguments in favour of the view that the second copper atom (Cu_{a_3}) is in the immediate vicinity of the high-spin heme group a_3 and that they are linked by a strong anti-ferromagnetic interaction: (1) the absence of the EPR signal from cytochrome a_3 and from the second copper atom; (2) data on the temperature dependence of magnetic susceptibility (the magnitude of the exchange integral $I = 200\ cm^{-1}$), the spin state $S = 2$; (3) Mössbauer spectroscopy data [350, 351].

The most detailed data on the active-site structure have been recently obtained by EXAFS [68, 662], which made it possible to determine the distance between the central atoms and the ligand atoms. The following values of the distances have been obtained (in nm): $Fe_{a_3}-Cu_{a_3}$ (0.375), $Fe_{a_3}-S$ (0.260), $Fe_{a_3}-N$ (imisazole in position 5, 0.214), $Fe_{a_3}-N$ (heme, 0.201), $Cu_{a_3}-S$ (bridge, 0.218), $Cu_{a_3}-S$ (0.228), $Cu_{a_3}-N$ (0.97), Cu_a-S (0.227), Cu_a-N (0.197), Fe_a-N (0.199). These data indicate that the Fe_{a_3} and Cu_{a_3} atoms are linked by a bridging sulphur. By the nature of the environment, cyt a_3 is similar to cyt *c*, cyt *a* is similar to cyt *b* and the copper atoms Cu_{a_3} and Cu_a are similar to the atoms in blue copper-containing proteins, e.g. in stellacyanin. The scheme that takes these data into account is shown in Fig. 9.3.

Upon interaction with O_2 the bridge is cleaved and O_2 becomes linked to a_3. The $Fe_{a_3}-O$ distance is 0.175 nm. In the oxidized state $a^{3+}Cu_a^{2+}Cu_{a_3}^{2+}a_3^{2+}O_2$ an EPR signal appears which is ascribed to $Cu_{a_3}^{2+}$ [353]. The other distances in the complexes ar $R_{Cu_aFe_a} \sim 1$ nm and $R_{Fe_aFe_{a_3}} \sim 1.4$ nm [652].

A number of data point to the functional relationship between hemes a and a_3 and between Cu_a and Cu_{a_3}, which is evidently an indication of their spatial disposition, thus allowing their cooperative binding [658]. For example, binding of CO by cytochrome a_3 causes distinct changes in the absorption spectra of

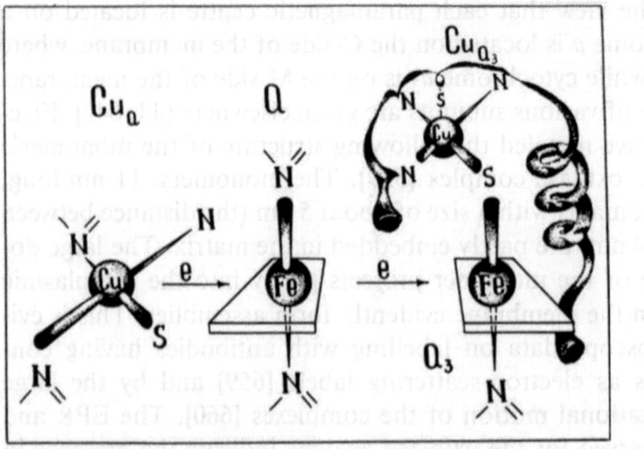

Fig. 9.3. The active site of cytochrome *c* oxidase. Constructed according to data reported elsewhere [68, 662]

heme a [663, 664]. Simultaneously, the EPR method records the shift of the heme of cytochrome *a* to the low-spin form [350, 661]. On the other hand, the reduction of heme a contributes to the conversion of a_3 to a centre accessible for complex formation. Resonance Raman spectroscopy has revealed the possibility of CO transfer from Fe_{a_3} to Cu_{a_3}. The rate of electron transfer from the reduced "primary" acceptors a and $Cu_a(I)$ to the secondary acceptors a_3 and Cu_{a_3} is not high and increases sharply upon complex formation with O_2. This property is evidently of great physiological significance, i.e., it prevents a "blank" electron transfer. The incorporation of O_2 into the coordination sphere of Fe_{a_3} increases the oxidizing ability of the complex. Electron transfer from cytochrome *c* to the enzyme converts it to a very active state. Two binding sites for cyt *c* have been detected, one with a high ($K = 10^{-8}$ M) and the other with a low ($K = 10^{-6}$ M) affinity. Binding of cyt *c* in the second site activates the electron transfer [656]. The primary acceptor a may be in two conformations with different E_0^0 [666].

Upon reduction of cytochrome *c* oxidase new properties appear [350, 351]. For a unit composed of paramagnetic atoms to be completely reduced, two electrons are required; the iron of hemes a and a_3 is converted to the low-spin ($S = O$) and the high-spin ($S = 2$) forms respectively, and the two copper atoms are converted to the univalent state. The ability to bind oxygen, CO, NO and other ligands via heme a_3 appears in the reduced state. Carbon monoxide stabilizes the reduced state of heme a_3. The rate of O_2 interaction with the reduced enzyme is very high ($k_I = 10^8\ M^{-1}\ s^{-1}$). If the reaction is carried out at a low temperature ($-130\,°C$), the oxy-form can be recorded, which is converted to a compound of the peroxide type at $-100\,°C$ and then to water at $-60\,°C$.

An intermediate product has been recorded [664] for the reduction of oxygen in the coordination sphere of a_3, which is a derivative of the peroxide type. It has been demonstrated in ingenious experiments, in which the state of the cytochrome *c* oxidase unit with two electrons was obtained by photodissociation of the corresponding carbonyl complex of a_3, that the doubly reduced state of the enzyme is capable of forming a complex with O_2 and an intermediate transition state. Another two electrons have to be supplied step by step for further reduction with formation of the ferri-intermediate [658]. Accounting of these results and also of the obvious one-electron mechanism of the reduction of the enzyme by cytochrome *c* makes it possible to assume the following mechanism of electron transfer via cytochrome *c* oxidase in terms of the symbols proposed in Chapter 1:

$$[1(4) \rightarrow 1(4) \rightarrow 2(1) \rightarrow 1(2)] \; .$$

The active-site structure of cytochrome *c* oxidase and its localization in the membrane are shown schematically in Figs. 9.2 and 9.3. The transfer of two electrons to O_2 occurs at $k_e \sim 2 \times 10^4 \, s^{-1}$; the further transfer and protonation take place at $k_e \sim 2 \times 10^3$ and $8 \times 10^2 \, s^{-1}$ [663].

Thus, the mitochondrial transport system accomplishes the switchover of the mechanism from the two-electron transport (upon oxidation of NADH and succinate by the flavin fragment) to the one-electron transfer within the chain. The mechanism is reversed (the switchover from the one-electron to the two-electron transfer) at the portion of the terminal oxidation with the aid of a heme $a_3 - Cu_{a_3}$ pair. The most important event in the overall process is the synthesis of ATP molecules. The general scheme of electron transport may be written in the following form:

$$2 \rightarrow 1 \rightarrow 1 \longrightarrow 1 \rightarrow 1 \rightarrow 1 \rightarrow 1 \rightarrow 1 \rightarrow 1 \rightarrow 1 \rightarrow 1 \rightarrow 2(2) \, .$$

$$2 \rightarrow 1 \rightarrow 1 \rightarrow 1 \rightarrow 1 \rightarrow 1 \rightarrow 1 \cdot$$

The electron transfer within the chain is effected with the aid of a number of one-electron carriers, the role of which may be played either by iron-containing heme and non-heme centres, which are rigidly embedded in protein complexes, or molecules of coenzyme Q and cytochrome *c*, which are more or less free to move inside or along the membrane, performing a shuttle motion between the complexes. The electron transfer through the membrane is accompanied by the buildup of a proton concentration or electric potential gradient across the membrane. This fact is of great physiological importance and probably has an impact on the synthesis of ATP from ADP and free phosphate. However, the detailed mechanism of the synthesis of ATP, just as the problem of the necessity of the membrane for its realization, remains unclear.

9.2 Electron-Transfer and Water-Oxidation Systems in Photosynthesis

The main outcome of photosynthesis is the oxidation of water and the synthesis of glucose from carbon dioxide at the expense of sunlight energy. The immensity of this process on the scale of the Earth is evidenced by the fact that annually about 200 billion tons of carbon is bound from carbon dioxide into forms that provide energy and structural material for all living organisms on Earth. The general scheme of photosynthesis is given in Fig. 1.2.

The enormous problem of photosynthesis embraces practically all aspects of modern biochemistry and molecular biology; we shall briefly consider only two aspects which seem to be of fundamental importance: (1) the structure and mechanism of action of the system of conversion of light energy into chemical energy in the primary charge photoseparation in bacterial and plant photosynthesis; and (2) the possible mechanisms of the participation of polynuclear manganese systems upon the photo-oxidation of water. The first system is a remarkable example of the appearance of qualitatively new properties upon combination of the centres into a system of the third type (see Chap. 1) and the second accomplishes one of the most surprising reactions that occur in nature, the production of strong reducing agents from water due to the energy of relatively soft light quanta.

9.2.1 The Primary Photoelectron-Transport During Bacterial and Plant Photosynthesis

The first result of the conversion of light energy into chemical energy in bacterial photosynthesis is the production of relatively strong reducing agents (reduced ferredoxin, NADH, molecular hydrogen) from weak mineral reductants, such as hydrogen sulphide, divalent iron, etc. Another result is the synthesis of ATP in cyclic and non-cyclic phosphorylation systems. There is every ground for believing that some of the parts of systems I and II of plant photosynthesis are similar in functional and structural properties to the bacterial system [665–667].

The energy and kinetic features of electron-phototransport will be discussed in Chapter 11 in connection with the problem of light energy storage. Here, we shall only note that the most important problem is the mechanism of photoseparation of charges, D_I^+ and A_I^-, where D_I and A_I are the primary electron donor and acceptor respectively, in which their immediate recombination is prevented:

$$[D_I A_I] \xrightarrow{h\nu} [D_I^* A_I] \rightarrow [D_I^+ A_I^-] \ .$$

The particles D_I^+ and A_I^- are a strong oxidant and a strong reductant respectively, and hence, accumulate light absorption energy.

The primary photophysical processes of photosynthesis take place within small membrane-bound complexes of pigments and proteins, which have been given the name "reaction centres" (RC). The pigments of these complexes differ in their structural and functional features from the other pigments of photosynthetic membranes [665–667]. An important advance in the understanding of the key steps in the energy transformation process of short-lived states (10^{-11} s), generated upon absorption of a light quantum to the energy of long-lived (10^{-1} – 10 s) or stable compounds (ATP, NADPH), has been achieved due to the isolation and purification of reaction centres.

It has been established that the basic low molecular components of the reaction centres are the pigments, bacteriochlorophyll (Bchl) and bacteriopheophytin (Bph). The latter differs from Bchl in that the central Mg atom in the tetrapyrrole ring is replaced by hydrogen atoms. One mole of RC contains 4 mol Bchl and 2 mol Bph. The pigments in photoactive reaction centres give absorption bands λ_{max} at 870, 800 and 600 nm and also absorption band λ_{max} at 760 and 535 nm and a common band at 370 nm. Apart from the pigments, the reaction centres contain one non-heme Fe atom per mole of RC and 1 – 2 mol ubiquinone (CoQ).

The protein component of RC from *Rhodopseudomonas viridis* consists of three polypeptide chains having sharply pronounced hydrophobic properties. The molecular masses of the polypeptides are 37571 (L), 35902 (M) and 28345 (H) daltons [676]. The subunit (H) does not carry pigments and the presence of the complex of subunits L and M is sufficient for the photochemical activity.

Some achievements have been recently made in establishing the basic photophysical steps and their time sequence and in assigning these steps to structures of a definite chemical nature [668, 669].

A series of excellent studies based on the use of subpicosecond and nanosecond methods have been devoted to the measurement of the kinetics of the principal electronic transitions during the separation of charges [666 – 671]. The light quanta are accumulated by so-called antenna proteins, which are chlorophyll-protein complexes [673]. Each protein subunit contains chlorophyll molecules, which are arranged, according to X-ray diffraction data, almost in parallel to one another at an average distance of about 1.2 nm. The minimal distance between the molecules in different subunits is about 2.4 nm. These distances are sufficient for an effective migration of energy by an inductive-resonance mechanism (Sect. 3.2.2). The singlet electronic excitation migrates along the "antenna" and enters the primary acceptor P_{870}, the dimer of bacteriochlorophyll $(Bchl)_2$, which also passes over into the singlet state.

This is followed by a chain of events (Fig. 9.4). During a time of the order of a picosecond, an electron is transferred to a molecule of bacteriochlorophyll $(Bchl)_1$ and then, in 5 – 10 ps, to bacteriopheophytin Bph and, in 200 ps, to the primary acceptor A_I (Q_A). These reverse transfer occurs at rates in the millisecond range. During this time the electron from the secondary donor, e.g. type *c* cytochrome, has the chance to be transferred to the cation radical of the chlorophyll dimer or A_I reduces the secondary acceptor A_2. As a result, the energy of a solar quantum is transformed to the energy of photoseparated charges.

The main changes in the physical properties upon electron transfer include the bleaching of the band at 870 nm in $(Bchl)_2$ with the appearance of an EPR signal with $g = 2.002$ and a width of 0.95 mT, the spectral changes of Bph and $(Bchl)_1$ and also the appearance of the EPR signal from Bph and a very broad EPR signal from the primary acceptor with $g \approx 1.8$ observed at 10 K. The primary acceptor is presumed to contain an ubiquinone molecule Q_A, near which there is an iron atom [670]. Mössbauer experiments have established that the iron atom is in the high-spin Fe^{2+} state and does not change valency upon electron transfer to the primary acceptor [671].

The spatial localization of donor and acceptor in centres has been studied [672, 674]. The experimental approach, which consists of the determination of the distance between the cation radical $(Bchl)_2^{+}$ and the paramagnetic state of the reduced acceptor, a complex of Fe with uniquinone $(Q \cdots Fe)$, is based on the method described in Section 3.1.3. The first series of experiments were concerned with the determination of the relaxation time of the EPR spectrum of the cation radical P_{870}^{+} of chromatophores from *Rhodospirillum rubrum* under conditions

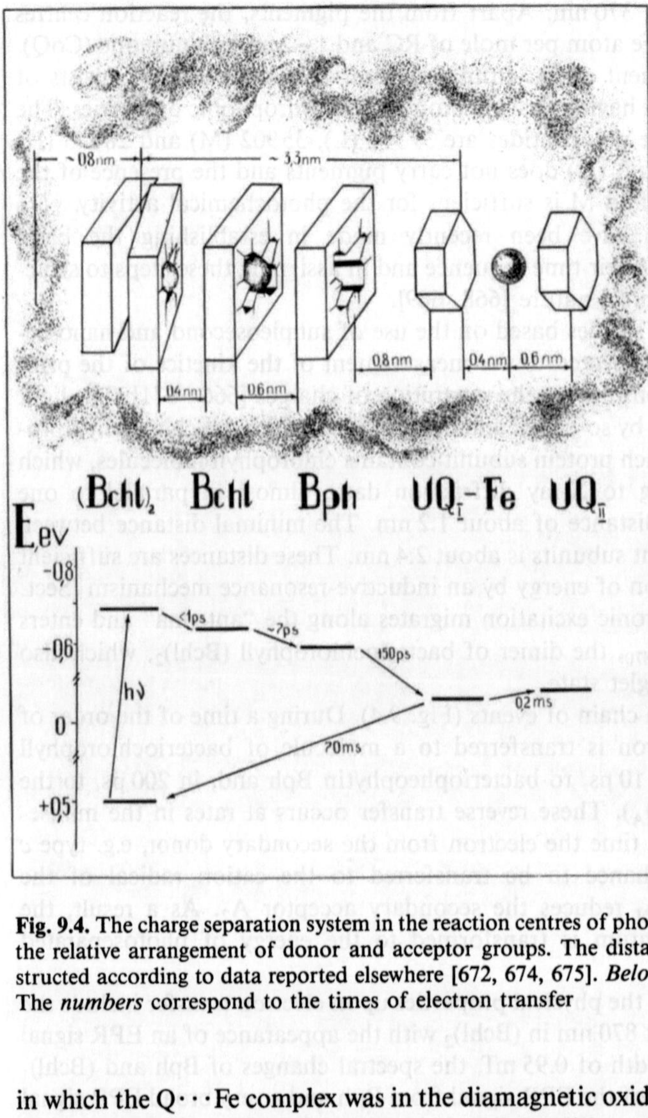

Fig. 9.4. The charge separation system in the reaction centres of photosynthetic bacteria. *Above:* the relative arrangement of donor and acceptor groups. The distances are given in nm. Constructed according to data reported elsewhere [672, 674, 675]. *Below:* the energy-level diagram. The *numbers* correspond to the times of electron transfer

in which the Q···Fe complex was in the diamagnetic oxidized state (the oxidation of P_{700} by iodine or the adaptation in the light with a photoelectron transfer to the secondary acceptors). In the second series of experiments P_{870} and Q···Fe were converted simultaneously into the paramagnetic state by photoreduction upon illumination at a temperature at which the EPR signal was recorded (77 °K).

A comparison of the temperature dependence of the rate of spin-lattice relaxation, $1/T_1$, in both series of experiments has made it possible to separate the paramagnetic contribution ($\Delta(1(T_1))$) to the relaxation of P^+_{870} from the electron spin of $Q^{(-)}$···Fe. This has, in turn, enabled the calculation of the distance between the paramagnetic centres by formula (3.12) which was found to be equal to 3.4–3.6 nm (Fig. 9.4).

New data have recently been obtained on the presence of spin-spin and electrostatic interactions between the components of RC and on the incorporation into the electron-transfer chain of another intermediate acceptor, a molecule of bacteriochlorophyll [295, 672, 674, 675]. Analysis of these data permit one to suggest the following scheme of electron transfer in RC:

$$\underset{\text{I}}{(\text{Bchl})_2} \rightarrow \underset{\text{II}}{(\text{Bchl})_1} \rightarrow \underset{\text{III}}{\text{Bph}} \rightarrow \underset{\text{IV}}{\text{Q}\cdots\text{Fe}} \rightarrow \text{A}_2 \ .$$

The distance between the components has been estimated using the calibration dependence of the exchange integral I on the distance between the centres (Fig. 11.3).

As can be seen from Figs. 9.4 and 9.5, the model proposed on the basis of a physico-chemical investigation agrees well with the X-ray model suggested recently [675]. According to this model of the reaction centre from *R. viridis* (resolution 0.3 nm), the pigments and other centres are located at the following distances (in nm): Mg-Mg in $(Bchl)_2$ (0.7), the angle between the tetrapyrrole planes being about 15°; the Mg-Mg between bacteriochlorophylls $(Bchl)_2$ and $Bchl_1$ (1.3), the angle between the planes being 70°; the distance between the $Bchl_1$ centre and the Bph centre is 1.1 nm, the angle between the plances being 64°; the Fe-Q distance is about 0.7 nm. The protein ligands of bacteriochlorophyll groups and non-heme iron are histidine residues. Attached to the reaction centre is cytochrome c_3, which contains four hemes (Fig. 9.5) arranged in pairs. Inside the pair the Fe-Fe distance is 1.4 nm, between the neighbouring pairs the Fe-Fe distance is 1.6 nm. The fifth iron ligand is histidine imidazole and the sixth ligands are histidine and threonine residues. Cytochrome c_3 is bound to the plateau of the reaction centre formed by the L- and M-chains from the side of the matrix. The H-subunit is located on the cytoplasmic side of the membrane.

A controversy has recently been carried on [226, 667, 680] concerning the direct participation of the bacteriochlorophyll molecule in the electron transfer of the anion radical $Bchl_1$ to pheophytin. The formation of this radical has been experimentally proved [673, 680] and disproved [680]. Note that in any case, with the known fixed $(Bchl)_2$-Bph distance equal to about 1.4 nm, the transfer with $k \sim 10^{12}\,s^{-1}$ seems to be improbable without the direct (via the formation of an intermediate anion radical) or indirect (via the corresponding orbital overlap, which increases V) participation of the conjugated system of the bacteriochlorophyll molecule.

The reaction centres in photosystems I and II of green plants have not yet been revealed. However, the application of physical methods has made it possible to establish in general outline the nature of the pigments and the kinetics of photochemical conversions in the reaction centres in preparations enriched with photosystems I and II [667, 677, 678, 682, 683, 887].

In photosystem I of green plants the primary donor is the chlorophyll dimer (P700) and the first intermediate acceptor is the chlorophyll monomer 695 (Chl 695). During the primary photoseparation with the formation of $P700^+$, Chl 695^- the energy of the system decreases by a factor of 0.4 eV (Fig. 9.6). The electron is transferred for 200 ps to the iron-sulphur protein X and then to the neighbouring iron-sulphur protein P-430. The state $P700^+ \cdot Chl\ 695 \cdot X \cdot P430^-$ is relatively stable ($\tau \sim 40$ ms) due to the considerable spatial separation of charges.

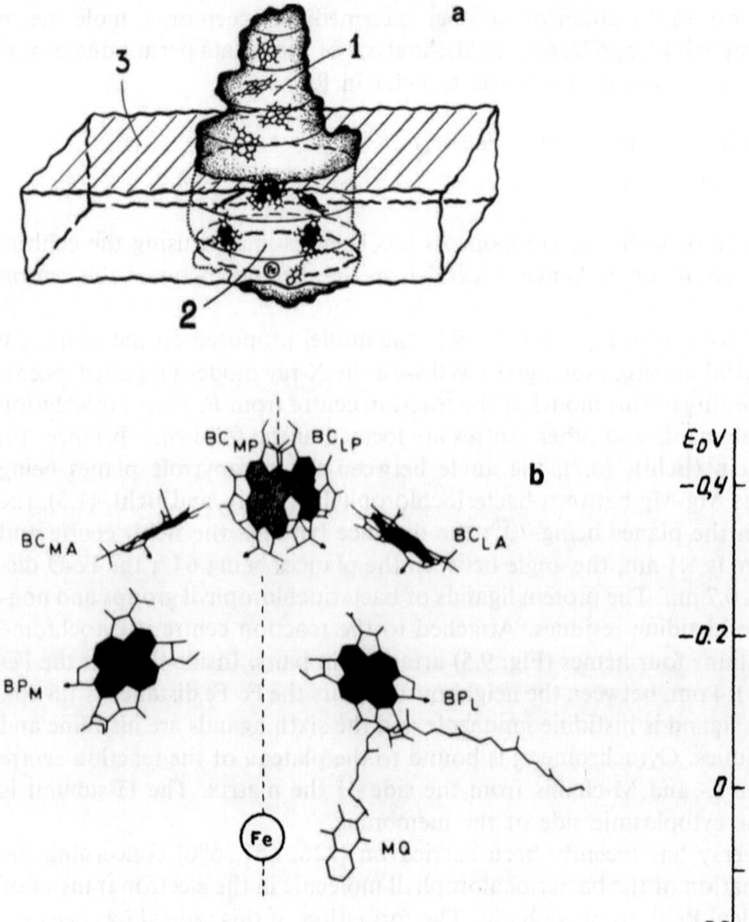

Fig. 9.5 a–c. Model of the reaction centre (RC) of the photosynthetic bacterium *Rhodopseudomonas viridis:* **a** General scheme of arrangement of the complex RC (*2*) with cytochrome *c* (*1*) in the membrane (*3*) (constructed on the basis of X-ray data [676]. In cytochrome *c* the arrangement of heme groups is shown; shown in RC is the location of the pigments and Fe. **b** Molecular organization of pigments in RC; *BC* bacteriochlorophyll b; *BP* bacteriopheophytin; *MQ* menaquinone. The subscripts *L* and *M* indicate the proximity to the L and M protein subunits respectively; *P* and *A* denote the special pair and the accessory of Bchl respectively [822]. **c** Enlarged model of the special pair [822]

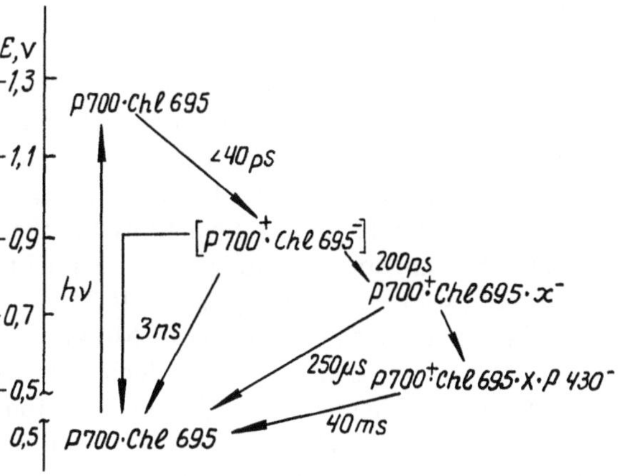

Fig. 9.6. The energy and kinetic schemes of the processes in the reaction centres of photosystem I of plant photosynthesis [667]). For designations, see text

P430 is a sufficiently strong reductant (E = −0.55 V), which is capable of reducing ferredoxin and $NAPH^+$.

The kinetics of charge photoseparation on the reaction centres of photosystem II of green plants and their structure are generally close to those of the above described reactive centres of photosynthetic bacteria (Fig. 9.7). The specificity of the system in question is that the oxidized primary donor $P680^+$ is reduced by the water oxidation centre S via two intermediate derivatives of plastoquinone, D_1 and D_2 [679].

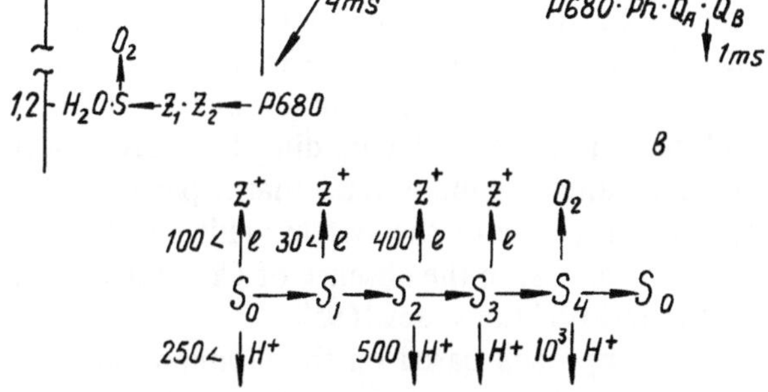

Fig. 9.7 a, b. Energy and kinetic schemes of the processes in the reaction centres of photosystem II of plant photosynthesis (on the basis of data reported elsewhere [667, 684]. For designations, see text. **a** Photoseparation of charges; **b** electron and proton transfer processes in the centre of oxygen evolution from water. The values of the times of the corresponding transitions are given in μs

9.2.2 The Role of Manganese in the Photo-Oxidation of Water to O_2

The photosystem II has a unit which releases O_2 under the action of light and which is adjacent to the reaction centre of the photosystem [681–687]. The reaction centres contain molecules of chlorophyll, pheophytin and primary acceptor. In his classical works Koch has shown that for O_2 to be evolved four light flashes are required [692]. As well be shown in Chapter 11, this is dictated by the thermodynamic specificity of water oxidation. Manganese has long been assumed to play a role in the photosynthetic decomposition of water by plants [688]. This supposition is confirmed by the correlations between the content of Mn and the rate of O_2 evolution in chloroplasts. At present, there are data which indicate that the thyllakoid membranes of photosystem II in plants incorporate about four manganese atoms per one reaction centre (per 400 chlorophyll molecules), the presence of which is necessary for the normal functioning of the system of O_2 evolution from H_2O under the action of light. The removal of at least one of the four Mn atoms leads to the malfunctioning of the entire system. Between Mn and RC there is one more carrier, which is presumed to be plastoquinone [684, 685].

A number of proteins have been isolated from chloroplasts with a molecular weight of 18 to 34 kD [684]. A correlation was observed between the ability of the preparations to evolve O_2 and the content of the inner protein with 34 kD. Japanese authors [686, 687] have reported the isolation of a complex from chloroplasts of photosystem II and cyanobacteria, which they called the "water: plastoquinone oxidoreductase" complex. The complex contains a group of five peptides of molecular mass 47, 40, 35, 30 and 9 kD and was found to be capable of evolving O_2 upon illumination in the presence of ferricyanide as an acceptor with an efficiency of up to 850 μmol O_2 mg^{-1} chlorophyll h^{-1}.

In the currently adopted terminology, the redox state of the O_2 evolution system is designated as S_0, S_1, S_2, S_3, S_4 in accordance with the number of electrons transferred from the complex. The evolution of O_2 is realized in the presence of inorganic ligands, which are arranged in the following series according to their efficiency: $Cl^- > Br^- \gg N_3^- > I^-$ [684]. Hydroxylamine is the inhibitor of the process. On the basis of the nature of the effect of the substrate structure and inhibitors on the kinetics of the process it has been concluded that the size of the active-centre gap is 4×2.5 Å. Apart from water, the system oxidizes in the light such ligands as hydrozine and H_2O_2, which are simultaneously the inhibitors of H_2O oxidation and which hinder the incorporation of Cl^- and other anions [694]. Presumably, the chloride ion is not a ligand directly involved in the coordination structure of manganese and it is more likely that it participates in the electron transfer from the reaction centre of the system. This conclusion is based on the observation of states S_2 and S_3 in the absence of Cl^-; the chloride ion, however, has an effect on the ratio of the states [695].

The data obtained by using the approach based on the illumination of the chloroplast systems by successive short flashes with a duration of the order of microseconds followed by EPR observation of the illumination results and by measuring the rate of the magnetic proton relaxation of water indicate the participation of an assembly of metal atoms in the photodecomposition of water [681, 684, 685, 698, 700–702].

In particular, the appearance of broad EPR signals has been reported upon illumination of chloroplasts with g = 2.79 [702] and g = 4.1 [685], whose intensity varied from flash to flash. Depending on the number of flashes spectral changes were observed in the region of 280–380 nm. According to data [700], these changes were monotonic and indicated the transition $Mn^{3+} \rightarrow Mn^{4+}$. According to other authors [701], the transitions $S_0 \rightarrow S_1$ and $S_2 \rightarrow S_3$ are not accompanied by spectral changes; the transition $S_1 \rightarrow S_2$ manifests itself distinctly in the spectra. The reduction time of the acceptor $P680^+$ increases with increasing degree of oxidation of the oxygen-evolving centre and is 23, 50, and 260 ns for reactions with the states $S_{0,1}$, S_2 and S_3 respectively. This process involves the release of protons into the medium: $S_0 \rightarrow S_1$ and $S_2 \rightarrow S_3$ (one proton) and $S_3 \rightarrow S_0$ (two protons) [685]. According to EPR spectra, new evidence has recently been obtained in favour of the functioning of a Mn-containing cluster in the reaction centres of photosystem II. An EPR signal has been detected at 35 K, which corresponds to several interacting Mn atoms with $I = 19\,cm^{-1}$ surrounded by nitrogen ligands [685]. Analysis of the hyperfine structure of the spectrum containing 20 lines has led to the suggestion that a binuclear or tetranuclear $Mn^{4+}Mn^{3+}$ cluster exists. In the S_2 state the intensity of the EPR signal increases sharply at the first light flash, decreases at the second flash and again increases somewhat at the fifth flash. All these effects disappear upon removal of Mn by dithionite or chelates. The incorporation of $[^{17}O]$ into H_2O leads to a broadening of the spectrum corresponding to a complex with H_2O_2 [699].

The cluster nature of the manganese complex has been convincingly demonstrated by Kulikov and co-workers [703]. Manganese was first removed from a preparation of photosystem II. In the system conditions were created for the appearance of an EPR signal from the anion radical of pheophytin $Ph^{\dot{-}}$ (a reduced medium, illumination); then Mn was gradually introduced. The state of the manganese complex was traced by the extent of saturation of the EPR signal of the anion radical, and the functional activity of the system by delayed fluorescence. As has been shown by experiments, the incorporation of one Mn atom into the centre changes sharply the saturation curve parameters of the EPR signal of $Ph^{\dot{-}}$; the introduction of a second Mn atom partly removes this effect; a third Mn atom again affects the saturation curves and, finally, the incorporation of further Mn atoms into the medium has no effect on the EPR spectrum of $Ph^{\dot{-}}$. Upon preliminary addition of Mg only two Mn atoms were incorporated into one centre. The introduction of four Mn atoms or two Mn atoms and two Mg atoms into the centre completely regenerates delayed fluorescence, which is indicative of the normal functioning of the donor portion of photosystem II.

The presence of four Mn atoms in the water oxidation centre of photosystem II of chloroplasts has been confirmed by Yocum and co-workers [691]. These authors removed Mn from the system and traced the results by the saturation of the EPR signal from the ion radical with g = 2.0 (the $Z^{\cdot +}$ or D^+ centre) at room temperature. This signal is ascribed to the product of plastoquinone oxidation by the radical cation of chlorophyll P^+_{680}. The relationship between the EPR signal of $Z^{\cdot +}$ with rapid Mn oxidation ($\tau \approx 10^{-6}$ s) and O_2 evolution has been shown in a number of works [691, 698]. It has been found that upon removal of one or three Mn atoms the signal of $Z^{\cdot +}$ becomes saturated with difficulty

($H_m \approx 40$ mW); upon removal of two atoms the magnitude of the saturation power decreases by a factor of about two. Upon transition from $S_0 - S_4$ the spin-relaxation time falls off monotonously. The use of the EXAFS method [681] has revealed the presence of at least two Mn atoms in the complex separated by distance of 0.27 nm. The ligands for Mn are N and O (r = 0.181 and 0.215 nm) According to the opinions of the authors, the most probable state in the dark is Mn(II) Mn(III). According to the data of the analysis of the EPR spectra, the complex has a structure of the cubane Mn_4O_4 with S = 3/2 [697].

All these data indicate the participation of an assembly of Mn atoms in the reaction of water elimination in biological systems. The model systems of water oxidation to O_2 also incorporate assemblies of interacting atoms. For example, in the photo-oxidation of water under the action of a Tris-bipyridyl complex of Ph(III) as the catalyst is a clustered form of manganese in the form of MnO_2 [704]. The possible elementary mechanisms of water oxidation in clusters will be discussed in Chapter 11.

9.3 The Microsomal Oxidation System

The liver microsomes, the various animal tissues, the mitochondria and microsomes of adrenal glands, the plant and yeast membranes and other microorganisms are sites where the substrate hydroxylation system is functioning.

A distinctive feature of this system is its surprising universality in the oxidation of substrates. The substrates may be natural compounds: saturated fatty acids, steroids, bile acids, cholesterol, metabolites of prostagladin synthesis, and also a large number of compounds that find their way into the organism from the surroundings, such as aromatic and saturated hydrocarbons, compounds with oxidative side chains, amines, esters, thioethers, phenols, amines, etc., more than a thousand compounds. The only requirements that the substrates must meet are the ability to add on hydroxyl in place of a hydrogen atom or to accept an oxygen atom (for example, to form an epoxide from a double bond, to form NO or SO oxides, etc.) and the presence of a hydrophobic section to ensure interaction with the active site. The electron donors used are NADPH or NADH; each of these reducing agents has its specific group of substrates.

The biological role of the microsomal system consists of the oxidation and neutralization of numerous foreign compounds and also of the participation in the synthesis of important biological compounds, e.g. prostaglandins, steroids, etc. In a number of microorganisms the most important function of the biological hydroxylation system consists in the primary oxidation of hydrocarbons, in particular, paraffins, which are converted to primary alcohols.

The vital importance of the hydroxylation system in animals and bacterial cells, the practical value of catalyzed reactions and the great scientific significance of the problem of the mild oxidation of saturated hydrocarbons attract the attention of many investigators. Here, we shall only be concerned with the problems related to the topic of this monograph, namely: the localization and mobility of the main components of the microsomal hydroxylation system of cytochromes P-450 and b_5, flavoproteins I and II (FP_1 and FP_2) and their redox properties, we will also consider the mechanisms of electron transport. In the mitochondria of adrenal glands and in a soluble system from the microorganism *P. putida* the electron transport from FP to cytochrome P-450 is effected with the aid of polynuclear iron-containing proteins, adrenodoxin and putidaredoxin respectively.

The primary proteins that supply electrons from reduced dinucleotides for the hydroxylase reaction in liver, adrenal glands and in microorganisms are flavoprotein reductases [407–411, 705]. However, in some cases the electron transfer to P-450 is not effected directly but via the corresponding carriers. It has been suggested that for the microsomal system of liver there is one more yet unidentified carrier X. It has been suggested [407] that the role of X may be played by cytochrome b_5.

NADPH, a specific flavoprotein of liver microsomes, FP(1), is made up of two identical subunits having a molecular mass of 40000, each of which contains one FAD molecule. The enzyme specific to NADH has a molecular mass of 43000 and also contains one FAD molecule. The enzyme is composed of two fragments separated by proteolysis: a hydrophilic peptide, which carriers FAD (33000) and a hydrophobic peptide buried in the membrane (10300). Cytochrome b_5 has an analogous structure (Chap. 6).

The function of the flavoprotein is to transfer electrons from NADH to cytochrome b_5, the amount of which in the membrane exceeds by almost ten times the amount of FP_2. On the basis of a number of indirect kinetic data, it has been concluded [407] that flavoproteins and cytochrome b_5 diffuse freely along the outer surface of the microsomal membrane with the aid of their hydrophobic sections and that the electron transfer in the hydrophobic sections occurs upon encounter of the reagents. The maximum rate of electron transport is about 500 acts per second per mole of enzyme. The hydride electron transfer from NADH to FAD is thermodynamically preferable (Fig. 9.8).

An important problem is the mechanism of electron transfer from cytochrome b_5 and other carriers to the enzyme cytochrome P-450. Such an electron transfer is effected twice: to a cytochrome P-450 substrate complex $[Fe^{3+}\cdots HR]$ and to a ternary complex $[Fe^{2+}O_2\cdots HR]$. Since the values of E_0 for these complexes are -200 to -400 mV and the E_0 value for cytochrome b_5 is 25 mV, this electron transfer is thermodynamically unfavourable and, according to correlational equations of the types (5.9) and (5.10), it must occur very slowly if no preliminary complex formation of redox reagents takes place. However, if the lifetime of such a complex is long, the probability of the process increases sharply (Sedt. 11.1.1). Crude estimates by formulas (5.9) and (5.10) show that in the case of kinetically favourable conditions (preliminary complex formation and orientation) with changes in ΔE_0 within the range 350–300 mV, the value of k_{tr} for the complex in an adiabatic process is $10^{+1}\,s^{-1}$. The experimental value of k_{tr} is $3.0-10^3\,s^{-1}$. This means that for electron transfer to be effective, the following conditions are required: first, preliminary complex formation of cytochromes b_5 and P-450 and, second, close contact between heme groups which will provide a relatively high value of the non-adiabaticity coefficient $\varkappa$ (see Chaps. 5 and 11). The formation of such complexes has been reported [708–710].

Since the heme group of cytochrome b_5 is localized in the hydrophilic portion, which projects above the membrane plane, the second condition indicated

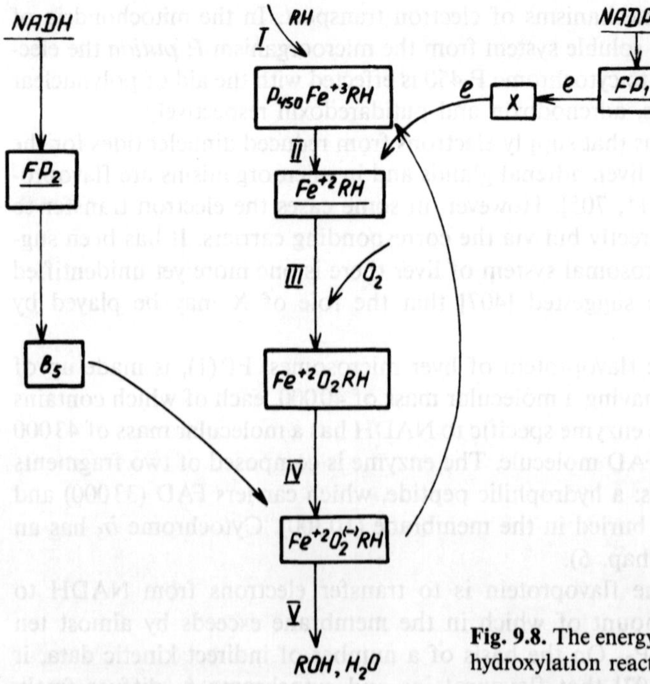

Fig. 9.8. The energy profile of the microsomal hydroxylation reaction

above can be fulfilled only if the heme portion of cytochrome P-450 is buried in the aqueous phase. Taking into account the presence of hydrophilic and hydrophobic sections in cytochrome P-450 [412] and also data on the non-accessibility of the heme group of the enzyme to an external nitroxide probe (spin-labelled lysosome) and to a triplet probe (anthracene butyrate), which diffuses freely in the surface layers of the microsomal membrane [712], it may be presumed that the heme group in cytochrome P-450 is localized near the boundary of the contact between the hydrophilic and the hydrophobic portions of the membrane and is buried in the protein globule at a depth of more than 0.8 nm [52, 711, 712, 715].

The problem of the distribution of cytochrome P-450 macromolecules in the microsomal membrane has been solved experimentally using electron-scattering mercarbide labels [142]. Each molecule of the enzyme was labelled along the free SH groups by two to four labels and could be observed on electron micrographs in the form of granules about 25 nm in size distributed statistically over the membrane surface (Fig. 9.9). The number of such granules increases appreciably in induced preparations with an increased content of cytochrome P-450. In the membrane the rotational diffusion of the enzyme is considerably reduced [708, 714, 886]. In isolated, labelled samples of cytochrome P-450 aggregates composed of five to six enzyme molecules were observed. The mechanisms of electron transport and hydroxylation will be considered in Chapter 11. It has been found through the use of the EPR method [711, 712] that the heme group is parallel to the plane of the membrane.

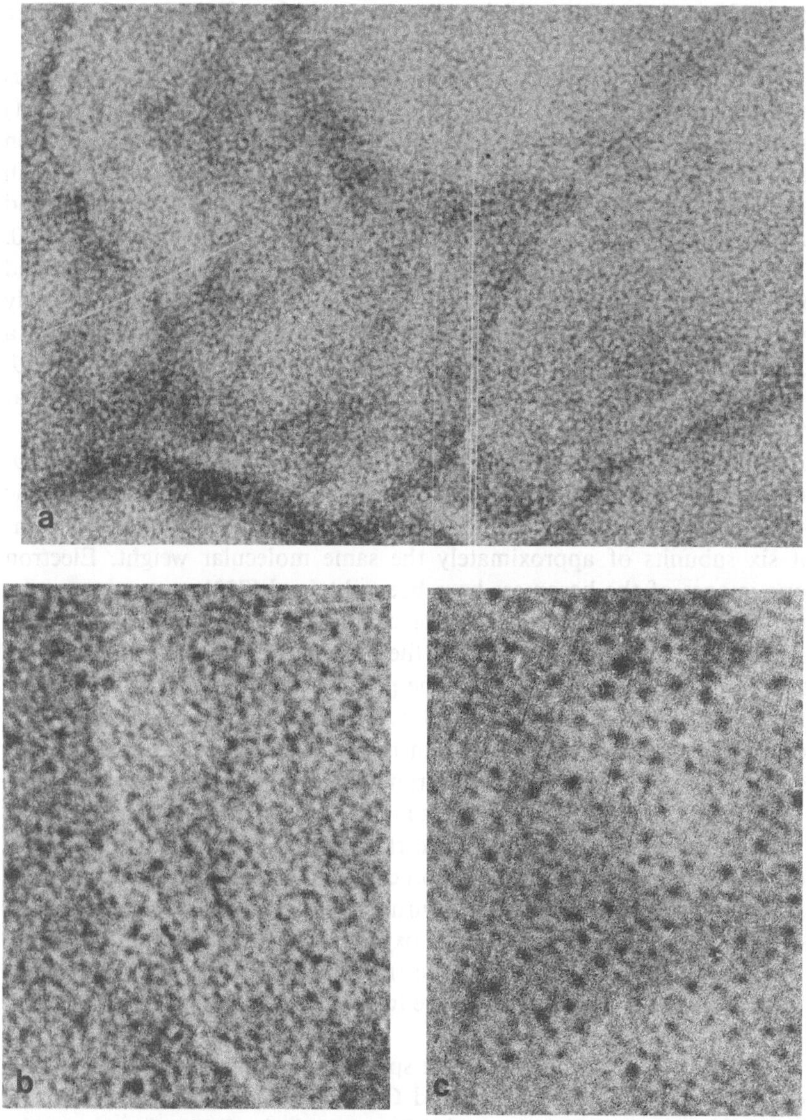

Fig. 9.9a–c. Electron micrographs of microsomal membrane shadows (rat liver) modified across the sulphhydryl groups by electron-scattering mercarbide labels: **a** initial membranes; **b** membranes modified by labels; **c** induced membranes with an increased content of cytochrome P-450. Magnification ×700000 [142]

9.4 The Methane-Hydroxylating System

In *Methylococcus* microorganisms utilizing methane a methane monooxidase (MOx) system has been detected, which catalyzes the reaction:

$$CH_4 + O_2 + DH_2 \rightarrow CH_3OH + D + H_2O \ .$$

The electron donor is NADH, which can be replaced by NADPH and ascorbic acid [706, 707].

The soluble enzyme has been isolated from two organisms *M. capsulatus* (MO_c) and *Trichosporum* (MO_t). MO_c consists of three protein components: (1) a protein having a molecular mass of 210000, which contains two non-heme iron atoms, and which consisted of two α (54000), two β (42000) and two γ (17000) subunits; (2) a protein with a molecular mass of 38000, which contains FAD and a cluster of the Fe_2S_2 type; and (3) a protein with a molecular mass of 16000. The first of these proteins is thought to perform the hydroxylation function and the second the reductase function. MO_t is bound to the membrane but is easily solubilized with the liberation of three protein components: (1) a protein with a molecular mass of 50000 per copper atom; (2) type *c* cytochrome with binds CO; and (3) a protein with a molecular mass of 9000. It has been found that cytochrome *c* does not affect the rate of the oxygenase reaction [719].

A preparation of methane monooxidase from the membranes of *M. capsulatus* (strain M) has recently been isolated and studied [722, 723]. The enzyme contains two iron atoms per molecule of molecular weight 180000, and is composed of six subunits of approximately the same molecular weight. Electron microscope images of the hexamers have been obtained [723].

Of fundamental importance for the understanding of the structure of the active site of the enzyme is the work of [707], the authors of which studied a MOx preparation from *M. capsulatus* (Bath). The protein with $M_r = 210000$ is characterized by an EPR signal with $g = 1.95, 1.88, 1.78$. The signal parameters were close to the corresponding parameters for semi-methemerythrin (Sect. 6.2.3; Fig. 6.11). The latter contains a binuclear complex of iron with histidine and carboxyl ligands. Oxygen adds only to one of the iron atoms and is converted to the peroxide state. A point that is also worth mentioning is a certain similarity between the iron-containing centre MOx and the centre of another iron-containing hydroxylating enzyme putidamonoxin [724]. The latter in turn is close in its physico-chemical parameters to model complexes of iron with oxygen ligands of the type EDTA.

For the enzyme MO_c in the membrane fraction $K_m = 5 \times 10^{-5}$ M, the maximum rate under optimal conditions is three to six methane molecules per second [722].

The enzyme possesses a high substrate specificity and catalyzes the reaction of oxygen with C−H, $\rangle C{=}C\langle$, N−H and C=O bonds with the formation of alcohols, epoxides, hydroxylamines and CO_2 respectively. Ethane and propane are oxidized at lower rates than methane. As has been shown [717−719, 724], the methane-monooxidase reaction has a number of specific features, which distinguish it from the "traditional" microsomal, mitochondrial and microbial monooxidase systems: (1) the value of K_M increases with increasing temperature; (2) carbon monoxide is both a substrate and a competitive inhibitor of methane oxidation; (3) CF_4 does not inhibit the oxidation of methane; (4) CH_4 and O_2 are bound in a cooperative manner [717]. All these factors are difficult to account for by the hydrophobic character of substrate binding alone. Therefore, it has been presumed [718, 719] that the activation of methane is possible in the coordination sphere of the metal by analogy with the activation in model systems on platinum [720, 721].

The second function of the active MO site is to activate oxygen. Hydrogen peroxide and organic peroxides are not active in the oxidation of methane and other substrates. Upon hydroxylation of methane and ethane an isotope effect ($k_H/k_D = 5$) is observed [725].

The study of the structure and mechanism of oxidation of the methane-hydroxylating system is undoubtedly one of the most important practical and theoretical trends.

Let us summarize the basic results of the study of the specificity of functioning of metalloenzymes and carriers in biological membranes.

The combination of mono- and binuclear protein macromolecules in biomembranes into systems of the third type and the "simple" incorporation of enzymes into the membranes make it possible to obtain protein-membrane complexes having qualitatively new properties, namely: (1) the ability to transport electrons over considerable distances by way of organization of a chain of centres in electron-transport systems; (2) to effect a spatial separation of charges upon phototransfer, which prevents their recombination (bacterial photosynthesis, systems I and II of plant photosynthesis); (3) to provide a shuttle electron transfer with the aid of proteins of the type of cytochrome *c* or cytochrome b_5, which slide on the membrane surface, or of intramembrane carriers of the ubiquinone type (biological oxidation and hydroxylation by O_2); (4) to form interaction centres between two and more enzymes an space.

The lipid environment of proteins in membranes is an evolutionally selected optimal medium for their functioning. The membranes favour the coupling of oxidation and phosphorylation reactions and also the localization of certain enzymatic processes in membrane sections. The specific and non-specific effects in membranes during the change of their lipid and chemical composition are powerful factors in the regulation of the activity and permeability of membranes. The systems of iron- and copper-containing active sites of metalloenzymes and carriers are specific devices for the switchover of the reaction mechanism from one-electron to multi-electron transport.

PART III

CHAPTER 10

Energy, Entropy and Molecular-Dynamic Relationships in Enzyme Catalysis

The high efficiency of enzymes as catalysts and their ability to effect a fine control of catalytic activity are due to a number of specific properties, which distinguish them from the majority of known chemical heterogeneous and homogeneous catalysts. Enzymes are first of all polyfunctional catalysts, whose active sites consist, as a rule, of several groups. Some of the groups bind and orientate substrates and others take part in concurrent or consecutive elementary chemical acts. As a result, the energy and entropy barriers in reaction steps are substantially lowered and the chemical process is directed a long the most favourable reaction path (Chap. 5).

However, the advantages of polyfunctional catalysis would not have been realized if the enzyme macromolecules had not possessed a combination of a sufficiently rigid and compact structure and an optimal molecular-dynamic mobility and, hence, the ability to change their conformation. The structural dynamics and conformational changes are requisites for the dynamic adjustment of the shape of the enzyme macromolecule to the steric and charge structure of the reaction complex which undergoes a change during the chemical reaction. The same factors are responsible for the ability of enzymes to display allosteric cooperative effects, when the entry of an inhibitor or activator molecule into the regulatory centres changes the state of spatially separated catalytic centres.

In this chapter we shall consider experimental data and theoretical conceptions pertaining to the effect of conformational changes and molecular dynamics of enzymes on their catalytic properties. From the standpoint of the chemical physics of enzyme catalysis, it is especially important to analyze the possible contribution of these effects to the experimentally determined kinetic and thermodynamic parameters and also to elucidate the relationship between these parameters and certain types of intramolecular motions in proteins and membranes.

10.1 Relationships Between the Energy and Entropy Parameters of Enzymatic Processes

10.1.1 Experimental Data

The enthalpy (E_a, ΔH_0) and entropy ($\Delta S^{\neq}$, ΔS_0) parameters can be determined experimentally through the use of the temperature dependencies of the rate and equilibrium constants (Eqs. (5.1) and (5.28)]. The thermodynamic parameters can also be determined by direct calorimetric measurements. While using these equations for enzymatic reactions (and for any real processes in the condensed phase, for that matter), one should, however, remember that, because of the complexity of these processes, the parameters that are determined from the temperature dependencies of the rate constants of the steps of substrate conversion (k_i), the affinity constants of the substrate (K_S), the inhibition constant (K_I), etc., are certain effective parameters, which jointly reflect the entire complexity of the occurrence of a particular macrostep on the enzyme (the presence of intermediate microsteps which do not manifest themselves externally, the participation of the protein matrix of the enzyme, the effect of the solvent particles, etc.). In particular, the discrepancy noted in some experiments [726–728] between the values of ΔH_0 and ΔS_0 obtained from Eq. (5.1) and directly by the calorimetric method for a complex-formation reaction involving proteins is probably accounted for by the fact that the steps, in which the structure of macromolecules of solvent is changed, make a contribution to the thermodynamic parameters but they do not influence the other characteristics of proteins, e.g. their spectral characteristics.

Most of the reactions catalyzed by enzymes differ from analogous non-enzymatic reactions by lower values of the apparent activation energy E_{app} [24, 99, 215–222, 726–729]. The entropy parameters of enzymatic reactions take on the various values ranging from -200 up to 200 eu, which corresponds to a difference of 10^{22} between the pre-exponential factors. The entropy parameter is often very sensitive to the structure of the substrate and enzyme and to the experimental temperature and pH. Many cases of anomalously high and anomalously low values of $\Delta S^{\neq}$ and ΔS_0 have been observed as compared with simple model reactions. The experimental dependencies of log k_i and log K on 1/T are rather often found to have a sharp break, which is an indication of a discontinuous variation of the apparent activation energy and entropy of the reactions over a narrow temperature range. The entropy parameters above or below this extreme temperature region lie, as a rule, in the regions of anomalous values. With regards to the values of ΔH_0 for the formation of enzyme-substrate and inhibitor complexes, they also vary within a rather wide range (from -60 to $+60\,kJ\,mol^{-1}$).

For the elementary acts of simple gas-phase reactions, the experimental values of $\Delta S^{\neq}$ vary from 0 to 50 eu (unimolecular reactions) and from -80 to -200 eu (bimolecular reactions of simple molecules uncomplicated by solvation). The entropy loss upon complex formation is $\Delta S_0 = -60$ to -300 eu. Gas-phase chemical reactions provide many examples of elementary processes [284]. The values of the entropy parameters of such reactions commonly lie well within the range calculated by methods of statistical physics on the basis of the activated-complex theory and may be regarded conditionally as the reference values for "normal" elementary chemical processes.

The value of $\Delta S_u^{\neq}$ for formally unimolecular enzymatic steps usually goes beyond the range of "normal" values. High negative values are especially characteristic of the hydrolytic enzymes α-chymotrypsin, pepsin, carboxypeptidase, etc. Since the theoretically predicted value of ΔS_0 lies within the range of -150 to -300 eu, the positive experimental values of ΔS_0 for a number of enzymes are undoubtedly evidence of a significant contribution of the concomitant processes involving the reorganization of the structure of the enzyme and solvent upon binding of the substrate on the enzyme surface. Anomalous values of $\Delta S^{\neq}$ are also characteristic of processes involving metalloenzymes, including electron-transfer reactions (Chap. 11).

Of special interest are data pertaining not to experiments with individual enzymes and substrates, but to series of experiments, in which the various parameters, such as the structure of substrates or inhibitors, reaction conditions (pH, the composition of the solvent, various temperature regions), undergo a regular change. The main advantage of this approach is that it allows one to elucidate the effect of the individual factors with the remaining factors being kept approximately constant.

Material on the kinetics and thermodynamics of enzymatic processes has been analyzed in review articles [24, 99, 726–729]. Cases were indicated, where for the same enzyme the change of the activation energy or enthalpy of the process caused by the variation of the chemical structure of the substrate, inhibitor or activator is accompanied by the change of the entropy of activation and entropy of a particular process respectively. The following dependencies are approximately satisfied:

$$E_a = A_c + \beta_c \Delta S^{\neq} ; \tag{10.1}$$

$$\Delta H_0 = A_d + \beta_d \Delta S_0 . \tag{10.2}$$

The difference in energy amounts to 80 kJ mol^{-1}, which, according to Eqs. (5.1) and (5.28), should have caused changes in the rate or equilibrium constants by 15 orders of magnitude, but these changes are compensated by significant (up to 240 eu) changes in entropy, and the real constants differ to a considerably smaller extent. The dependencies, such as Eqs. (10.1) and (10.2), are known as compensation effects. Analysis of experimental data has shown that the values of the coefficients β_d for the equations that describe the data on the Michaelis and inhibition constants are close to the average experimental temperatures (T_{av}). A similar relationship holds for the data obtained for the same enzyme in various temperature ranges. For the kinetic parameters corresponding to different substrates, activators or inhibitors, β_c is, as a rule, not equal to T_{av}. Examples of compensation effects for processes involving the binding of ligands to hemoglobin, the oxidation of alcohols by catalase and the hydroxylation of substrates by cytochrome P-450 are presented in Fig. 10.1 (see also [734]). An approximate correlation between these parameters has also been observed for an electron-transfer reaction involving metal carriers [24] (see Sect. 11.1.3).

The anomalies and compensation relations considered above are not only characteristic of enzymatic reactions. The compensation relationships between the experimental energies and entropies of protein denaturation are noted for enormous ranges of ΔE_a values (from -20 to 500 kJ mol^{-1}) and of $\Delta\Delta S^{\neq}$ values (from -40 to 1600 eu). These quantities are highly sensitive to external conditions (pH, additives, moisture content; etc.). Linear correlations have also been found for rotational diffusion processes of spin labels and probes introduced into the various portions of globular proteins [24, 99]. They have also been observed, though to a lesser extent, in various processes taking place in the con-

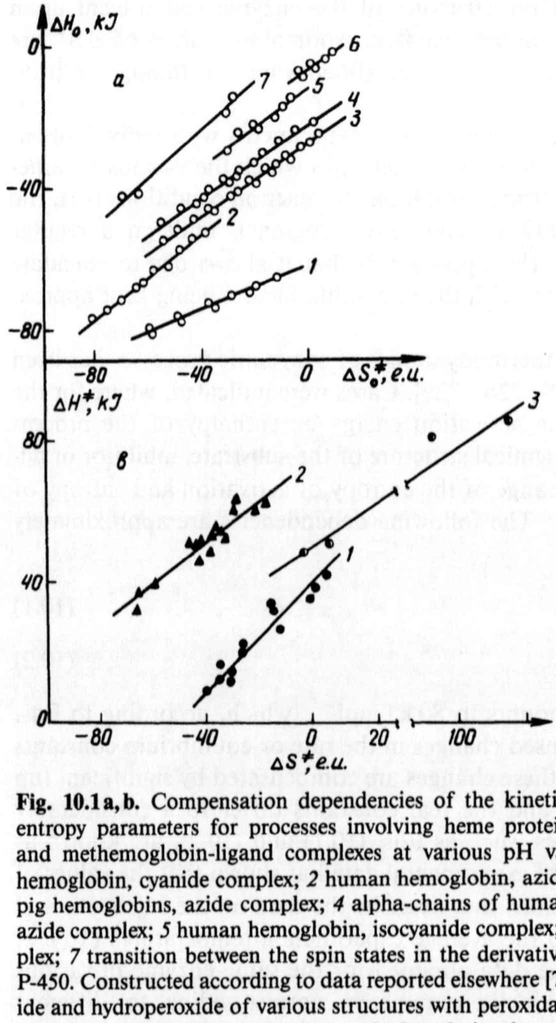

Fig. 10.1a, b. Compensation dependencies of the kinetic and thermodynamic enthalpy and entropy parameters for processes involving heme proteins. **a** The formation of myoglobin- and methemoglobin-ligand complexes at various pH values [726] (*curves 1–6*): *1* human hemoglobin, cyanide complex; *2* human hemoglobin, azide complex; *3* dog, pigeon and guinea pig hemoglobins, azide complex; *4* alpha-chains of human hemoglobin and whale myoglobin, azide complex; *5* human hemoglobin, isocyanide complex; *6* human hemoglobin, fluoride complex; *7* transition between the spin states in the derivatives of metmyoglobin and cytochrome P-450. Constructed according to data reported elsewhere [730]. **b** *1* Reactions of hydrogen peroxide and hydroperoxide of various structures with peroxidases, leading to the formation of compound I. *2, 3* Reactions of substrate hydroxylation by microsomal cytochrome P-450 [432]

densed phase (chemical reactions, diffusion, evaporation, electrical conduction, etc.). The main property of these systems, which differ from simple gas reactions, is the interaction of particles which gives rise to qualitatively new effects. One effect is the manifestation of the cooperative properties of particle assemblies, when the behaviour of one particle in the assembly depends to a considerable extent on the state of the other particles (Sect. 4.1).

10.1.2 Theoretical Treatment

The anomalies pointed out in the previous section in the thermodynamic and kinetic parameters of enzymatic processes, including compensation effects, may be accounted for in general on the basis of the assumption that the chemical elementary steps on the enzymes are accompanied by the rearrangement of the structure of protein globules and the surrounding water. The kinetic and thermodynamic parameters of such structural rearrangements make a contribution to the experimentally measured E_{app} and $\Delta S^{\neq}_{app}$, whose values reflect the cooperative properties of the water-protein matrix.

It has been shown [24, 99, 227–231] that if a chemical reaction with rate constant k_r and activation energy E_r and entropy $\Delta S^{\neq}_r$ is accompanied by fast and reversible rearrangements of protein units and water molecules (altogether, n particles), then at $T<T_c=\Delta H^0_0/\Delta S^0_0$ the equalities are fulfilled that follow from Eq. (4.3) with $k_i = k_r$:

$$E_{app} = E_r + n\Delta H^0_0 \quad \text{and} \quad \Delta S^{\neq}_{app} = \Delta S^{\neq}_r + n\Delta S^0_0 + A \ , \tag{10.3}$$

where ΔH^0_0 and ΔS^0_0 are the standard enthalpy and entropy of one elementary rearrangement respectively; the value of A depends on the manner in which the particles are packed.

From Eq. (10.3) it follows that in such systems anomalous values of E_{app} and $\Delta S^{\neq}_{app}$ may be observed if the number of particles or units n is sufficiently large. Since the quantities ΔH^0_0 and ΔS^0_0 may be either positive or negative, which correspond conditionally to the "melting" or and "crystallization" of the water-protein matrix, as a result of the rearrangement processes at a considerable value of n, the values of the parameters E_{app} and $\Delta S^{\neq}_{app}$ may be too high or too low as compared to the true chemical parameters E_r and $\Delta S^{\neq}_r$. In other words, the cooperative rearrangements are responsible for the kinetic and thermodynamic anomalies of enzymatic and chemical processes.

The compensation dependencies may also be accounted for within the scope of Eq. (10.3). For systems that differ in the number of particles in a cooperative unit, n, and have approximately equal ΔH^0_0, ΔS^0_0 and E_r, the compensation Eq. (10.1) must be satisfied with:

$$A_c = E_r - T\Delta S^{\neq}_r \quad \text{and} \quad \beta_c = \Delta H^0_0/\Delta S^0_0 = T_c \ , \tag{10.4}$$

where T_c is the "melting" temperature of the cooperative unit.

According to Eqs. (10.1) and (5.28), at the temperature of the experiment $T = T_c = \beta_c$ the systems that differ in the number n must have the same rate constant, $k_{app} = k_r$. Therefore, the quantity β_c is called the "isokinetic temperature". As the temperature continues to increase, the reaction series must be reversed. However, as has been shown in Section 4.1, within the framework of the model of the cooperative process under consideration at $T \approx T_c$, the temperature dependence of k_{app} changes sharply and upon further increase of the temperature $k_{app} = k_r$ and no inversion of the series takes place. The theoretical dependencies of k_{app} on T in the Arrhenius coordinates at different n are shown in Fig. 10.2a. The experimental data on the denaturation of proteins have confirmed the theoretically predicted temperature dependencies [231].

In a general case, the chemical reaction and the rearrangement process cannot be regarded as independent events. If the quantities E_r and $\Delta S^{\neq}_r$ are represented, to a first approximation, as linear functions of n with the coefficients being, respectively, equal to a_1 and b_1, then the isokinetic temperature will be given by:

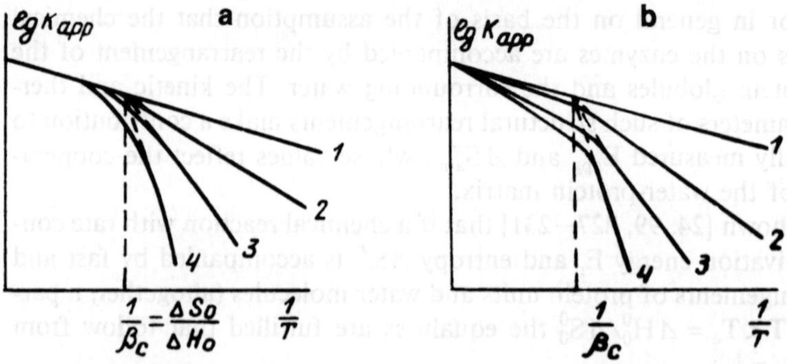

Fig. 10.2 a, b. Types of theoretical dependencies of log k_{app} on 1/T for various mechanisms of a reaction involving the compensation effect. Calculated by Eqs. (5.28), (10.3) and (10.5). **a** ΔH_0 and $\Delta S_0 > 0$, $a_1 = b_1 = 0$; **b** ΔH_0 and $\Delta S_0 > 0$, $a_1 > 0, b_1 = 0$. The *curves 1–4* correspond to an increase in the length of the cooperative chain n; β_c are isokinetic temperatures

$$\beta_c = (\Delta H_0^0 + a_1)/(\Delta S_0^0 + b_1) \ . \tag{10.5}$$

Thus, the experimental value of β_c may significantly deviate from the "melting" temperature of the matrix, T_c, depending on the sign and magnitude of the coefficients a_1 and b_1. The Arrhenius dependencies corresponding to various cases are shown in Fig. 10.2.

In conclusion, it should be noted that equations of types (10.1) and (10.2) are obtained in those cases where in the reaction series the values of E_r and $\Delta S_r^{\neq}$ are close to each other, but E_{app} is determined with an error of $\delta \Delta E_a$. It is easy to show that in such a case $\beta_c \approx T_{av}$ and the range of E_a values is equal to the maximum value of $\delta \Delta E_a$. Therefore, the criterion of the true compensation effect is a ray of straight lines that converge to the isokinetic temperature in the Arrhenius coordinates (Fig. 10.2).

The above treatment is based on the simplest model of the cooperative process. In real cases, not all the portions of the cooperative unit are identical; the rates of the elementary conversions are different and the "initiation" processes may occur simultaneously at various points of the unit, etc. Such cases require special consideration. Nevertheless, in my opinion, equations of types (10.3), (10.4) and (10.5) disclose, in principle, the physical meaning of the experimentally determined activation energy, and entropy parameters of enzymatic processes, including those which involve metalloenzymes and carriers.

10.2 Molecular-Dynamic Properties of Proteins and Enzymes

10.2.1 General

The present-day conception of the molecular dynamics of proteins is based on three hypotheses advanced in the late 1950s and the early 1960s.

Lumry and Eyring assumed in their early work [732] that the interaction of substrates with enzymes is accompanied by a certain distortion of the structure (the conformation) of enzymes. The process is accompanied by changes in the energetic state of the system which contribute to the favourable occurrence of the particular reaction. Later, Lumry and Biltonen [733] developed these conceptions into the concept of the "complementarity of free energies" of chemical and conformational processes. According to this concept, the energy redistribution upon binding of the substrate into the conformational energy occurs in such a way that the energy profile of the process as a whole is eventually flattened. In other words, all the intermediate products of the chemical conversion on the enzyme have nearly the same free energy.

Another basis for the modern conceptions of the molecular dynamics of proteins is the Linderström-Lang hypothesis of the mobile fluctuating state of the protein macromolecule [735]. One of the manifestations of structure fluctuations of protein macromolecules is the ability of the inner peptide groups to exchange hydrogen atoms for the water protons, the process being effected in the open conformations of the protein globule.

The third conception is the Koshland induced-fit theory, according to which the structure of the enzyme can be induced to fit the structure of the substrate during their interaction [736]. According to his hypothesis, the binding of the substrate to the enzyme must be accompanied by a change in the conformation of the enzyme. The induced-fit hypothesis has been applied to metalloenzymes by Volkenshtein [223], who introduced the concept of electronic-conformational interaction. According to these conceptions, to each electronic state of the metal atom in the enzyme (valence, spin nature, etc.) a certain conformational structure of the protein globule corresponds. On the whole, in the equilibrium state the system may be characterized as a "conformon" (by analogy with a polaron, which characterizes the state of the electric charge and of the environment in crystals).

Though the early hypotheses were based on speculative and indirect considerations, they nevertheless had a strong impact on the development of the science of enzymes, since they provided a clear-cut formulation of the problem and, to a certain extent, outlined the paths for its solution.

At the next stage of the solution of the problem qualitative data were used on the isotopic exchange of hydrogen between the aqueous phase and the inner portions of protein globules [737–740], the kinetics of enzyme catalysis [726–729, 741], the reactivity of the functional groups buried in the protein globule [326] and on the proteolysis of proteins [739].

The analyses carried out by some authors [726, 729, 741] of the entropy and energy specificity of enzyme catalysis and denaturation processes have led to the conclusion that the elementary acts of chemical conversions in the active sites of enzymes are accompanied by cooperative changes in the structure of the surrounding water-protein matrix. Numerous data have been obtained on the change of the conformation during the functioning of proteins, which provide indirect evidence in favour of their intramolecular mobility [223, 727]. In particular, so-called transglobular conformational transitions have been discovered in enzymes consisting of a single protein subunit, which is evidence of their intramolecular nature [24, 99].

However, the details of the real picture of the intramolecular dynamics of proteins could be clarified only when more direct physical methods were used, which were capable of directly recording the mobility of the individual portions of the protein globule and of determining the concrete amplitude and frequency characteristics of this mobility. Such methods were NMR [742–747], luminscence methods [179, 748–751] and the methods of spin and Mössbauer labels [24, 99, 143, 146, 313, 752–756]. The investigations carried out by means of these methods have led to the elucidation of the general dynamic characteristics of mobility in various layers of the water-protein matrix and, in a number of cases, in the interior of the protein globules. These investigations have led to the very important conclusion that a physiological temperatures the protein globules exhibit internal mobility in the frequency range $10^7-10^8\ s^{-1}$ and at amplitudes higher than 0.03–0.05 nm. There are data indicating that this type of mobility is associated with the functional activity of proteins [24, 208–213, 756–759].

At present we are witnessing the intensive development of a new stage in the investigation of protein dynamics [117, 259, 728, 760–765]. A distinctive feature of this new stage is the application of the latest methods (particularly NMR, X-ray diffraction, dynamic spectral methods and physical labels) which make it possible to obtain detailed information on the mobility of the individual functional groups in proteins and to compare the kinetic and dynamic characteristics of enzymatic processes, and also the use of theoretical calculations.

In this section we shall be concerned with the present-day status of the problem of the intramolecular dynamics of globular proteins. Special emphasis is made on those aspects that are associated with the functional properties of this most important type of biopolymer and, in particular, with the ability to catalyze chemical reactions.

10.2.2 Methods of Studying the Intramolecular Dynamics of Proteins

Complex systems, such as globular proteins, undergo a number of dynamic processes, which have specific frequencies and amplitudes: (1) "ordinary" intramolecular nuclear vibrations ($\nu_c = 10^{11}$ to $10^{14}\ s^{-1}$, A = 0.001–0.01 nm; (2) reorientation (rotational) displacements of the side groups; (3) vibrational and rotational motions of relatively large blocks, e.g. helical portions; (4) rotational (segmental) mobility of the main polypeptide chains inside the blocks; (5) rotational and translational mobility of the medium's molecule, e.g. water molecules. The last four types of motion may, in principle, be characterized by a wide range of frequencies ($\nu_c = 10^{11}$ to $10^{-3}\ s^{-1}$) and amplitudes (0.01–1.0 nm).

The first type of intramolecular motion takes place in all systems, including solids at low temperatures. The intensity of the other types of motion is different in different systems.

A wide range of physico-chemical methods are used for the study of the intramolecular dynamics of protein globules and adjacent solvent molecules [24, 99, 207–213, 742, 749, 756–769]. The joint application of these methods makes it possible to study the mobility of proteins over an extensive range of characteristic correlation frequencies ($\nu_c = 10^{-3}$ to $10^{14}\ s^{-1}$) and over a wide range of amplitudes (A = 0.001 to 5.0 nm). The methods used for the study of the dynamic state of proteins are briefly described in Table 10.1. The theoretical and ex-

Table 10.1. Brief characterization of physico-chemical methods of investigation of protein dynamics

Method	Parameters measured	Information obtained[a]
Luminescence		
Fluorescence	Depolarization	$\tau_c^{-1} = 10^8 - 10^9 s^{-1}$
	Radiation autocorrelation	$\tau_c^{-1} = 10^8 - 10^9 s^{-1}$
	Relaxational shift	$\tau_c^{-1} = 10^8 - 10^9 s^{-1}$
	Quenching by reagents	$k_{tran} = 10^8 - 10^{10} M^{-1} s^{-1}$
Phosphorescence	Depolarization	$\tau_c^{-1} = 10^{-1} - 10^6 s^{-1}$
	Relaxational shift	$\tau_c^{-1} = 10^{-1} - 10^6 s^{-1}$
	Quenching by the exchange mechanism	$k_{tran} = 1 - 10^{10} M^{-1} s^{-1}$
Rayleigh scattering	Doppler shift of frequency	$\tau_c^{-1} = 10^9 - 10^{10} s^{-1}$
	Line broadening	$\tau_c^{-1} = 10 - 10^7 s^{-1}$
Spin labelling	Shape of absorption lines	$\tau_c^{-1} = 10^7 - 10^{10} s^{-1}$
	Saturation transfer parameters	$\tau_c^{-1} = 10^4 - 10^6 s^{-1}$
	Quenching by paramagnetics	$k_{tran} = 10^6 - 10^{10} M^{-1} s^{-1}$
Recombination-kinetic method	Kinetics of radical recombination	$k_{tran} = 10 - 10^8 M^{-1} s^{-1}$ $\tau_{tran}^{-1} = 10^4 - 10^{-1} s$
Mössbauer labels	Recoilless absorption	$\tau_c^{-1} = 10^7 s^{-1}$
	Diffusion line broadening	$\tau_c^{-1} = 10^7 s^{-1}$
Dielectric measurement	Absorption and dispersion of electromagnetic radiation	$\tau_c^{-1} = 10^3 - 10^{10} s^{-1}$
Ultrasonic methods	Sound absorption and dispersion	$\tau_c^{-1} = 10^5 - 10^9 s^{-1}$
NMR	Line splitting in high-resolution spectrum	$\tau_c^{-1} = 10 - 10^3 s^{-1}$
	Frequency and temperature dependences T_1 and T_2	$\tau_c^{-1} = 10^4 - 10^{10} s^{-1}$
	Relaxation in the magnetic field gradient	$\tau_c^{-1} = 10^7 - 10^{10} s^{-1}$
	Narrowing of broad lines	$\tau_c^{-1} = 10 - 10^4 s^{-1}$

[a] τ_c^{-1} is the correlation frequency; k_{tran} is the rate constant of collisions; τ_{tran}^{-1} is the translational jump time.

perimental aspects of this problem have been considered in a number of works review papers and monographs cited above.

Below, we shall briefly consider the applications of the methods of physical labels developed in recent years (Chap. 3) to the solution of the dynamic problems. The use of the label technique significantly widens the possibilities of obtaining detailed information on the dynamics of the individual portions of the water-protein matrix by physical methods.

In going from the quantities that are directly determined by experiment to the dynamic parameters of the medium, it is necessary to take into account the specific features of each of the methods, which should be considered separately.

The Spin-Label Technique. The variants of the EPR method allow one to determine the rotational mobility ($\nu_c = 10^4$ to $10^{11}\ s^{-1}$) and the rate constant of collisions ($k_{tran} = 10^7$ to $10^{10}\ m^{-1}\ s^{-1}$) of spin labels and probes. Apart from the intensity of the rotational mobility of radicals, a number of other factors considered in Section 3.1.2 have a considerable effect on the EPR parameters. Strict accounting of all these factors is a difficult task; in the case of protein systems this task has not yet been resolved. Therefore, at present, use is made of simplified model conceptions of the motion of the radicals and a comparison is made with the behaviour of nitroxide radicals in simpler systems with known molecular-dynamic characteristics.

In the frequency region of 5×10^6 to $5\times10^8\ s^{-1}$, the 2-mm EPR spectroscopy is most promising [162, 163]. The high resolving power of the method (see Sect. 3.1.3) makes it possible to obtain, directly from the analysis of the EPR spectrum of the spin label, values of the g-tensor and HFS tensor, which reflect the microenvironment of the spin label [770, 780]. Moreover, it has become possible to reveal the nature of the motion of the spin label relative to the rotation axes and, in particular, the presence or absence of anisotropic rapid motion with $\nu_c > 10^9\ s^{-1}$ [770, 771].

The first investigations of the dynamics of proteins with the use of the 2-mm spectroscopy [770, 771] demonstrated that this method enables one to effectively study the anisotropic motions of the nitroxide fragment of a spin label. The intensity of these motions, in turn, depends on the microrelief and dynamics of the environment.

Data on the mobility of nitroxide radicals in model systems may be summarized as follows [24, 99]. In liquids, the apparent activation parameters of the mobility of the radicals are close to the corresponding parameters of the molecules of the medium. In polymers, at temperatures higher than the glass-transition temperature T_g, the mobility of spin probes is mainly determined by the segmental mobility of polymeric chains. Below T_g, the motion of the probes is modulated by relatively low-amplitude motions of the matrix walls, which surround the probe.

The motion of the nitroxide fragment of the label or probe is determined to a considerable extent by the microrelief of the protein in the region of incorporation. Under steric constraints the motion will be essentially anisotropic and may be regarded as the motion inside a cone with a specified angle. The EPR parameters substantially depend in this case on the size of the cone, the extent of its reorientation and on the velocity of motion of the nitroxide fragment inside the cone [157–161]. If these motions are similar to isotropic motions (the long and flexible "legs" of labels), then the value of ν_c can be calculated using formulas such as (3.1) and (3.2). For the estimation of ν_c in comparative experiments these formulas may also be used for systems with the motion of nitroxide fragments in the cone, except for cases where formula (3.2) is used for a very rapid rotation of the fragment in the cone with $\nu_c \gtrsim 2\times10^9\ s^{-1}$ [157].

In the frequency range 10^4 to $10^6\ s^{-1}$ the value of ν_c is determined with the aid of saturation-transfer methods [160]. Unfortunately, the development level of the theory does not permit the precise calculation of ν_c for anisotropic rotations. Without a special theoretical treatment it is only possible to estimate ν_c within the limits of one order of magnitude in the region where $\nu_c^{-1} \approx T_I$, where T_I is the rate of spin-lattice relaxation (Sect. 3.1.2).

Fluorescent and Phosphorescent Labels. As has been shown in Section 3.2.1 for the simplest model, in which the rearrangement of electrical dipoles may be characterized by a single characteristic relaxation frequency ν_r, formula (3.18), which allows one to determine the value of ν_r, is valid. However, in real systems deviations from the simplest physical picture are observed, which are caused by the effect of the electric field of the excited chromophore group on the specific phenomena of environment and also by the intrinsic microheterogeneity of the samples. If we assume a Gaussian distribution over the free activation energies ($\Delta F^{\neq}$) of the reorientation of chromatophores, it is possible to find an expression that would relate the experimentally determined, apparent activation energy (E_{app}) to the most probable activation energy (E_{max}) and the second moment of the distribution curve (ΔF_0^2) [125, 756, 772]:

$$E_{app} = E_{max} - \Delta F_0^2/RT \ . \tag{10.6}$$

This expression is valid for cases where the measurements are performed in a narrow range of frequencies (ν_r, i.e. within the framework of one method (fluorescence or phosphorescence). In a joint analysis of the data of these methods, which differ in characteristic frequency by six and more orders of magnitude, the value is $E_{app} \approx E_{max}$. For example, it has been shown [756] that the values of ν_r calculated from Eq. (3.18) are much (up to several orders of magnitude) lower than the values of ν_D obtained by dielectric measurements in the range of temperatures at which $\nu_r \ll \nu^*_{f,ph}$. In the region of $\nu_r \gg \nu^*_{f,ph}$, the value is $\nu_r \approx \nu_D$ with an accuracy of one order of magnitude.

Another method that reveals the microheterogeneity of the medium is an approach based on the simultaneous determination of a set of fluorescence or phosphorescence parameters: the intensity I of the shift $\Delta\lambda_{max}$ and the lifetime $\tau^*_{f,ph}$. The presence of several phases, which differ appreciably in these parameters, gives rise to a substantial difference between the corresponding temperature dependencies. For example, the appearance of a new phase with strong quenching of the phosphorescence of the chromophore by the solute oxygen molecules will bring about a sharp decrease in the parameter I measured with invariable experimental values of $\Delta\lambda_{max}$ and $\tau^*_{f,ph}$ for the phase observed.

Mössbauer Spectroscopy. According to theory, the probability of recoilless resonance absorption of a γ-quantum (f′), which determines the intensity of the Mössbauer spectrum, is given by the relation (2.3). If the decrease of f′ is caused by an increase in amplitude $\langle \Delta x \rangle_c$ due to conformational mobility, the quantity f′ is a quantitative measure of the motion of the protein portion rigidly bound to a Mössbauer atom. Another dynamic parameter of the spectrum is the diffusion broadening ($\Delta\Gamma$) [formula (2.4)]. The experimentally measured dependencies of the fall of f′ on temperature and moisture content may be described in terms of three groups of factors [756, 763, 764]: (1) a monotonic increase of $\langle \Delta x \rangle_c$; (2) the oscillation of the Mössbauer atom between several states of different f′, these states being characterized by the measurable value of f′; (3) the practically complete disappearance of the spectrum due to fractions characterized by critical values of $\langle \Delta x \rangle_{cr}$, which are so high that the contribution of these fractions to the measured values of f′ is negligibly small.

It should also be noted that even an insignificant dynamic microheterogeneity of the medium must have an effect on the experimental dependence f′(T) [756]. Indeed, if it is assumed that a Gaussian distribution occurs in the sample among the vibrational energies of Mössbauer atoms, $E_v \sim \langle \Delta x \rangle^2$, then the experimental value of f′ will be equal to:

$$f' = \exp\left[\beta_M\left(1 - \beta_M [\langle \Delta x \rangle]_0^2 / 2 \langle \Delta x \rangle^2_{max}\right)\right] ,$$

where $\beta_M = 4\pi^2 \langle \Delta x \rangle^2_{max} / \lambda^2$, and $\langle \Delta x \rangle_{max}$ and $[\langle \Delta x \rangle]_0^2$ are the most probable amplitude and the second moment of the amplitude distribution respectively.

As the temperature increases, the value of f′ falls due to at least two factors: an increase in $\langle \Delta x \rangle_{max}$ and in the population of states with $\langle \Delta x \rangle > \langle \Delta x \rangle_{cr}$ and frequencies $\nu_c \gtrsim 10^7\ s^{-1}$. The latter states will no longer make an appreciable contribution to the measured value of f′.

Radical Pairs. Ultra-slow motions in a condensed medium at correlation frequencies $\nu_c \leqslant 1 - 10^2$ can be recorded by studying the recombination of radical pairs with the aid of the EPR method [165]. Radical pairs can be easily produced by introducing certain compounds, e.g. tetraphenylhydrazine, into the samples under study with a subsequent dissociation of these compounds by radiation. Methods have been proposed [165] for the treatment of experimental data on the kinetics of recombination of radical pairs with a broad distribution over recombination rate constants.

The capabilities of the above considered methods of physical labels are listed in Table 10.1, and the strategy of using physical labels is given in Fig. 10.3. Each

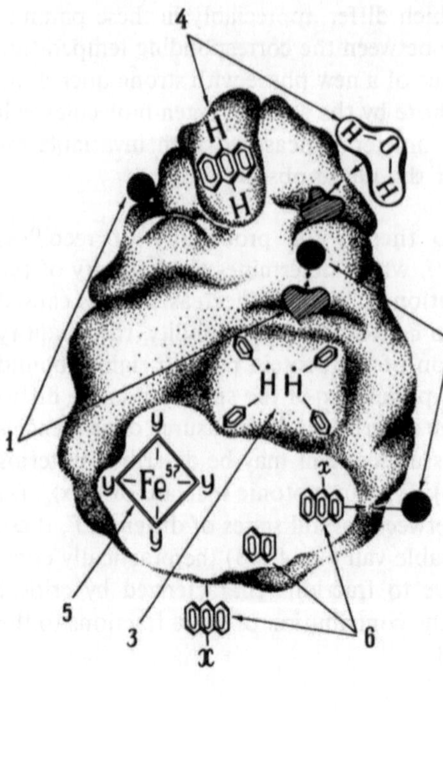

Fig. 10.3. The principles of the introduction of physical labels and probes into a protein globule: *1* spin labels; *2* spin probes; *3* radical pairs; *4* NMR probes; *5* Mössbauer labels; *6* luminescent labels and probes

of the methods used separately is sensitive to motions over a relatively narrow range of correlation frequencies. As we have seen above, there are difficulties and uncertainties in the precise determination of frequencies. But the joint use of these methods makes it possible to obtain insight into the motions over a very wide frequency range and there is ground for believing that the general dynamic picture obtained is sufficiently reliable.

10.2.3 Experimental Data on the Intramolecular Dynamics of Proteins

In this section we shall be concerned with experimental data pertaining mainly to metalloproteins. But we shall first consider data on serum albumins that have been studied relatively well.

Bovine and Human Serum Albumins (BSA and HSA). In 1950 Karush [773], while analyzing data on the binding of organic molecules to albumins, arrived at a conclusion on the flexibility of the binding-site structure. It was later found that the binding of a number of substrates in the hydrophobic pocket of serum albumins causes a change in the conformation of the protein globule. The example, a change was observed in the rotational mobility of a spin label located at a distance greater than 1.6 nm from the hydrophobic pocket, into which steroid molecules were introduced [774]. This was accompanied by a change in the efficiency of the dipole-dipole interaction of the labels with the water molecules, which was recorded from a change in the rate of magnetic relaxation by NMR.

The mobility of various portions of the water-protein matrix of SA has been studied by many methods [756, 759]. These investigations were concerned, in particular, with the separation of the internal motions from the motions of the protein globule as a whole. To this end, the experiments were carried out under conditions in which the rotation of SA macromolecules was retarded by increasing the viscosity of the medium (glycerine or sucrose is added, measurements are carried out in suspensions or in dry and moistened samples, etc.). These experiments are illustrated in Fig. 10.4, which shows the dependencies of the dynamic parameters of internal and external physical labels on SA in the frequency range of $\nu_c = 10^7$ to $10^9\,s^{-1}$ on temperature. As can be seen from Fig. 10.4, the independent approaches, such as the methods of spin, luminescent and Mössbauer labels and NMR, record distinctly the appearance of molecular mobility in the water-protein matrix of SA at $T \approx 200$ K. It should be emphasized that this mobility is practically absent in dry preparations. As shown by experiments, a certain critical, relative moisture content ($P/P_2 = 0.6$) is required for the appearance of molecular dynamics with $\nu_c = 10^7$ to $10^9\,s^{-1}$. The data obtained by other methods (NMR, dielectric constant) also point to the presence of mobility with the frequency ranges indicated in the aqueous phase surrounding the protein.

The study of the mobility of BSA and HSA over a wide temperature range (30–330 K) by the joint use of methods that enable one to study the motions at frequencies of $\nu_c = 10^{-4}$ to $10^{12}\,s^{-1}$, has revealed a general picture of dynamic effects in these proteins (Fig. 10.5). It can be seen from Fig. 10.5 that the general tendency is as follows: the lower the value of the characteristic frequency of the

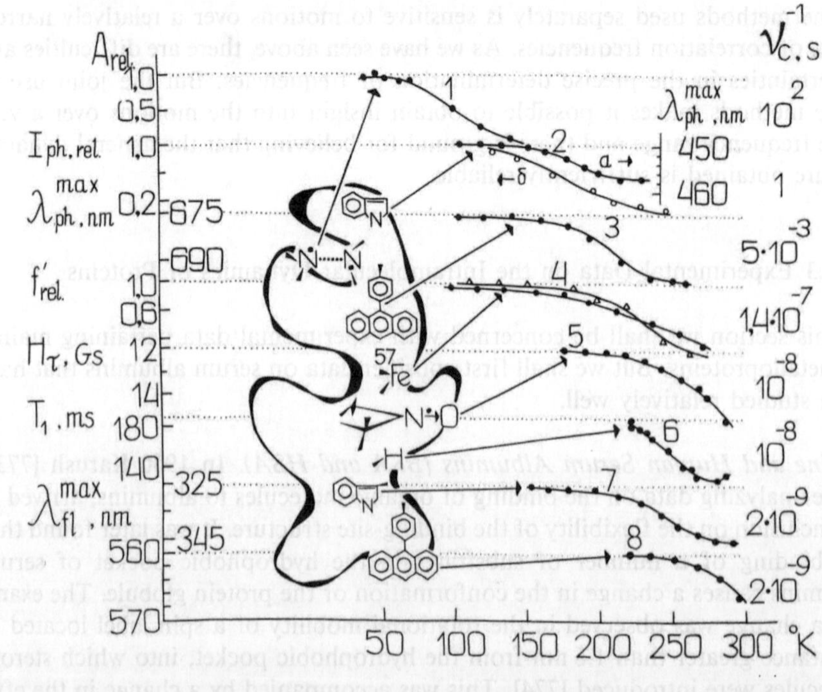

Fig. 10.4. Temperature dependencies of physical labels on dynamic parameters in serum albumins [748]. *Left:* conventional diagram of the label locations; *right*: characteristic time of appropriate method. *1* Radical pair recombination; *2* tryptophan phosphorescence; *3* eosin label (phosphorescence); *4* Mössbauer label; *5* spin label, T-parameter; *6* PMR; *7* tryptophan fluorescence; *8* eosin label (fluorescence)

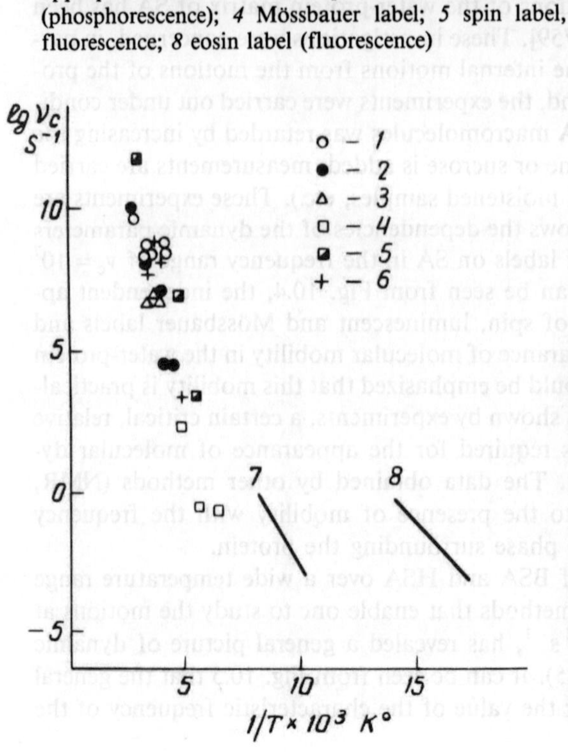

Fig. 10.5. Data on the correlation frequencies of the mobility of physical labels and the environment on bovine and human serum albumins in the Arhenius coordinates [790]: *1* fluorescent labels; *2* spin labels and probes; *3* Mössbauer labels; *4* fluorescent labels; *5* translational jumps of O_2 in the surface layers; *6* PMR of protein and water protons; *7, 8* Arrhenius dependencies calculated form experimental data on the recombination of radical pairs (diphenylamine radicals) for fractions with maximum (*7*) and minimum (*8*) activation energy of recombination

method (ν_c), the lower the temperature at which the label mobility can be recorded. In a wide temperature range (130–300 K) the experimental data for surface labels lie about the Arrhenius straight line with $E_{app} \approx 70\ \mathrm{kJ\ mol^{-1}}$ and $\Delta S^{\neq}_{app} \approx 160$ eu. Thus, the mobility recorded is the result of a gradual softening of the water-protein matrix rather than of individual phase transitions. The mobility of the spin label and diphenylamine radicals in the hydrophobic pocket of HSA has been described by the following parameters: $E_{app} \approx 20\ \mathrm{kJ\ mol^{-1}}$, $\Delta S^{\neq}_{app} \approx -23$ eu in the temperature range 80–280 K.

However, it is evident that the general picture given above is approximate. A more detailed measurement has revealed appreciable differences in the dynamic properties between various portions of the protein globule which are hidden when studied in a wide range of temperatures and correlation frequencies. The side groups exposed to the solvent move at physiological temperatures with higher frequencies and the groups that are closely packed into peptide blocks move with lower frequencies than those expected from the averaged temperature dependence shown in Fig. 10.5. It should be noted that the radical-pair and luminescence methods disclose a substantial heterogeneity in dynamic parameters which is characterized by an activation-energy spread of $4-6\ \mathrm{kJ\ mol^{-1}}$ even in those cases where the labels are localized in the same portion of the protein macromolecule. The heterogeneity of the macromolecule of serum albumin is evidenced by the recently discovered phenomenon of the long wavelength shift in the fluorescence spectra of the only tryptophan in HSA upon edge long-wavelength excitation [775].

According to the localization of physical labels and probes, three zones can be distinguished: (1) the peripheral aqueous phase, which does not differ from the buld water at a distance of 1.4–1.6 nm; (2) the zone formed by the functional groups of the protein and the surrounding water and, possibly, also by water molecules in wide "grooves" of the protein globule; the molecular dynamic characteristics of the zone correspond to a "glycerinelike" state, i.e. the state of viscous amorphous liquids and soft natural and synthetic polymers above the glass-transition temperature; (3) the zone, which contains narrow grooves and hydrophobic pockets on the surface and also polypeptide chains; the motion in the zone is characterized by $\nu_c \approx 10^7\ \mathrm{s^{-1}}$ and a weak temperature dependence of ν_c. It may be presumed that the mobility of the hydrophobic pocket in serum albumins at physiological temperatures is functionally important, allowing the pocket walls to be adjusted to the structure of the corresponding substrate.

Myoglobin and Hemoglobin. Heme-containing proteins, oxygen carriers, are convenient objects for the study of intramolecular dynamics due to the availability of these proteins and the existence of X-ray diffraction models. It was suggested in early works (see, for example, [732]) that the hemoglobin macromolecule undergoes a change during the binding of ligands by the heme group. In the classical works of Perutz and colleagues these conformational changes in hemoglobin crystals were studied in detail (Chap. 6). It has been demonstrated with the aid of the spin-label technique that the substantial rearrangements in the disposition of the hemoglobin subunits occur also in solutions [99]. The results of experiments on isotopic exchange and proteolytic degradation have confirmed these inferences [739]. In myoglobin, whose structue is similar to the structure of the hemoglobin subunit, conformational changes have also been recorded. For example, the spin labels added to the histidine groups of sperm whale myoglobin at a distance of at least 1.7 nm from the heme group have distinctly traced transmolecular conformational transitions upon entropy of cyanide into the active site of the protein [99]. The occurrence of such transitions has also been confirmed by a change in the mobility of the $^{13}CH_3$ groups of surface-methylated methionines 55 and 131. The presence of some flexibility of the cavity of the active sites of hemoproteins and other portions of the globule is evidenced by data on the entry of large molecules into the active site [326] and also by data on isotopic exchange of hydrogen between the internal portions of the protein globule and water [739].

The first direct data on the intramolecular mobility of the hemoprotein globules were obtained by the Mössbauer-label method [143–146]. The experiments were carried out on dry and moistened powders and in myoglobin single crystals, which excluded the motion of the macromolecule as a whole. [^{57}Fe] atoms were incorporated into the heme group of hemoglobin and myoglobin. Thus, is became possible to trace the displacement of the [^{57}Fe] atom tightly bound to the porphyrin ring relative to the protein globule. Since the heme ring is bound to the protein matrix by numerous hydrophobic contacts, it is evident that the motion of the heme must be accompanied by the displacement of the helical polypeptide chains. As has been shown by experiments, in dry and moistened samples at temperatures below 200 K, the vibrations of the [^{57}Fe] correspond to "ordinary" solid-state vibrations with $\langle \Delta x \rangle_c$ being equal to 0.001 nm. However, as the relative moisture content exceeds a certain critical value ($P/P_S > 0.65$), a new type of motion arises with $\langle \Delta x \rangle_c = 0.02$ to 0.05 nm and a frequency of $\nu_c \gtrsim 10^7\ s^{-1}$. These values characterize the dynamics of intramolecular vibrations. The increase in mobility recorded by gamma resonance spectroscopy correlates well with data on the isotopic exchange in myoglobin [146]. The vibration of the heme group begins at the same moisture content as the exchange in the inner regions of the globule. Under conditions in which the exchange involved the portions directly adjacent to the heme group, the latter becomes involved in a diffusive motion with $\nu_c \gtrsim 10^7\ s^{-1}$. The latest investigations have confirmed the basic inferences made in the works cited above [208, 211, 212, 776].

The flexibility of the cavity of the myoglobin active site is evidenced by the mobility of a rather bulky spin label, a derivative of isocyanide attached to the heme group in the single crystal. In the range of room temperatures the mobility parameters, $\nu_c = 10^7$ to $10^8\ s^{-1}$, $E_{app} = 12\ kJ\ mol^{-1}$, $\Delta S^{\neq}_{app} = -60$ eu [99].

The values of mean displacements ($\langle \Delta x \rangle = A$) have been calculated [768] for separate myoglobin atoms on the basis of X-ray diffraction data. These results are in good agreement with the findings obtained by the physical-label technique and with those obtained in subsequent works devoted to the Mössbauer spectroscopy of myoglobin. For example, according to the authors of the works cited, $A = 0.022$ nm for the atoms of the heme group and $A = 0.02-0.03$ nm for the motion of most of the atoms in the polypeptide chain. For the more mobile end of the globule and the cavity of the active site, $A = 0.035$ to 0.040 nm and for the

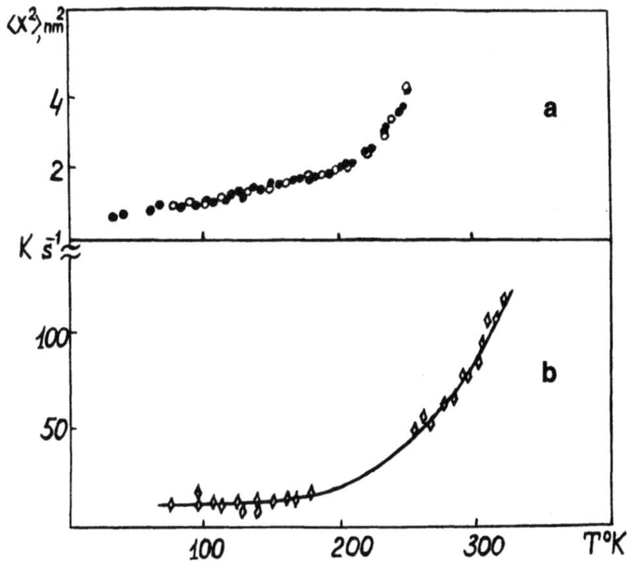

Fig. 10.6. **a** $\langle x^2 \rangle$ of human deoxyhemoglobin as a function of temperature: [^{57}Fe]-enriched in the α- and β-chains [779]. **b** Temperature dependence of the electron-transfer rate k_1 in the [α(Fe''(H_2O); β(Zn)] hybrid hemoglobin [781]

sheath of the globule, A = 0.045 nm. The temperature dependencies of the quantity A and $\langle \Delta x \rangle_c$ and the rate constant of electron transfer between the hemes of Hb have also been found to be similar [208, 779, 781] (Fig. 10.6).

The difference in the mobility of the various parts of the myoglobin protein globule is evidenced by further data [742], whereby the rotation of the heme CH_3 group was studied by the PMR method. The following values of the dynamic parameters have been obtained: $\nu_c = 3 \times 10^2\ s^{-1}$ at 30 °C, $E_{app} = 90\ kJ\ mol^{-1}$ and $\Delta S^{\neq}_{app} = -8.5$ eu.

The mobility of a spin label on the 151-Tyr group and the chromophore group at the N-terminal NH_2 group has been studied [755, 779] in the temperature range 77–300 K and in the range of characteristic correlation frequencies, $\nu_c = 10^2 - 10^9\ s^{-1}$ (Fig. 10.7). It has been found that the mobility of the myoglobin heme group with $\nu_c > 10^7\ s^{-1}$ and an amplitude exceeding 0.03 nm, which arises at T > 200 K (the data of Mössbauer spectroscopy), correlates with the dynamics of the water-protein matrix with $\nu_c \gtrsim 10^7\ s^{-1}$ recorded by physical-label methods (Fig. 10.7).

It is interesting to compare the data on the dynamic state of myoglobin with the results of kinetic studies of ligand binding in CO-deoxymyoglobin. As seen from Fig. 10.7, the following correlations are observed in water-glycerine solutions. The region of the motions of fluorescent and spin labels with $\nu_c \approx 10^7\ s^{-1}$ near 230 K coincides with the region of variation of the apparent activation energy of the association of CO with the heme group from 80 up to 12 kJ mol^{-1} [782]. Approximately in this temperature range the slope of the temperature dependencies N(t) changes in the Arrhenius coordinates, where ($N_{(t)}$) is the fraction of myoglobin molecules that has no time to be complexed in the reaction between the CO molecules in solution and the heme during the time 10^{-2} s. The rate-limiting step under these conditions is apparently the movement of CO from the solvent into the active-site cavity.

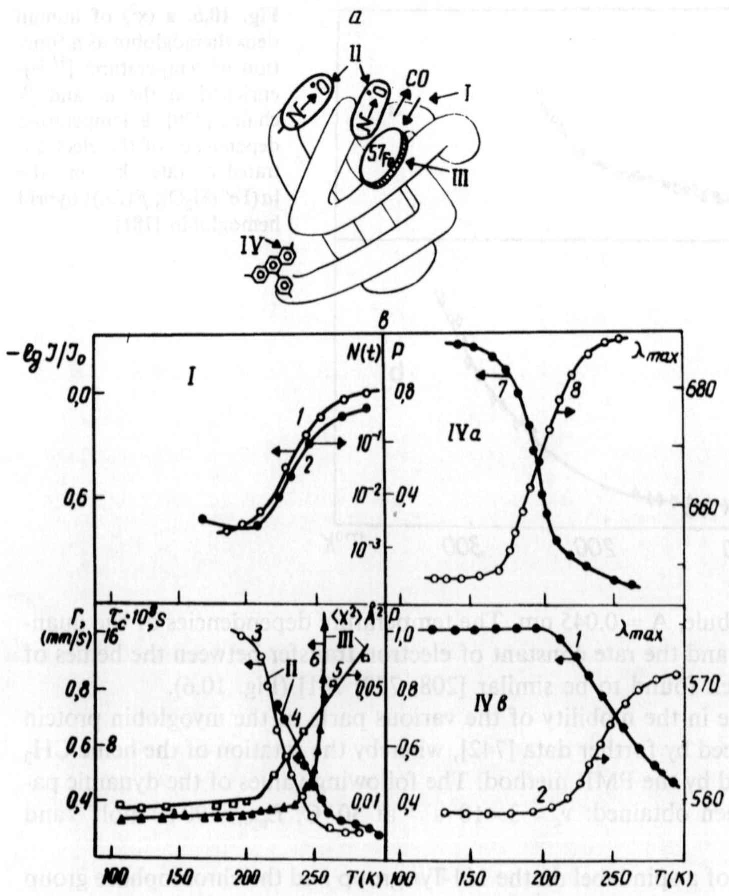

Fig. 10.7 a, b. Temperature dependence of the dynamic parameters of physical labels and of the kinetic parameters of the processes of carboxylation and decarboxylation on sperm whale myoglobin: **a** introduction of labels into the myoglobin macromolecule; **b** temperature dependencies of the parameters of physical labels on sperm whale myoglobin; *I* kinetic parameters of the photodissociation of MbCo; I/I_0 is the ratio of quantum yields of the photodissociation of MbCO at a given temperature and at high temperatures (*1*) [755]: $N(t)$ is the fraction of MbCO molecules left non-dissociated upon elapse of 10^{-2} s after a short light flash (*2*) [755]; *II* spin labels; *III* Mössbauer labels (τ_c is in the correlation time of rotational diffusion of spin labels; the label at the active site of the myoglobin is the monocrystalline state (*3*); the label on the Tyr 151 residue of the myoglobin is in solution (*4*) [779]; $\langle x^2 \rangle$ is the rms shift of the displacment of the Mössbauer nucleus [^{57}Fe] at the active site of the myoglobin in nm^2(*5*); Γ is the line width of the Mössbauer spectrum of the [^{57}Fe] atom at the active site of the myoglobin (in relative units) (*6*) [208]; *IVa* phosphorescent label, eosin isothiocyanate (EITC) on the terminal NH_2 group of thy myoglobin [*P* is the relative fluorescence intensity (*7*)]; λ_{max} is the position of the maximum of the fluorescence spectrum in nm (*8*); *IVb* fluorescent EITC label on the terminal NH_2 group of the myoglobin [*P* is the relative fluorescence intensity (*9*)]; λ_{max} is the position of the maximum of the fluorescence spectrum in nm (*10*)

The temperature dependence of the quantum yield of photodissociation of carboxymyoglobin, I_f, correlates with data on fluorescence and Mössbauer spectroscopy [756]. Below $T = 200$ K the quantum yield is close to zero. An increase in the quantum yield begins in the same temperature region in which I and f′ drop sharply, the value of $\Delta\lambda_{max}$ of fluorescence spectra increases and, hence, the conformational mobility of the heme increases, and the dynamics of the water-protein matrix is intensified. It may be presumed that the mobility of the matrix under these conditions makes the release of CO possible from the active-site cavity after the elementary photodissociation act is completed. In the temperature range 140–190 K the method of phosphorescent labels records several dynamic processes in the matrix: slow reorientations of electrical dipole assemblies around the excited chromophore ($\nu_c \approx 10^3\ s^{-1}$) and the motion of the O_2 molecule with ν_{tran}^{-1}tran $\approx 10^{-7}$ to 10^{-8} s. According to available data [782], in this region there occur a number of activation processes of CO association present in the cavity of the active site. The activation energy of these processes varies from 10 to 50 kJ mol^{-1} and the entropy of activation from 45 to 120 eu; there is a linear relationship between these quantities (compensation effect). This fact is consistent with the conceptions developed earlier [726, 729], according to which the presence of the compensation effect is evidence that rearrangements of the water-protein matrix have an impact on the kinetic parameters of the chemical process studied.

It may also be presumed that the physical processes responsible for the change from the temperature-independent tunnelling mechanism of reassociation of carboxymyoglobin to a temperature-dependent mechanism in the region of low temperatures [782] is associated with the dynamic state of the myoglobin macromolecule. The region of the transition at $T \approx 70$ K is close to the region of the increase of the parameter $\Delta H_{1/2}$, obtained in a study of the temperature dependencies of the saturation curves of the spin-label EPR spectra on human oxyhemoglobin. It has been shown [166] that the increase of $\Delta H_{1/2}$ correlates with the intensification of various motions: slow, high-amplitude displacements of radical pairs ($A = 0.5$ nm, $\nu_c = 10^{-2}\ s^{-1}$) and faster reorientations of electrical dipole assemblies with $\nu_c \approx 1\ s^{-1}$ and relatively rapid jumps of O_2 molecules with $\nu_c = 2\times10^{+4}\ s^{-1}$.

The regularities in the dynamic properties obtained for myoglobin preparations are confirmed in general for human hemoglobin. Thus, the mobility of the heme ring with $A = 0.02$ nm and $\nu_c \gtrsim 10^7\ s^{-1}$ has been recorded by means of the Mössbauer-label method at a certain critical moisture content ($P/P_S \gtrsim 0.73$) [146]. Under analogous conditions a rotational motion of the spin label on the surface arises. For spin labels tightly packed in the groove of the hemoglobin macromolecule in the region of the β-93 SH group, low amplitude, high-frequency librations have been detected, which reflect, in the opinion of some authors [783], the vibrations of the surrounding portions of the protein. The NMR data [746] are consistent with the theoretical models of protein rotation only under the assumption of the internal mobility of the hemoglobin protein globule with frequencies near the nanosecond region. Analysis of the temperature dependencies of the rotational diffusion of spin-labels in the surface layers of hemoglobin and myoglobin also confirms its glycerinelike nature [99].

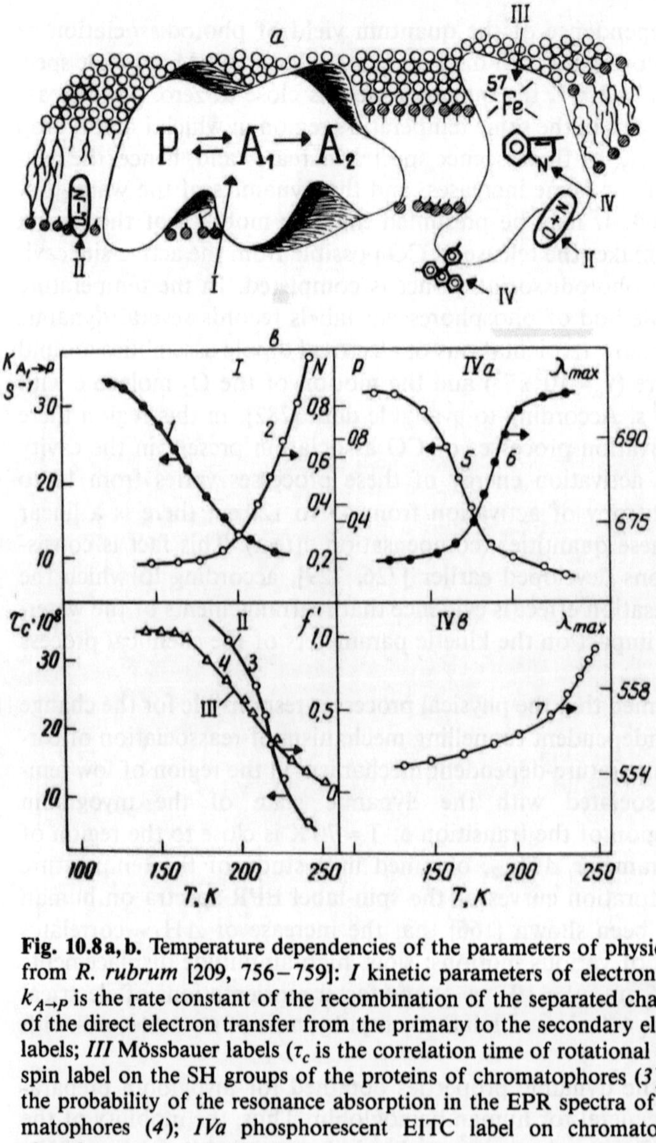

Fig. 10.8 a, b. Temperature dependencies of the parameters of physical labels on chromophores from *R. rubrum* [209, 756–759]: *I* kinetic parameters of electron transfer in chromophores; $k_{A \to P}$ is the rate constant of the recombination of the separated charges (*1*); *N* is the efficiency of the direct electron transfer from the primary to the secondary electron acceptor (*2*); *II* spin labels; *III* Mössbauer labels (τ_c is the correlation time of rotational diffusion of the labels. The spin label on the SH groups of the proteins of chromatophores (*3*); *f'* is the relative value of the probability of the resonance absorption in the EPR spectra of the [^{57}Fe] nucleus in chromatophores (*4*); *IVa* phosphorescent EITC label on chromatophores [*P* is the relative phosphorescence intensity (*5*)]; λ_{max} is the position of the maximum of the phosphorescence spectrum in nm (*6*); *IV* fluorescence of the EITC residue [λ_{max} is the maximum of the fluorescence spectrum in nm (*7*)]

Other Metalloproteins. Photosensitive systems are convenient objects for the analysis of possible correlations between the dynamic and functional properties of proteins and membranes. After a short light pulse it is possible to observe the transfer of an electron along the chain and to trace the dynamic state of the matrix with the aid of internal and external physical labels [209, 756–759]. Such an investigation has been carried out on the reaction centres (RC) extracted from *Rhodopseudomonas spheroidos* in the isolated state and in the composition of the membranes of chromatophores. Spin, Mössbauer, fluorescent and phosphorescent labels were introduced into the various portions of the protein matrix: they were covalently bound to the surface groups of RC and adsorbed by the hydrophobic segments of the protein and membrane and [^{57}Fe] atoms were introduced by way of biosynthesis into iron-containing proteins (Fig. 10.8). Then, the dependencies of the label mobility and the rate constants of electron transfer between the components of the photosynthetic chain on temperature and moisture content were measured in the same samples. The results of one of the experiments are illustrated in Fig. 10.8. It can be seen from Fig. 10.8 that the emergence of an electron from the primary photosynthetic cell, which is favourable for photosynthesis (the transport from the primary acceptor A_1 to the secondary acceptor A_2), takes place only under the conditions in which the labels record the mobility of the protein moiety of the membrane with $\nu_c \gtrsim 10^{-7}\ s^{-1}$. The rate of another undesirable kinetic process, the recombination of the primary products of the photoseparation of the charges of the reduced primary acceptor (A_1^-) and oxidized primary donor (D_1^+), falls from 10^2 to $10^3\ s^{-1}$ when dynamic processes occur (Fig. 10.8).

Thus, the transformation process of sunlight energy into chemical energy of spatially separated charges occurs only in a certain dynamic state of the water-protein matrix.

Data have been reported in the literature on the intramolecular mobility of various proteins. Although these data are scanty, they nevertheless confirm the general picture of dynamic phenomena that emerges from the material analyzed above. Let us consider some of these phenomena.

The internal mobility of [^{57}Fe] atoms with $\nu_c \gtrsim 10^7\ s^{-1}$ in the active site of spinach ferrodoxin, which depends on temperature and moisture content, has been revealed by means of the Mössbauer-label technique [146]. Such a mobility has been established for lysozyme, α-chymotrypsin and collagen.

In type *c* cytochromes at physiological temperatures a spectrum of dynamic characteristics has been observed, determined by the NMR method: slow rotations of tyrosine residues 75 and 97 in cytochrome *c* with $E_{app} = 100\ kJ\ mol^{-1}$ and $\Delta S^{\neq} = 105$ eu [743]; the absence of an appreciable rotation of tyrosine and phenylalanine residues in cytochrome c_{552}. Analysis of the decay curves for the spin echo of the protons of globular proteins in solutions of RNase, DNase, hexokinase, alcohol dehydrogenase and peroxidase has demonstrated the presence of the decay component due to the intramolecular motion of protons with a frequency of $10^7 - 10^8\ s^{-1}$ [747].

10.2.4 Theoretical Conception of the Intramolecular Dynamics of Proteins

The data on diffusion in condensed media are interpreted by using free-volume concepts [784]. Condensed media are essentially structures with maximally, closely packed molecules, which contribute to the minimization of the energy of the system as a whole. In real systems, however, there are practically always vacant sites, unsubstituted voids and other structural disturbances, i.e. defects which form a free volume (V_0). The defects are often dynamic in nature, i.e. they are moving in the medium. When the defect (a "bubble") reaches a given molecule or group, it is trapped for a certain period of time by the defect and has a chance to take part in a certain type of motion (rotation, libration, high-amplitude vibration, translational jump). The larger the free volume (V) required for a given type of motion, the lower the probability of its realization. According to theory, the value of the rate constant of the elementary act of motion is given by:

$$k \sim \exp(-V/V_0) \ . \qquad (10.7)$$

That the aperiodicity of the globular protein structure makes the existence possible of diverse static and dynamic defects is confirmed by X-ray diffraction analysis [728, 738]. On the other hand, closely packed portions (blocks) are seen distinctly in the protein structure. The data obtained by means of the isotopic exchange method also point to a high heterogeneity of proteins [737–740]. Taking into account the data given in the previous section, it should be emphasized that the mobility of surface layers of the water-protein matrix has a decisive effect on the intramolecular dynamics of proteins. In dry powders the protein globules are tightly packed by the forces of intermolecular interaction and by electrostatic charges not shielded by water, while the internal portions are held together by intramolecular forces. The free volume is not large. The inevitable thermal fluctuations lead to an instantaneous opening of the structure, this being followed by its rapid collapse. The first portions of the water are bound by polar groups and have no substantial effect on protein dynamics. However, beginning from a certain critical amount, the water molecules shield the charges, enter the cavities formed, saturate the broken contacts, preventing their collapse, increase the free volume of the system and, eventually, intensify the dynamic processes in the medium, especially in the physiologically important high frequency region.

The above given experimental data on proteins may be treated from the standpoint of cooperative rearrangements (Sect. 10.1). These data indicate the presence of two types of motion. One type of motion is characterized by relatively high values of activation energy ($E_{app} = 24-80$ kJ mol^{-1}) and activation entropy ($\Delta S^{\neq}_{app} = 40-360$ eu) with a linear correlation between these quantities. Within the framework of the model considered in Section 10.1, this means that the motion is associated with the "melting" of the cooperative portion, i.e. ΔH^0_0 and ΔS^0_0 are positive and n is sufficiently large. The second type of motion is characterized by the following values: $E_{app} = 5-16$ kJ mol^{-1} and $\Delta S^{\neq}_{app} = -90$ to -65 eu. The latter type of motion, which is realized in the hydrophobic segments and in the folds of polypeptide chains, is presumably associated with relatively low-amplitude (0.02–0.05 nm) vibrations of the protein matrix walls. In such cases the separation of the walls in this portion is accompanied by the effective ordered "crystallization" of other groups, e.g. water molecules, around the hydrophobic contacts exposed as a result of motion. Intermediate cases are also encountered.

The third approach is based on the thermodynamic theory of fluctuations [728]. According to this theory, the fluctuations in energy and volume are associated with the values of the specific heat capacity C_P, which in turn depends on the number of degrees of freedom of the particles moving in the system. The largest contributions to the heat capacity come from the translational, rotational and low-frequency vibrational motions. The rms fluctuation of the energy is given by

$$\langle \Delta H \rangle = T(C_p R)^{1/2} \ . \tag{10.8}$$

For example, with a conformational contribution to the heat capacity C_p = 400 kJ mol^{-1} and at T = 300 K, the fluctuation of the conformational enthalpy $\langle \Delta H \rangle = 16.0$ kJ mol^{-1}. This value is not high considering the large size of protein macromolecules. If the fluctuations are localized in small portions, e.g. at the boundary between the blocks or in the region of the active site, they may become significant.

The approaches described above are concerned with different aspects of the same phenomenon and there is an obvious relationship between them. For example, upon phase transitions, which are typical cooperative processes with the number of particles $n = 10^{12}$ (according to the estimates given in [728]), a precise compensation is observed, $\Delta H_0 = T \Delta S_0$; C_P, $\langle \Delta H \rangle \to \infty$, and the number of interacting defects that make up the free volume sharply increases [784].

Stochastic models have lately been proposed for describing the conformational dynamics of proteins, which take into account the viscoeleastic properties of the internal portions of protein globules. In particular, a model of the conformational transition in a viscous medium between two states separated by a wide static barrier of height E_A has been reported [316]. The average frequency of the conformational transition in this case is determined by Eq. (5.54) in Table 5.1 with the value of $l \sim 0.01$ nm (the width of the local minimum of the initial state) and $L = 0.1 - 1$ nm (the length of the barrier). The viscosity of the protein is estimated at about 10 poise and $\nu_{conf} = 10^7 - 10^9 \ s^{-1}$. Consideration of the elastic properties for extended α-helices leads to values of the displacement amplitude $0.1 - 0.3$ nm and frequencies of the order of $10^6 \ s^{-1}$; the corresponding values for β-structures are $0.25 - 0.7$ nm and $\sim 10^6 \ s^{-1}$. When conformational fluctuations with an amplitude of the order of 0.1 nm are considered, use is made of the Brownian oscillator model limited by a parabolic potential and of the model of the viscous rectangular potential box [763].

According to other authors [739], the local trans-conformations in proteins have a common nature with the “wandering” equilibrium point defects (vacancies) in the lattice points and extra atoms in the interstices of solid crystals. These local displacements are realized due to the elastic deformation of the surrounding matrix. Concepts on the role of mobile defects in proteins for the realization of their functional properties have been developed [728, 738].

Successful attempts have been made in recent years at a direct theoretical calculation of the kinetics of conformational transitions and averaged fluctuations in concrete proteins [259, 777]. In these attempts, the inital state is taken to be the positions of atoms determined from X-ray diffraction data. The dynamics of the displacement of the protein was calculated by proceeding from the corresponding values of the atom-atom potentials. For the pancreatic inhibitor chymotrypsin the calculation was performed with a time interval of 10^{-15} s. According to the calculation, the displacement of the polypeptide chains amounting to 0.05 nm is achieved during a time of the order of 10^{-10} s. This value differs appreciably from the experimental value (10^{-8} s), which is typical of proteins, probably because the theory does not take into account the effect of the medium on the dynamics of the macromolecule. The mean deviations of the nuclear positions in cytochrome *c* have been calculated [259]: 0.07 nm for the α-carbon atoms of the main chain, 0.085 nm for other heavy atoms and 0.051 nm for the heme group. These calculations have confirmed the theoretical inference made by Lifshits that under certain conditions the free polymeric chain folds into a globule with a tightly packed core and a loose "edge" [766]. For example, for cytochrome *c* the mean deviations vary from 0.066 to 0.164 nm in going from a core of radius 0.6 nm to the "edge" of radius 2.2 nm.

Thus, the conceptions of the intramolecular dynamics of protein globules are based on a sound theoretical foundation. However, much work is required for the study of the dynamics of concrete proteins and for the development of a general quantitative theory for these systems.

10.2.5 Intramolecular Dynamics and Functional Properties of Proteins

The analysis of data on the dynamics of proteins and chemical conversions in the active sites of enzymes should be based on the conceptions of chemical kinetics with account taken of the specific features of enzyme catalysis. According to these conceptions, molecular dynamics plays a key role in the realization of the elementary act of a chemical conversion in the condensed phase. This role boils down to the creation of conditions for a fit between electrons and nuclei (an electron-nuclear complementarity) during the elementary act [see Chap. 5, Eqs. (5.4), (5.5), (5.13), (5.21), (5.26), (5.29), (5.34) and Table 5.1]. The energy of the "reaction centre-medium" system depends to a considerable extent on two types of interaction between the electronic and nuclear systems: an electron-vibronic and an electronic-orientation interaction. In the first type of interaction a state of equilibrium between the electronic and nuclear systems is achieved during a single vibration period ($\nu^{-1} = 10^{-12}$ to 10^{-14} s). The orientation of electrical dipoles of the environment can occur over a wide range of times (10^{-12} to 10^{3} s). Evidently, the electronic-nuclear complementarity required for the minimization of the energy in the reaction can be realized only if the medium exhibits some molecular mobility. Many chemical processes in polymers and other condensed systems are limited by the molecular mobility of the medium or depend significantly on the arrangement of molecules and units in the solvent cage.

The most specific feature of enzymes is their multi-centre nature. It is this feature that provides the realization of the main advantages of enzymes, such as the binding and orientation of substrates, synchronous elementary acts and the possibility of occurrence of a multi-step process with an optimal rate for each step [24, 215–222, 728]. However, a detailed analysis of concrete reactions shows that these advantages cannot be realized in absolutely rigid structures. Indeed, even such a "simple" process at the transfer of an electron from one centre to another must be completed with the reorganization of structures, in which an electronic-conformational interaction takes place. Chemical enzymatic processes are, as a rule, accompanied by a significant change in the size of the reaction complex (by 0.05 to 0.2 nm). Some bonds are broken, new ones are formed and the covalent

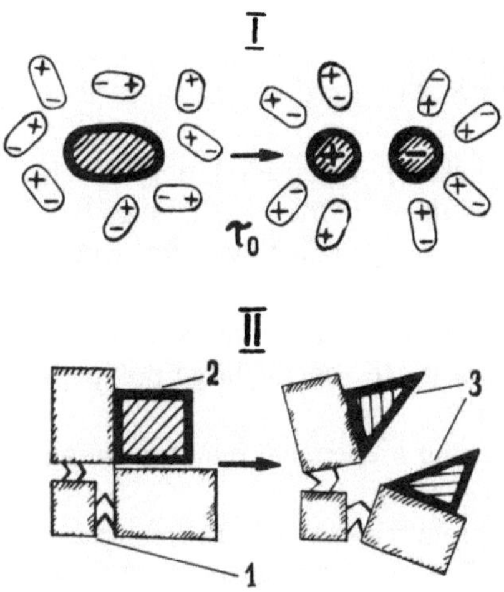

Fig. 10.9. Examples illustrating the principle of the dynamic adaptation of the environment of the surrounding matrix to the structure of the reaction complex. *I* Adaptation by way of orientation of the dipole system around the charged states upon ionization. *II* Block adaptation of the shape of the enzyme to the shape of the reaction complex at the active site; *1* blocks; *2* substrate; *3* reaction product of substrate cleavage

contacts change to van der Waals contacts and vice versa. It is obvious that the structure of the protein matrix must fit the varying shape of the reaction complex in all the process steps. This general property of the protein matrix may be defined as dynamic adaptability (Fig. 10.9) [24, 741].

It is evident that the processes involving proteins, which are accompanied by a change in their conformation, such as complex formation (for example, the formation of an antigen-antibody complex), allosteric transitions and mechanical transformations, are also realized only due to the ability of proteins to change their conformation rapidly and reversible.

The experimental frequency values for relatively large-scale conformational changes in enzymes are $\nu_c = 10^2$ to $10^4\,s^{-1}$ [727]. Slow transitions in complex proteins occur in the second and minute ranges. At the same time, as has been shown for a number of processes (the binding of ligands, electron transfer, the hydrolysis of an enzyme-substrate complex), for chemical processes to be realized a high-frequency is required ($\nu_c = 10^{-7} - 10^8\,s^{-1}$) and relatively low-amplitude ($A = 0.03 - 0.06$ nm) dynamics. Hence, the overall conformational transition consists of a large number of small-scale motions.

At present, there is every ground for believing that the high-frequency and low-amplitude intramolecular mobility of protein globules is associated to a considerable extent with the displacements of relatively large polypeptide blocks rather than with the motion of the major portion of atoms inside the blocks [24, 145, 198, 213, 798]. This assumption is confirmed by the following arguments:

1. The temperature heat-capacity coefficient ($0.92 \times 10^{-3}\,J\,g^{-1}\,K^{-2}$) depends weakly on temperature and moisture content [787, 788] under the conditions of a sharp intensification of the intramolecular motion registered by direct physical methods. This means that the motions under consideration are associated with the appearance of a relatively small number of degrees of freedom, whose contribution is insignificant as compared with the increase of heat capacity brought about by the oscillatory motion of a large number of atoms in the protein globule. Indeed, the intramolecular dynamics of proteins is characterized by a frequency

of $10^8\,s^{-1}$, each vibration must make a contribution to the heat capacity amounting to about $8\,J\,mol^{-1}\,K^{-1}$. For an independent motion of each of several hundreds of peptide and other rigid fragments in a protein, the contribution would be enormous.

2. Many globular proteins consist of closely packed polypeptide blocks which are bound relatively weakly [765]. The compressibility of the α-helices and β-structures in proteins is several times lower than the average taken over the protein globule [790].

3. The conformational changes of a number of globular proteins upon interaction with substrates are associated with the displacement of large polypeptide blocks upon variation of temperature and pH.

4. The displacement of polypeptide blocks during a chemical reaction on an enzyme must be accompanied by minimal entropy losses.

The dynamic behaviour of the protein macromolecule may be approximately described in terms of the following model [790]. The protein globule is made up of rigid blocks (α-helical portions, β-structures, closely packed side groups) and a softer, amorphous region (disordered, loose portions and water molecules). The blocks are bound by relatively flexible hinges and sufficiently weak hydrogen, hydrophobic and polar interactions at the contacts between the blocks and the amorphous regions. Thus, the block system is embedded in the viscous amorphous medium. The protein globule is surrounded by a glycerinelike "sheath" composed of water molecules and exposed surface protein groups. The globule may be in a state in which the blocks are closely packed (the closed state) and in an open state. The number of open states is relatively small. In an open state the blocks may have a looser structure than in the closed state. This may manifest itself, for example, in the ability to undergo a hydrogen exchange between the peptide protons and water. Like many other amorphous bodies, the external "sheath" is characterized by high temperature viscosity coefficient. As the temperature increases, the sheath is substantially softened and the motion of the blocks is determined by the interactions in the system composed of blocks and an internal amorphous region. The dynamics of the amorphous region is characterized by a low temperature coefficient.

The concepts of proteins as systems possessing the properties of amorphous bodies have recently been worked out on the basis of data on the temperature dependence of the heat capacity in the region of helium temperatures [789]. Analogous concepts have been developed for protein globules and rubberlike systems [728]. The viscous properties of protein bodies have been taken into account in a number of theoretical works, in which the act of the chemical reaction is treated as the result of the combination of the thermal-disorder-activated motion of the nuclei of the reaction complex along the reaction coordinate at a thermal velocity and the diffusive motion of the nuclei system of the surrounding medium [see, in particular, Sect. 5.11 and Eqs. (5.53) and (5.54) in Table 5.1]. From this standpoint it is possible to account for the effect of the solvent viscosity on the rate of enzymatic reactions [792] and on the intramolecular dynamics of proteins recorded by Mössbauer spectroscopy [793].

The numerous entropy and enthalpy anomalies in the kinetic and thermodynamic parameters of enzymatic processes pointed out already in the 1960s

[726, 729] find their natural explanation (see Sect. 10.1) within the framework of the concept of conformational and solvation changes that accompany elementary chemical reactions.

In conclusion, it should be noted that the above described conceptions of the block nature of the intramolecular mobility of proteins are to a certain extent similar to the concept according to which protein globules are treated as systems having elements characteristic of machines, i.e. systems having a few specific degrees of freedom [264, 727, 764]. It is essential, however, that the motions over these degrees of freedom take place in a viscous, amorphous medium and are presumably associated to a considerable extent with water-protein interactions and, in particular, with the rearrangements of water molecules in the surface layers of the globule. It is evident that the models under discussion are similar, in part, to the model that describes the intramolecular dynamics of the protein as a diffusional motion in a potential box [763].

The above described model of the reversible equilibrium, dynamic adjustment of the enzyme shape to the structure of the reaction complex furnishes a consistent picture of a wide range of experimental phenomena. However, at present one must not ignore the possibility of other, more complex effects of the intramolecular dynamics of proteins on the course of the enzymatic process. In particular, the conceptions of the complementarity of the free energies of the chemical and conformational subsystem of enzymes [733], the relaxation model of the elementary act of the enzymatic reaction [727] and the concept of the role of fluctuations of the charge and electrical dipoles [760] deserve serious consideration. The theoretical conceptions on the nature of the intramolecular dynamics of proteins and its role in enzyme catalysis together with the corresponding experimental data have been described elsewhere [793–800]. Some of the quantitative aspects of the problem will be discussed in Section 11.1 for one-electron processes.

Further experimental and theoretical investigations of the molecular dynamics of proteins and enzymes and its correlation with the functional properties are undoubtedly promising and intriguing areas in the chemical physics of enzyme catalysis.

CHAPTER 11

Mechanisms of the Elementary Acts of Redox and Coupled Processes Involving Metalloenzymes and Carriers

An enzymatic process involves a certain sequence of chemical and conformational steps. For the reaction to occur effectively each of the elementary steps must proceed at a high velocity and the spatial and time organization of the chemical process as a whole must be favourable. Studies on a large number of concrete cases have convincingly demonstrated the unique ability of enzyme protein macromolecules to bind and orient with precision the reaction groups in the active site and to provide an entire set of chemically active groups for the organization of an optimal sequence of chemical elementary acts [216–219, 221–223]. Interesting conceptions have been worked out concerning the ability of enzymes to convert substrate molecules into reactive forms, whose concentrations are low under conditions of non-enzymatic reactions [312]. Thus, it is evident that enzymes are excellent "organizers" and "conductors" of chemical processes.

However, the central problem of the mechanism of the elementary steps in the theory of enzyme catalysis has not yet been solved. What is the cause of the sharp decrease in the energy activation barriers in enzymatic reactions as compared with chemical reactions? Why is the transfer involving metal carriers so effective? What is the mechanism of the activation of inert molecules, such as N_2 and H_2O, whose conversion into NH_3 and O_2 respectively, has not been accomplished until recently under mild conditions by a non-enzymatic route? These and many other questions require a fundamental solution.

Physical concepts have been formulated that account for the excellent catalytic properties of enzymes and their ability to utilize the energy released in a chemical reaction for lowering the energy barriers. The possible mechanisms of storage of this energy that have been proposed are the mechanisms of its transformation to conformational energy or the energy of elastic vibrations of protein macromolecules [791]. The "rack" concept based on the assumption of the deforming action of the enzyme protein globules on the substrate fixing groups is discussed in the literature [732]. While giving credit to the conceptions advanced in the literature, especially the latter one, it should be noted that they are formalized abstractions which are not supported by direct experimental data or by analogies in model chemical systems.

On the other hand, modern chemical kinetics has provided extensive and concrete material for a theoretical analysis of the elementary mechanisms of enzymatic processes and for the formulation of experimental approaches to their investigation. An important criterion for the correctness of the mechanism of the enzymatic step is the possibility of its chemical modelling under mild conditions close to physiological conditions.

In dealing with multi-step chemical reactions use is ordinarily made of the concept of the rate-limiting step, which is the slowest step, whose velocity determines the rate of the overall process. But this is insufficient for describing enzymatic processes and, especially, for comparing them with the corresponding model chemical reactions. For example, the steps that can be rate-limiting in an enzymatic reaction (diffusion into the groove of the active site, protonation, a conformational transition, an electron transfer) either occur very rapidly or do not take place at all in model systems. Comparison of the rate-limiting steps is, therefore, often meaningless.

In order to reveal the principal factors that determine the high efficiency of enzymatic processes, it is expedient to introduce the concept of the decisive step of the process. The decisive step of an enzymatic process may be defined as the step, whose occurrence in the corresponding chemical process is either impeded or practically impossible under mild conditions. Examples of the decisive steps in the reduction of N_2 and O_2, oxidation of H_2O and in other processes will be considered below.

The present chapter is devoted to the analysis of the elementary mechanisms of redox enzymatic processes from the viewpoint of chemical kinetics. Simpler one-electron mechanisms will be treated within the framework of the theories of outer-sphere electron transfer. Unfortunately, the present-day level of development of the theory of inner-sphere reactions in the coordination shell of transition metals and, especially, of multi-electron processes permits only a qualitative treatment of possible mechanisms. This chapter concludes with a discussion of the chemical mechanisms of the most important biological process of coupling of ATP hydrolysis with electron transport and with other chemical and mechanical processes.

11.1 One-Electron Transfer

11.1.1 Theoretical Concepts. General

The theory of one-electron transfer is based on the general conceptions of the mechanisms of chemical reactions outlined in Chapter 5. According to these conceptions, the rate of transfer between the donor (D) and acceptor (A) molecules is primarily determined by the following factors: (1) the number of collisions between the molecules; (2) the lifetime of the complex DA before and after the electron transfer; (3) the thermodynamic preferableness of the transfer; (4) the degree of overlap of the wave functions of the system before and after the transfer; (5) the efficiency of the adjustment of the nuclear environment to the new electronic configuration of the donor and the acceptor (the electronic-nuclear complementarity); (6) the presence or absence of spin forbiddenness.

The role of the first two factors has been considered in Section 5.3. Here, we will consider in more detail the process of electron transfer in a complex with a fixed arrangement of donor and acceptor centres. The following simplified physical picture may be visualized. Suppose that as a result of a thermal fluctuation or other factors, the nuclear configuration of the system has attained a transition state corresponding to the equality of the energies of the system in the initial (i) and the final (f) state (Fig. 5.4). Then, the electron density will oscillate between the D and A centres with a frequency ν_e, which depends on the degree of overlap of the wave functions and, in particular, on the distance between the centres [Eq. (5.36)]. The electron-transfer act will terminate only after an electron-nuclear fit is achieved in the reaction products as a result of the nuclear adjustment. The probability of the adjustment process is determined by the efficiency of the rapid vibrational relaxation of the nuclei ($\tau = 10^{-12}$ to 10^{-14}s) and by the relatively slow orientational relaxation, including the conformational relaxation ($\tau = 10^{-10}$ to 10^{-2}s). The quantitative aspects of the relaxation processes have been considered in Section 5.5.4.

Let us consider the possible mechanisms of the attainment of an electronic-conformational fit for the case of an electron transfer between the transition-metal ions M_D^{+n} and M_A^{+m} present in the protein environment. Let square and brace brackets denote the presence of an electronic-conformational fit in the reduced (red) and oxidized (ox) states respectively; the absence of such a fit in the conformation-strained forms is denoted by an asterisk. The electron-transfer process in the complex may then be described by the following scheme:

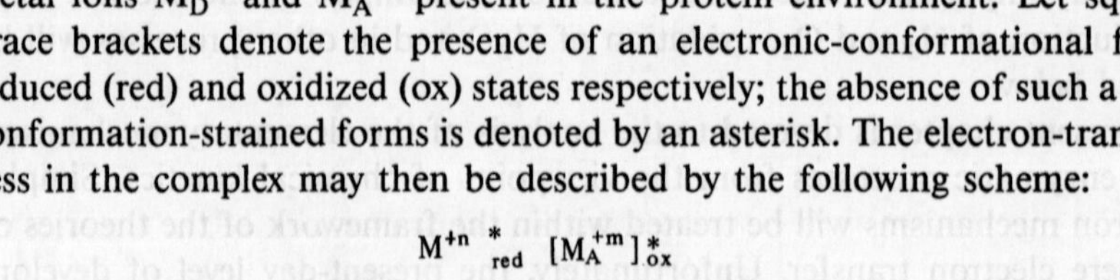

(III)

Mechanism I involves first an equilibrium adiabatic conformational adjustment of the protein and of the surrounding water to the configuration of the final state f, this being followed by an electron-transfer process (Fig. 11.1 a). In mechanism II, which in fact corresponds to the Marcus model (Sect. 5.5.3), the electron transfer occurs at the moment of attainment of the transition-state configuration, which is denoted by a double asterisk (Q_{tr}, Fig. 5.4). According to mechanism III, first a rapid electron transfer takes place from M_D^{+n} to M_A^{+m}, the two centres remaining in their "old" conformational forms. The subsequent conformational transition completes the process.

Evidently, with small differences in free energy between the initial and the final state (ΔF_0), mechanism II, which involves the formation of a transition state, is the most effective. Thus, at $\Delta F_0 = 0$ the free activation energy $\Delta F_r^{\neq}$ for this mechanism is four times lower than the free energy of reorganization ΔF_r [Eq. (5.29)]. The latter energy is close to the activation energy in mechanisms I and III. In real protein systems the conformational transitions involve many nuclei and, according to the principle of optimal motion, the rearrangement is a

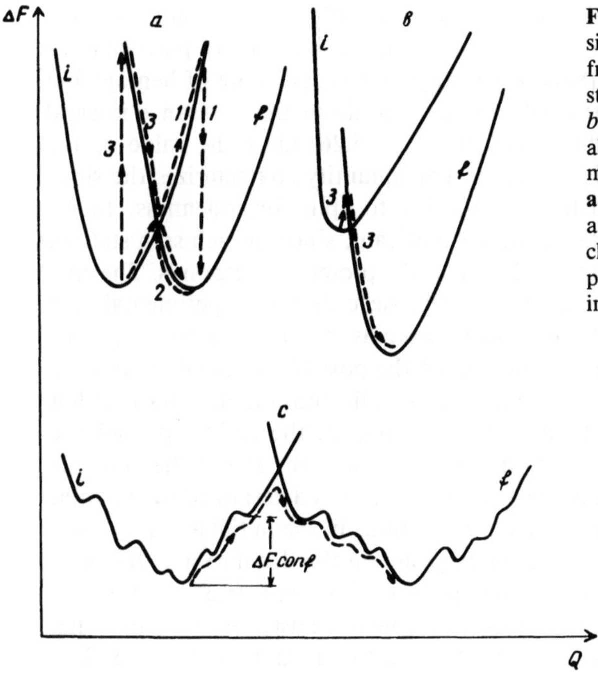

Fig. 11.1 a–c. Possible ways of transition of an electron-nuclear system from the initial (*i*) to the final (*f*) state upon electron transfer. The *broken lines with arrows* show the alternative routes; *1–3* correspond to mechanisms I, II, III (see text); **a** $\Delta F_0 = 0$; **b** $\Delta F_0 > \Delta F_r$; **c** route with a multi-stage conformational change; ΔF_{conf} is the free energy of preliminary conformational matching

cooperative multi-step process (Sects. 5.7 and 10.1). The actual electron-transfer processes most likely occur by an optimal route: this route consists in part of a multi-step conformational adjustment with a limited number of nuclei taking part in each rapid and reversible intermediate step and the reaction terminates in an electron transfer through the transition state by mechanism II. Taking the above factors into account and also formulas (5.5), (5.29) and (5.57), one can write the following expression for the rate constant of the electron transfer

$$k_{tr} = \nu_0 \varkappa a_{syn} \exp -(\Delta F_r^{\neq} + \Delta F_{conf})/RT \ , \qquad (11.1)$$

where ν_0, $\varkappa$ and a_{syn} are the frequency factor, the transmission coefficient and the synchronization factor respectively; ΔF_{conf} is the free energy of the preliminary conformational transition (Fig. 11.1 c). In a general case the quantity $\varkappa$ takes into account vibronic as well as electronic quantum effects.

The dependence of the rate of electron transfer on the distance between the centres, r_{DA}, is given by formulas (5.35) and (5.36) and, in particular, by the dependence $\varkappa \sim |V|^2$ in formula (11.1). If the electron-nuclear fit is attained due to zero vibrations of the nuclear system, the quantity k_{tr} is independent of temperature and the process can occur even at absolute zero [Eq. (5.42)]. As has been shown in Section 5.5.4, [Eq. (5.36)], a pure electron tunnelling can be efficient over sufficiently large distances (up to 1.8 nm) even in the absence of special electron-conducting channels. An important factor that limits the long-range mechanisms of electron transfer is the decrease of the probability of transfers upon attainment of an electron-nuclear fit.

A further factor that has an effect on the rate of electron transfer is the requirement of spin conservation. As has been pointed out in Section 4.3, the spin forbiddenness is not absolute. There is a finite probability of formally spin-forbidden reactions due to spin-orbital and spin-spin interactions. In particular, as has been shown by Redi et al. [801], the change in the spin state of hemoglobin from $S = 0$ to $S = 2$ is associated with the magnitude of the resonance integral $V \approx 10^{-3}$ eV. Substitution of this value into Eq. (5.26) gives the value of the transmission coefficient $\varkappa = 10^{-2}$ to 10^{-3}. This quantity characterizes the slowing down of the electron-transfer process due to spin forbiddenness. In experiments the slowing down may be more significant, since to each spin state of the active site a certain conformational state of the protein corresponds. The conformational transition may be the slowest step. According to experimental data on such conformational transitions in hemoproteins, $\tau = 10^{-3}$ to 10^{-5} s [727].

Of considerable interest in the treatment of the possible mechanisms of electron transfer and other chemical reactions, especially those taking place at low temperatures, is the problem of nuclear tunnelling, i.e. the ability of nuclei to adopt the configuration of the reaction products before the top of the potential barrier is reached [803]. The probability of this process is determined by the value of the overlap integral for the nuclear wave functions in the initial and final state $\langle \psi_i^{nuc} \psi_f^{nuc} \rangle$. The dependence of the tunnelling rate on the distance may be represented approximately by formulas similar to Eqs. (5.35) and (5.36) with an exponential damping of the rate with distance. Estimates based on these formulas with account taken of the particle's mass (M) and the magnitude of energy barriers (ΔE) [$a_B \sim (M\Delta E)^{-1/2}$] leads to the following approximate data, which characterize the dependence of the tunnelling time (τ_t) on the tunnelling distance (Δr_t). For $\Delta r_t \approx 0.1$ nm, i.e. the distance travelled by nuclei in chemical reactions, the time $\tau_t \approx 1$ s for the atom [^{1}H] and $\tau \approx 10^{+37}$ s for the atom [^{18}O]; for distances of 0.04 nm the corresponding values for τ_t are equal to 10^{-8} and 10^{+7} s, i.e. the tunnelling of protons is a realistic event. At the same time, direct tunnelling of heavy nuclei is evidently an unreal process, except for very short distances of the order of 0.01 nm.

It is, however, quite possible that temperature-independent, synchronous processes occur along the reaction coordinates with zero or low barriers. Let us illustrate one of the possible mechanisms of processes of this type with a hypothetical example (Fig. 11.2). Suppose that in a pair of molecules DA, which are in direct contact with each other, a complete or partial electron transfer with the formation of the state D^+A^- can occur when the barrier (ΔE) is overcome. Suppose also that upon attainment of this state the particles D^+ and A^- move towards each other under the influence of electrostatic attraction and enter into a chemical reaction without surmounting the barrier with the formation of DA. On the whole, the process will occur with an activation energy of $E_a = \Delta E$. However, if the pair DA is in an environment containing electric dipoles or polarizable molecules, it is probable that, as a result of zero vibrations or rotational motion, the dipole assembly will synchronously adopt the configuration, whose electrostatic interaction with the pair D^+A^- will compensate for the energy consumed for the formation of this pair. As a result, a new barrierless reaction route will appear (Fig. 11.2). The reaction time for this route (τ_{syn}) can be roughly estimated by means of formula (5.56). For example, if the synchronous movement of the surrounding nuclei can be described by 40 degrees of freedom, then $\tau_{syn} \approx 1$ s.

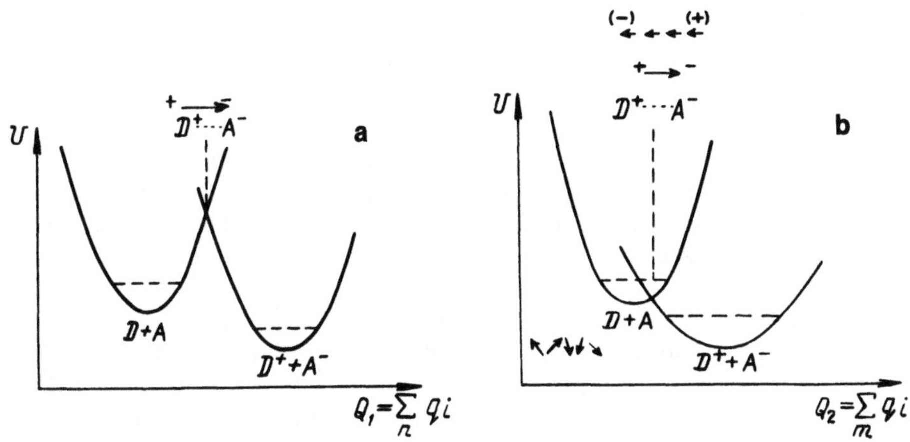

Fig. 11.2 a, b. The possibility of barrier-free chemical reaction at the reaction site of DA due to the formation of the compensating electric dipole moment upon synchronous motion of the nuclei of the environment. **a** Dependence of the energy of the system on the generalized coordinates of the reaction site, Q_1. The *long arrow* shows the variation of the dipole moment in the transient state. **b** Dependence of the energy of the system on the generalized coordinates of the reaction site and the medium, Q_2. The *short arrows* show the possible change of the reorientation of the dipoles of the medium

The theoretical conceptions outline in this section will be used further in the text in the analysis of experimental data on metalloproteins. But first we will consider the important problem of the relationship between electron-transfer and spin-exchange processes.

11.1.2 Electron Transfer and Spin Exchange

It has been repeatedly pointed out in the literature that there is a detailed analogy between electron-transfer and spin-exchange processes [24, 174, 196, 802–805]. Spin exchange is a change in the spin state of paramagnetic particles caused by an exchange interaction which occurs when the orbitals of their unpaired electrons overlap. The intensity of spin exchange is characterized by the exchange integral I_{AB}. The relationship between the exchange integral I_{AB} and the overlap integral is given by the equation:

$$I_{AB} = I_0 S_{AB}^n , \tag{11.2}$$

where $n = 2.2-2.5$. This equation is analogous to Eq. (5.35) for the square of the resonance integral, V^2, the value of which determines the probability of electron transfer. It therefore seems that the approach that includes an independent determination of S_{AB} from spin-exchange data is very promising for evaluation of V^2. Such an approach permits the assessment of the role of purely electronic factors in a complex electron-transfer process.

The probability of spin exchange as a result of an exchange interaction is given by [174]:

$$P(t) = \sin^2 (It/2) , \qquad (11.3)$$

where t is the interaction time. At $t = \pi/I$ thew signs of the spin densities of the partners A and B change simultaneously (the flip-flop process). At sufficiently large t the spin densities at the centres A and B oscillate with a frequency of $\nu_{ex} = \tau_{ex}^{-1} = I/2$. These oscillations are similar to the electron-density oscillations between two molecules with a fixed nuclear configuration corresponding to the transition state [Eq. (5.36)] that occurs upon electron transfer. The spin exchange my be treated as an idealized model of electron transfer with the nuclear coordinates being fixed.

Two types of spin exchange are distinguished: direct and indirect. The first type of exchange is associated with a direct overlap of orbitals in space. Indirect exchange takes place as a result of the delocalization of the spin density over the orbitals of the atoms residing between the paramagnetic atoms and also due to spin polarization. The latter phenomenon takes place as a result of the interaction of unpaired electrons with the electrons of filled shells; a splitting of the latter molecular orbitals occurs. This interaction occurs differently with electrons, whose spins are differently oriented with respect to the spin of an unpaired electron. Here, the electrons of filled shells become partly unpaired and an extra spin density appears on the interacting paramagnetic atoms. Indirect exchange between spatially separated atoms usually predominates over the direct exchange.

The exchange integral may be represented approximately in the form [806]:

$$I_{AB} \approx I^0_{AB} \varrho_A \varrho_B / n_A n_B , \qquad (11.4)$$

where I^0_{AB} is the exchange integral at the van der Waals distance; ϱ_A and ϱ_B are the spin densities on the directly contacting atoms; n_A and n_B are the numbers of unpaired electrons A and B. The values of $I^0_{AB} \sim 10^{16} s^{-1}$.

It should be noted that apart from the similarities between electron transfer and spin exchange, there are significant differences between these processes: (1) exchange interactions and electron-transfer processes may be determined by the overlap of different orbitals; (2) spin polarization can introduce an appreciable contribution to the value of the exchange integral; (3) nuclear relaxation processes play an important role in electron transfer.

Exchange interactions manifest themselves in a number of physical phenomena. They determine the energy of covalent bonds, the anti-ferromagnetic and ferromagnetic properties of polynuclear complexes and manifest themselves in the spectral and relaxation characteristics of the EPR and NMR signals of polyradicals and complexes and also in triplet-triplet transitions (Chap. 2). As can be seen from Fig. 11.3, the experimental data on the dependence of the quantitative exchange parameters, the exchange integrals (I) and rate constants (k_{ex}) of triplet-triplet transitions and triplet-singlet conversion (intersystem crossing), on the distance r between the centres lie on two, approximately exponential curves described by the following equation [24, 196, 675]:

$$I,\ k_{ex} \approx 10^{16} \exp(-L'r) . \qquad (11.5)$$

For systems, in which the exchanging centres are separated by a non-conducting medium (molecules or groups with saturated chemical bonds) $L' \approx 26\ nm^{-1}$.

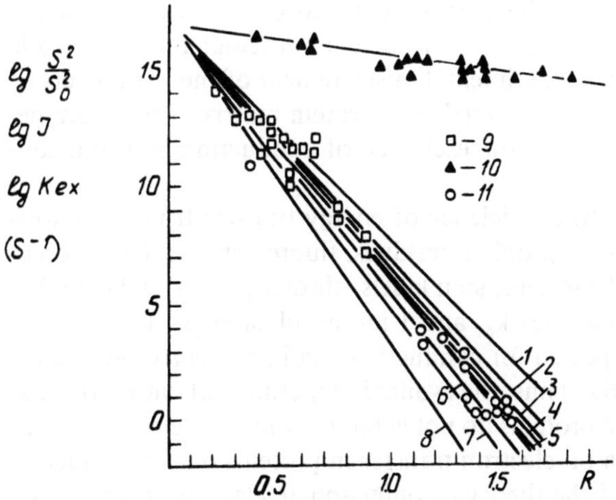

Fig. 11.3. Dependence of the parameters of spin exchange and overlap integrals (S) on the distance between the centres: *1–8*-dependencies S^2/S_0^2, where S and S_0 are, respectively, the overlap integrals at a distance R (in nm) and the distances of van der Waals contacts for various atomic pairs. The electronic configurations of the atoms are given in parentheses: *1* = C (^{1}D)-C (^{1}D); *2* = C (^{3}P)-C (^{3}P); *3* = C (^{3}P)-C (^{2}D); *4* = H-C (^{3}P); *5* = H–N (^{2}D), N (^{2}D)-N (^{2}D), N (^{2}D)-O (^{3}P); *6* = C (^{3}P)-N (^{4}S); C (^{3}P)-O (^{3}P); *7* = H-O (^{3}P); *8* = N (^{4}S)-N (^{4}N); *9* = O (^{3}P)-O (^{3}P), N ^{4}S)-O (^{3}P). *9–11* Experimental data on exchange interactions: *9* exchange integral (J), spin exchange between nitroxide radicals, the radicals and atoms of paramagnetic metals in complexes and between metal atoms in complexes in the case of a non-conducting medium between the centres; *10* relative exchange integral (J′) of spin exchange between nitroxide radicals linked via conjugated bridges; *11* rate constants (k_{ex}) of exchange quenching of phosphorescence by chromatophores and paramagnetic centres

For systems in which the radical centres are linked by conducting conjugated bonds, $L' = 3\ nm^{-1}$. In non-conducting systems the exchange intensity is damped by about ten times when the centres are at a distance of 0.1 nm from each other; in conducting systems, the effect can be reduced by ten times when the centres are separated by a distance of 0.7 nm.

As can be seen from Fig. 11.3, the experimental data on exchange interactions correlate fairly well with the theoretical dependencies of the square of the overlap integrals S^2 on the distance. This confirms the validity of the spin-exchange concepts according to which spin exchange is an idealized electron-transfer model. The empirical Eq. (11.5) offers interesting possibilities of using exchange data for the solution of a number of theoretical and structural problems of metalloenzyme catalysis. They can be used to determine the distance between the centres, the depth of the active sites of enzymes and the transmission coefficients and to study the ability of protein macromolecules to conduct spin and electron density, etc. The value of the ratio of the rates of electron transfer (k_{tr}) and exchange (k_{ex}), $\beta = k_{tr}/k_{ex}$, may serve as a parameter which characterizes the effect of electron-nuclear interaction (reorganization of the medium, vibrational relaxation, etc.) on the rate of electron transfer.

To illustrate, we will describe examples of the use of spin-exchange concepts for the experimental study of the electronic conductivity of proteins. The problem of the anomalous electronic conductivity of proteins has often been discussed in the literature (see, for example, [816]). For instance, an attempt has been made [196] to solve this problem by an experimental measurement of the parameters of spin exchange between the centres separated by a protein matrix. These parameters were compared with the data on the exchange of conducting and non-conducting model systems (Fig. 11.3).

The experimental results on the efficiency of energy transfer from triplex-excited chromophores in proteins (tryptophan residues, fluorescent labels) to acceptor groups (the heme groups of proteins, spin labels, chromophores) indicate that the experimental measured quantities k_{ex} and I are in all cases 5–13 orders of magnitude lower than those expected for conducting media and correspond most closely to non-conducting media. Thus, the available experimental data show that in all the cases studied [196] the proteins do not exhibit anomalous electronic conductivity. In analyzing the data on electron transfer in proteins one can evidently make use of the conceptions of the theory of outer-sphere electron transfer and, in particular, of the Eqs. (11.6–11.11) presented in Table 11.1.

Other examples of the use of the spin-exchange method will be considered in the subsequent sections.

11.1.3 Methods of Quantitative Estimation of the Electron-Transfer Parameters in Proteins

According to the theories described in Sections 5.1–5.5, the rate of outer-sphere electron transfer between a donor and an acceptor is determined by the following factors: the equilibrium constant for the formation of a non-specific collision (K_{nsp}) or a specific (K_{sp}) complex; the equilibrium constant for the formation of an active conformation (K_{conf}); the steric geometric factor (f_g); the electrostatic factor (α_{el}); the pre-exponential term (ν_0); the non-adiabaticity coefficient ($\varkappa$); the free energy of the process (ΔF_0); and the free energy of reorganization of the medium (ΔF_r).

For reactions involving protein molecules in solution the quantity K_{nsp} is estimated from formulas of types (5.15)–(5.17) and is of the order of unity. The values of the specific constants K_{sp} range over wide limits: $K_{sp} = 10^{-3}$ to 10^{-7} M.

It is known that proteins are characterized by an intricate mosaic of surface and internal charges [807–811, 818], which substantially complicate the theoretical computation of the electrostatic factor α_{el}. For this reason, it is most expedient to consider data obtained at a high ionic strength of the solution when $\alpha_{el} \sim 1$. The dependence of the equilibrium constant of the electron transfer reaction, spin exchange or exchange quenching of triplet states on the ionic strength of the solution allows one to estimate the magnitude and sign of local charges and the values of the electrostatic factor by using Eqs. (5.18–5.20). In experiments with different ionic strengths it is necessary to take into account the possibility of specific interactions between salts and proteins and also their effect on the conformation of proteins.

The values of the geometric steric factor (f_g) for proteins are estimated by using well-known X-ray structural models. Metal-containing active centres are

Table 11.1 Basic formulas for the rate constants of electron transfer (k_{tr}), exchange-resonance energy transfer (k_{ex}) and the parameter $\beta = k_{tr}/k_{ex}$ [a]

Nature of energy exchange and electron transfer	k_{ex}	Equation No.	k_{tr}	Equation No.	$\beta = k_{tr}/k_{ex}$	Equation No.
Weak exchange	$k_d f_g \alpha_{el} \tau \cdot \nu_0 \exp(-\alpha r)$	(11.6)			$\dfrac{\nu_e \varkappa \exp(-\Delta F^{\neq}/RT}{\nu_0 \exp(-\alpha r)}$	(11.10)
Kinetic regime			$k_d f_g \cdot \alpha_{el} \nu_e \tau \varkappa \exp(-\Delta F^{\neq}/RT)$	(11.7)		
Strong exchange	$k_d f_g^{1/2} \alpha_{el}$	(11.8)			$f_g^{1/2} \tau \nu_e \varkappa \exp(-\Delta F^{\neq}/RT)$	(11.11)
Kinetic regime			$k_d f_g \alpha_{el} \tau \nu_e \varkappa \exp(-\Delta F^{\neq}/RT)$	(11.9)		

[a] f_g and α_{el} are the geometric and the electrostatic factor respectively; τ is the lifetime of the collision complex. Weak exchange corresponds to the case $k_{ex}\ \tau < 1$ and strong exchange to $k_{ex}\ \tau > 1$; the kinetic regime corresponds to the case $k_{tr}\ \tau < 1$.

localized, as a rule, near the surface of the macromolecule (for example, cytochromes *c* and *b*, azurin, plastocyanin, ferrodoxins). According to estimates, $f_g = 2$ to 4×10^{-2} [812]. Close values have been obtained experimentally from the efficiency of exchange processes (spin exchange, quenching of triplet states) involving heme-containing metalloenzymes [125, 712, 813–815].

Serious difficulties are encounterd in calculations of the energy of reorganization of the medium (ΔF_r) in complex inhomogeneous systems, such as proteins. According to theory [812], $\Delta F_r = \Delta F_r^i + \Delta F_r^s$, where ΔF_r^i and ΔF_r^s are the energy change due to a change in bond length between reagents inside the reaction complex (Δq_i) and due to the reorientation of groups of proteins and solvent respectively. Calculation of these quantities is performed using the following formulas:

$$\Delta F_r^i = \sum_j f_r^i f_j^f (f_j^i + f_j^f)^{-1} \cdot (\Delta q_i)^2 \; ; \tag{11.12}$$

$$\Delta F_r^s = (\Delta e)^2 [\tfrac{1}{2}r_1 + \tfrac{1}{2}r_2 - 1/(r_1+r_2)] (1/D_{op} - 1/D_s) \; ; \tag{11.13}$$

where f_j^i and f_j^f are the force constants of the jth linkage in the initial and final states; Δe is the charge of the change; r_1 and r_2 are the radii of spherical reagents in nm; D_{op} and D_s are the optical and static constants of the microenvironment. Taking account of the non-sphericity of the reaction complex and the difference between the dielectric constant inside the complex (D_{loc}), the bulk dielectric constant lowers the value of ΔF_r^s in the limiting case $D_{loc} \rightarrow D_{op}$ by 25% [807].

Several approaches are employed when attempts are made to quantitatively evaluate ΔF_r in proteins. In one approach, simplifying assumptions are made concerning the quantities D_s inside and outside the protein and the quantity ΔF_r^i is either neglected [807] or is found under the assumption of the participation of a certain vibration in electron transfer [812]. For example [807], using values of D equal to 4 and 81 for the protein and water respectively, values of ΔF_r^s were obtained which varied from 60 to 150 kJ mol^{-1}, depending on the distance between donor and acceptor centres and on their localization with respect to the globule surface and the radius of the globule. In another approach, use is made of X-ray structural data on the nuclear coordinates of proteins in oxidized and reduced states. Such an estimate performed for cytochrome *c* [800, 810] led to $\Delta F_r^i \sim 4$ kJ mol^{-1} and $\Delta F_r^s = 22$ kJ mol^{-1}. A semi-empirical approach to the evaluation of ΔF_r in proteins will be described in the next section.

Another problem consists in the choice of parameters that are required for the calculation of the frequency factor (ν_0) and the non-adiabaticity coefficient $\varkappa$. Strictly speaking, the latter parameter includes the quantity $\varkappa'$, which reflects the probability of electron tunnelling and σ, which characterizes quantitatively the tunnelling of nuclei. According to Eqs. (5.28) and (5.40), the value of the pre-exponential factor in the absence of nuclear tunnelling is given by:

$$k_0 = (\pi^2 h^{-2} k T^{-1} \Delta F_r^{-1})^{1/2} |V|^2 \; .$$

We represent the resonance integral in the form $V = V_0 \exp(-\beta r)$, where V_0 is the value of the resonance integral at the point of direct contact of the reagents; r is the distance between the edges of the molecules when they move apart; then, after substitution of the estimated values of V_0 and ΔF_r, we find that [812]:

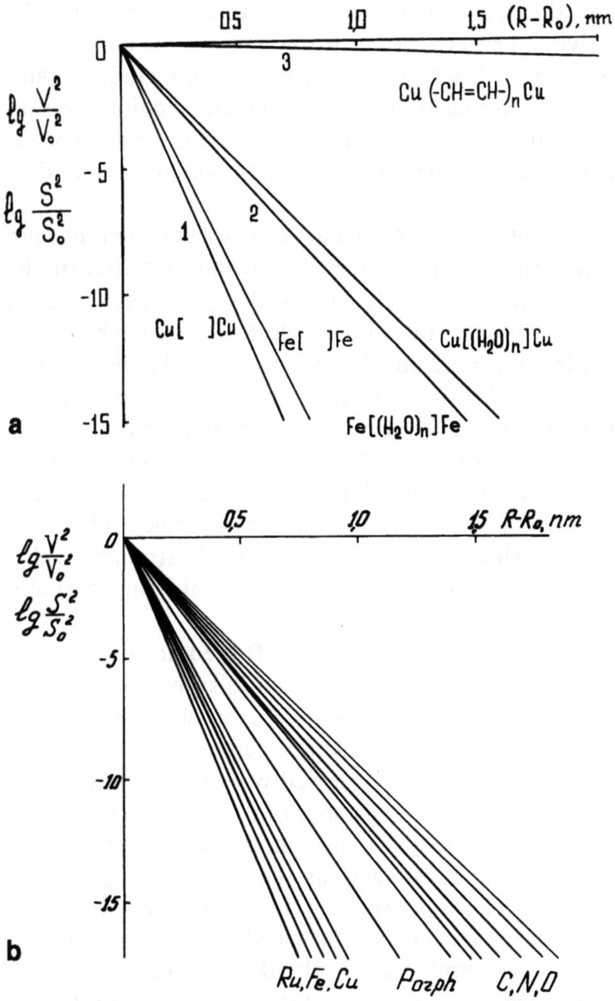

Fig. 11.4a, b. Dependence of theoretically calculated exchange (S) and resonance (V) integrals on the distance (R) (plotted on the basis of data reported elsewhere [804, 834, 835]. **a** Effect of the structure of the interacting atoms; **b** effect of the medium between the atoms

$$k_0 \sim 10^{13} \exp(-\beta r) . \tag{11.14}$$

The quantity β characterizes quantitatively the decrease in the degree of overlap of the orbitals with increasing distance between the centres.

Figures 11.3 and 11.4a show the theoretical dependence of V/V_0 and S/S_0 for various pairs of atoms. Figure 11.4b demonstrates the effect of the medium between the atoms. As can be seen from Fig. 11.4b, the involvement of the orbitals of the medium's molecules in the interaction contributes to an increase in the effective overlap and the quantity V/V_0 falls more slowly with distance than in a vacuum. In the presence of a conducting medium, e.g. a system of conjugated bonds, this decrease is insignificant. The effect of peptide α-helical portions in proteins on electron transfer has been considered theoretically [816]. Of interest is the fact that the experimental dependance of the exchange process parameters [the exchange integral of spin-spin interaction (I) and the rate constants for a triplet-triplet transition] on the distance corresponds approximately to the theoretical curves for V/V_0 with the values of β being equal to 26 nm^{-1} for a conducting medium and to 2.8 nm^{-1} for non-conducting media, including water and proteins (see Sect. 11.1.2.). In connection with the similarity pointed out in Section 11.1.2 between electron transfer and exchange processes, it seems to be expedient to use that above indicated values of β in corresponding theoretical estimations.

The quantities V and S for certain pairs or reagents can be calculated through the use of Eq. (11.4) from the experimental values of the spin densities of the directly interacting atoms [804]. Another, purely empirical approach is based on the experimental dependence of the rate constants of electron transfer k_e on the distance. As has been shown [817], $\beta = 12\ nm^{-1}$ corresponds to the dependence $k_e(r)$ for aromatic, anion radicals linked by a rigid system of saturated bonds to aromatic acceptors in frozen solutions. The direct use of this quantity for proteins, however, does not seem to be justifiable. Anion radicals are reactive compounds, for which the barrier of electron tunnelling is not high; they can interact according to a charge-transfer mechanism with the solvent molecules, whose orbitals can thus contribute to tunnelling. The rigid mechanically strained bridge systems that have been used [817] may exhibit the properties of being partially unsaturated.

The probability of long-distance electron transfer is affected not only by the distance between donor (D) and acceptor (A) centres, but also by their relative orientation. According to theoretical calculations [819], for oblate spheroids, which simulate the porphyrin rings at fixed distances between the edges, the highest values of the resonance integral V_{DA} is observed with a face-to-face configuration. For example, for σ-interactions at a distance of 0.5 nm upon transition from this configuration to an edge-to-edge configuration, the calculated value of V_{DA} drops from 10^{-2} to 10^{-3} eV. For π-interactions the value of V_{DA} falls to zero. With the specified distances between the centres, the quantity V_{DA} increases only slightly upon transition from the face-to-face to the edge-to-edge configuration (by about three times at $r_{DA} = 1$ nm) because the decrease of V_{DA} with a change in the orientation is compensated for by an increase in this quantity due to the approach of the edges of the centres.

A point worth mentioning is that with intermediate orientations, which are, as a rule, characteristic for both fixed biological centres and static systems, e.g. chromophores in solid solutions, the distance between the centres has a stronger effect on the efficiency of electron transfer than the relative orientation. As has been shown above, when the centres move apart by 0.1 nm, the probability of electron transfer falls by about one order of magnitude, whereas a significant change in relative orientation changes this quantity to within one order of magnitude.

In accordance with the formulas of Table 5.1, the rate of electron transfer is proportional to the square of the resonance integral $|V|^2$. The value of the latter reflects, in a general case, both the direct interaction of the orbitals of the centres with the quantity V^0_{AB} and the indirect interaction with the participation of the orbitals of the medium's molecules [293]. In the general case:

$$V_{AB} = V^0_{AB} + \sum_i V_{AC_i} V_{C_iB} / (E_{C_i} - E_B) , \tag{11.15}$$

where V_{AC_i} and V_{C_iB} are the resonance integrals, which characterize the interactions of the orbitals of the initial and final states with the intermediate ith C-center respectively; $E_{C_i} - E_B$ is the difference between the corresponding energy levels. From formula (11.15) it follows that under certain conditions the medium can contribute to long-distance electron transfers.

In cases where the electron transfer is accompanied by a change in the high-frequency vibrations of the surrounding nuclei with $h\omega > k_BT = 200\,cm^{-1}$ at room temperature, it is necessary to take into account the nuclear tunnelling factor (σ) [Eq. (5.38); Table 5.1]. According to theoretical estimates [804, 812], for the electron transfer between various transition-metal complexes in aqueous medium the quantity σ varies from 1 to 0.15.

The values of the standard free energies of electron transfer ΔF_0 are calculated from experimental values of the standard redox potentials (E_0^0). A priori calculations of the equilibrium constants for the transition to the active state (K_{conf}) are not possible at present. However, because of the entropy-enthalpy compensation effect, the value of K_{conf} is presumably close to unity.

For reactions of low molecular complexes a number of theoretical predictions prescribed by formulas (11.12) and (11.13) are realized, e.g. a certain dependence of k_e on the dielectric constant of the medium, the change the coordinates of the ligands and on the free energy of the reaction (ΔF_0) [812]. In a number of cases it is possible to achieve an agreement between experimental and theoretical values of k_e [804, 812].

11.1.4 Experimental Data on One-Electron Transfer in Metalloenzymes

A large amount of experimental material has accumulated on the measurement of the kinetics of electron exchange between proteins and between proteins and low molecular weight compounds by using the stopped-flow, temperature-jump, NMR and pulse radiolysis methods [812, 830]. The investigations carried out have revealed a large variety of kinetic properties of the systems studied. Here, we will briefly consider the most important of these properties (Tables 11.2 and 11.3).

Comparison of experimental data with theoretical conceptions is supposed to provide answers, in particular, to a number of fundamental questions: (1) does an electron transfer occur directly during the collisions of reagents (non-specifically) or does it initially form a specific complex? (2) Does the process occur via the equilibrium or non-equilibrium conformational forms of the protein? (3) Do the reagents undergo first a conformational change? (4) What is the particular mechanism of the electron transfer and at what values of the parameters do the corresponding electron-transfer theories agree with experimental data?

An appreciable effect on the rate of electron transfer is exerted by the ionic strength and pH, which points to the substantial role of electrostaic effects in the interactions of the reagents. The influence of these factors does not, however, agree quantitatively with the data on the total charge of the reagents [825].

For a number of objects studied it has been shown that electron transfer is preceded by the complex formation of the reactants. For an azurin-ferricyanide pair [826] it has been demonstrated that there is a slight change in the free energy

Table 11.2. The rate constants of electron self-exchange in proteins

Protein	k_{11}, $M^{-1}s^{-1}$	T, K	I	Medium	Reference
1. Cyt *c* (horse)[a]	5.5×10^3	330	1	Phosphate, pH = 7, KCL	[821]
2. Cyt c_{551}[a]	7.5×10^6	325	0.05	Phosphate, pH = 7	[840]
3. Rd (*C. pasteurianum*)[b]	10^8-10^9	298	0.1	Tris-HCl, pH = 7	[822]
4. Fd (spinach)[b]	1.7×10^{-3}	298	0.1	Tris-NaCl, pH = 7.0	[822]
5. HiPIP (*Chromatium*)[b]	1.3×10^{-2}	298	0.1	Tris-NaCl, pH = 7.0	[822]
6. Cyt_{553}[c]	2.8×10^8	298	0.1	d	[829]
7. Cyt_{551}[c]	4.6×10^7	298	0.1	d	[829]
8. Azurin (Ps)[c]	9.9×10^5	298	0.1	d	[829]
9. Plastocyanin[c]	6.6×10^2	298	0.1	d	[829]
10. Cyt *c*[c]	1.5×10^2	298	0.1	d	[829]

[a] Experimentally measured values.
[b] Calculated by the Marcus formula for reactions involving low molecular weight reagents.
[c] Calculated by the Marcus formula for reactions involving proteins.
[d] Phosphate buffer, pH = 7, 10 μm EDTA.

at various steps, which is typical of enzymatic processes. The formation of short-lived complexes has been observed in a reaction between a high-potential iron protein (HiPIP) and type *c* cytochromes [827].

In a number of cases it has been shown that specific complexes are formed between reagents with dissociation constants of $10^3-10^6\,M^{-1}$ [708–710, 744, 831, 832]. For example, complexes of ctochrome *c* with $[Fe(CN)_6]$ and [Fe(EDTA) H_2O], complexes of cytochrome *c* with cytochrome *c* peroxidase and complexes of cytochrome P-450 with andrenodoxin have been recorded. In other cases, e.g. in reactions of stellacyanin with cytochromes *c* and c_{551}, the formation such complexes have not been found [828].

An analysis of quantitative data on the kinetics of electron exchange is customarily carried out within the framework of the Marcus equation [829]:

$$k_{12} = (k_{11}k_{22}K_{12}f_{12})^{1/2} , \qquad (11.16)$$

where

$$\log f_{12} = (\log k_{12})^2/4 \log (k_{11}k_{22}/k_d^2) ,$$

and k_{11} and k_{22} are the rate constants for the electron transfer between identical reactants; k_D is the number of collisions. Equation (11.16) has been derived under the assumption of the additivity of the free activation energies and logarithms of transmission coefficients for the reactions considered and is thus based on the conceptions of the absence of specific effects in collisions and electron transport.

Examples have been reported in the literature which illustrate the validity of Eq. (11.16) (cytochromes *c* and c_{551}, azurins, plastocyanins), but these are excep-

Table 11.3. The rate of electron cross-exchange in proteins [812]

Protein	Complex	K_{12}, $M^{-1}s^{-1}$
Azurin	$Co(phen)^{3+}$	4.4×10^3
R. aeruginosa	$Ru(NH_3)_5Py^{3+}$	2.0×10^3
	$Fe(EDTA)^{2-}$	1.3×10^3
	$Co(OX)_3^{3-}$	2.9×10^{-2}
Plastocyanin	$Co(phen)_3^{3+}$	1×10^3
spinach	$Ru(NH_{35}Py^{3+}$	7.1×10^3
	$Fe(EDTA)^{2-}$	8.2×10^4
	$Co(OX)_3^{3-}$	2.4×10^{-1}
Stellacyanin	$Co(phen)_3^{3+}$	1.8×10^5
R. vernicifera	$Ru(NH_3)_3Py$	1.9×10^5
	$Fe(EDTA)^{2-}$	4.3×10^5
	$Co(OX)_3^{3-}$	7.3×10^2
Cytochrome *c*	$Co(phen)_3^{3+}$	1.5×10^3
horse	$Ru(NH_3)_6^{2+}$	3.8×10^4
	$Ru(NH_3)_5Py^{3+}$	9.6×10^3
	$RuBzIm^{2+}$	5.8×10^4
	$Fe(EDTA)^{2+}$	2.6×10^4
	$Co(OX)_3^{2-}$	
Cytochrome *c*-azurin		3.5×10^2
Plastocyanin-cyt *c*		10^6
Stellacyanin-cyt *c*		3.5×10^2
Azurin-cyt c_{551}		6×10^6
Plastocyanin-cyt c_{551}		7.5×10^5
Cyt *c*-*cyt c* peroxidase		5×10^8
Cyt *c*-cyt c_{551}		6.7×10^4
Cyt *c*-cyt b_5		3×10^7

tions to the general rule [828, 830]. For different pairs of reagents the value of k_{11} for metalloproteins differs significantly in many cases, depending on the second reagent 2 used (see Table 11.2).

Another methods employed to analyze experimental data is based on the use of the following equation (r, nm) [278]:

$$k_{11} = 5\times10^7 \exp(-14\ r),\ M^{-1}s^{-1}\ . \qquad (11.17)$$

This equation corresponds to Eq. (5.40) for a non-adiabatic electron transfer between the centres which are at a distance r from each other. Experimentation has shown that the values of r calculated from Eq. (11.17) depends on the nature of the second reagent, which also points to a substantial contribution of non-specific effects to the electron transfer involving proteins and also to the possible differences in the reorganization energies of the reagents.

The available experimental data and theoretical concepts allow one to arrive at certain conclusions concerning the electron-transfer mechanism. For example, it has been found [825] that the equilibrium constant K_{12} determined experimentally is equal to the constant calculated from the ratio of the rate constants of the forward and reverse reactions of HiPIP with ferricyanide. This is an indication of the absence of relaxationally non-equilibrium forms that result from a rapid electron transfer (Sect. 11.1.1, mechanism III). As can be seen from Fig. 11.6, the experimental values of the activation parameters $\Delta H^{\neq}_{app}$ and $\Delta S^{\neq}_{app}$ vary over a wide range [24]. It has been established that there is an approximate compensation relation between these parameters. As has been shown in Section 10.1, this is an indication that the electron transfer is accompanied by a change in the conformational state of the protein and water molecules. Thus, the most probable energy profile of the electron transfer is the one shown in Fig. 11.1c.

However, if solvated electrons or anion radicals, which typically have a high reducing potential ($E_0 = -0.4$ to -2.7 V), are used as the reducing agents, it is possible to record the conformationally strained forms of metal carriers [727, 833]. The energy profile of the process in this case is presumably similar to the one shown in Fig. 11.1b.

Figure 11.5 presents experimental values of the rate constants of electron transfer in proteins as a function of the distance between the centres. As can be seen, these data are concentrated mainly around the dependence of log k_{ex} on the r characteristic of the exchange processes in non-conducting media. It should be pointed out that the data referring to reactions involving low molecular complexes lie below this straight line and that the reactions of strong oxidants, such as a triplet-excited Zn-porphyrin in the composition of hemoglobin, lie above the straight line. Presumably, the typical situation for electron transfer in proteins is a low reorganization energy ΔF_r and substantial non-adiabaticity. Here, as has been noted above, the ability of the protein matrix to contribute to an effective overlap of the orbitals of redox centres does not differ substantially from "ordinary" non-conducting media. However, in the case of reactions involving strong oxidants or reductants intermediate chemical states are not ruled out.

In order to separate the factors that influence the rate of an outer-sphere electron transfer, spin-exchange data have been used [715]. The model system was an electron transfer between the oxidized and reduced forms of cytochrome *c*. The experimental value of k_{11} is $5.5 \times 10^3\ M^{-1} s^{-1}$ at a high ionic strength (1 M) [823]. The factors that determine the rate of the electron-transfer reaction were estimated in the following sequence. First, the steric factor (f_s) was determined from the value of the rate constant of the exchange quenching of the triplet state of eosin by cytochrome *c* using formula (5.13) for a strong exchange, in which case $f_s = \sqrt{f_g}$, where f_g is the geometric factor. The value of f_g obtained (1.5×10^{-1}) was then used to determine the products: $\varkappa \exp(-\Delta F_r^{\neq}/RT) = 2 \times 10^{-7}$ for $k_d = 5 \times 10^9\ M^{-1} s^{-1}$, $\tau_r = 3 \times 10^{-11}$ s, $a_{el} = 1$, $\nu_0 = 6 \times 10^{12} s^{-1}$ [see Eq. (11.7) and Table 11.1]. Estimation of the transmission coefficient $\varkappa$ for the collision of the heme edges carried out on the basis of spin density data [815] by means of Eqs. (11.2), (11.4) and (5.44) has led to the value of $\varkappa \approx 10^{-6}$. From this it follows that $\Delta F_r^{\neq} \sim 4$ kJ mol^{-1} and $\Delta F_r \approx 16$ kJ mol^{-1}. The value of $\Delta F_r^{\neq} < 6$ kJ mol^{-1} has recently been obtained on the basis of X-ray diffraction data [800, 810].

Thus, the electron transfer between molecules of cytochrome *c* is more likely characterized by substantial non-adiabaticity and a relatively slight reorganization of the medium. The validity of this inference for other type *c* cytochromes is also supported by slight differences in the properties of the heme group in native reduced cytochrome *c* and in a conformationally strained sample obtained by the

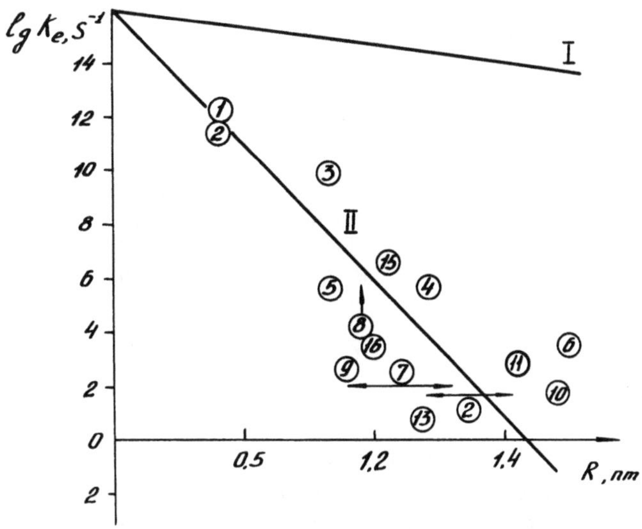

Fig. 11.5. Dependence of the rate constant of electron transfer k_e on the distance between the edges of the centres (R). *I* and *II* Theoretical dependencies for conducting and non-conducting systems respectively. The *arrows* indicate the direction of motion of electrons; *1* $(Bchl)_2 \to Bchl$ [667, 822]; *2* $Bchl^- \to Bph$ [667, 822]; *3* $Bph^- \to Q_A$ [667, 822]; *4* Cyt $c \to (Bchl)_2^+$ [294, 822]; *5* Cyt $b_5 \to$ Cyt c [812]; *6* Cyt $c \to$ Cyt c peroxidase [812]; *7* Cyt $a \to Cu_A$ (cytochrome c oxidase) [663, 874]; *8* $[Fe_4S_4] \to [Fe_4S_4]$, ferredoxin [481, 840]; *9* $Q_A^- \to Q_B$ [667, 882]; *10* $ZnHb^* \to FeHb$ [781]; *11* $Q_A^- \to (Bchl)_2^+$ [838, 822]; *12* Cyt*c*→Cyt*a* [812]; *13* $(NH_3)_5Ru^{2+} \cdot$hys-83 (azurin) $\to Cu^{2+}$ in azurin [836]; *14* $(NH_3)_5Ru^{2+} \cdot$hys-48 (Mb)$\to Fe^{3+}$Mb [836]; *15* $Cr^{3+}(bipy)_3^* \to Cu^{1+}$ plastocyanin [837]; *16* $[Fe(CN)_6]^{4-} \to Fe^{3+}$ cyt c [744]. (a) $(Bchl)_2$, Bchl and Bph = bacteriochlorophyll dimer, bacteriochlorophyll and bacteriopheophytin of the reaction centre of photosynthetic bacteria respectively; Q_A and Q_B are the primary and secondary quinone acceptors of the reaction centre respectively; (b) the distance was calculated from the data of Ref. [822] with account taken of the size of the non-conducting gaps between the pigments

reduction of oxidized cytochrome c solvated electrons at 77 K [833]. X-ray studies have shown that there are relatively small differences in the conformation of tuna cytochrome c in redox processes [362].

An estimate of the constant of electron transfer between the molecules of horse cytochrome c has also been performed [812] the following parameters were used: $f_g \sim 10^{-2}$, $\varkappa = 4 \times 10^{-4}$ [estimated from Eq. (11.14) with $\beta = 12\,nm^{-1}$ and $r = 0.64$ nm]; $\Delta F_r^s \simeq 80\,kJ\,mol^{-1}$, (calculated from formula (11.13) at $D_s = 10$); $\Delta F_r^i = 16\,kJ\,mol^{-1}$; $a_{el} = 0.25$. With such a set of parameters $k_e^{th} = 2 \times 10^3\,M^{-1}\,s^{-1}$ at $k_e^{obs} = 5 \times 10^3\,M^{-1}\,s^{-1}$ (at $\mu = 0.1$ M). In contrast to the results of the semi-empirical approach described above, according to further data [812], the electron transfer between molecules of cytochrome c is characterized by the reorganization energy close to ΔF_r for low molecular complexes and by relatively large values of the non-adiabaticity coefficient.

Thus, within the framework of a single theory the different approaches lead to somewhat different conclusions and the problem requires further study. However, it seems to us that the semi-empirical approach, which relies on experimental data and which is freer from arbitrary estimates, is more reliable. In particular,

Table 11.4. Values of the parameters that determine the rate of electron transfer between cytochrome *c* and the cation-radical of the chlorophyll dimer based on the date if different authors (the parameters are given in electron volts)

Energy of quantum vibration $\langle h\omega_0 \rangle$	Free energy ΔF_0	Free energy of reorganization ΔF_r	Resonance integral	Reference
0.025	0.05	0.5	4×10^{-4}	[265]
0.05 – 0.1	0.1	1	3×10^{-5}	[266]
0.2	0.45	~2.4	10^{-2} to 4×10^{-3}	[843]
–	0.17	0.58	$\sim2.5\times10^{-6}$	[812]
0.13		1.45	6.5×10^{-4}	[879]

the choice of the value of $D_s = 4$, which is typical of proteins, instead of $D_s = 10$ used elsewhere [812], leads to low values of ΔF_r ($\approx 12\ \mathrm{kJ\ mol^{-1}}$). The choice of the value of $\beta = 12\ \mathrm{nm}^{-1}$ also seems to be substantially arbitrary. As noted in Section 11.1.3, this value is typical for reactions involving anion radicals with low ionization potentials and for systems with increased electronic conductivity.

The problems that arise in a theoretical estimation of the electron transfer parameters are also illustrated by calculations referring to the electron transfer from cytochrome *c* to the cation radical of the chlorophyll dimer on the reaction centres of photosynthetic bacteria *Chromatium*. The experimental values of the rate constant of electron transfer in the temperature range of 4.5 – 100 K are $k \approx 4.3\times10^2\ \mathrm{s}^{-1}$; as the temperature is raised, the value of k_e increases with the activation energy $E_a = 0.14$ eV [842]. According to X-ray structural data [822], the distance between the edges of pigments is about 1.2 nm. The sets of theoretical parameters that describe the experiments discussed above on the basis of the data of different authors are shown in Table 11.4. As can be seen from Table 11.4, these parameters vary within wide limits: the vibrational energy $\langle h\omega_0 \rangle$ 0.025 – 0.2 eV, the free energy $\Delta F_0 = 0.05 - 0.45$ eV; the reorganization energy $\Delta F_r = 0.5 - 2.4$ eV; the resonance integral $V = 2.5\times10^{-6}$ to 2.5×10^{-2} eV. Unfortunately, it is difficult at present to formulate a criterion for the choice of a single correct set of these parameters. It should also be noted that the conclusion was obtained [841] on the basis of measurements of k_e and the spectral characteristics of cytochrome *c* in the reaction centres of *Ectothiorhodospira shaposhnikovii*, that the true value of k_e is generally temperature-independent. The observed weak temperature dependence of k_e^{app} was accounted for by the transition from an inactive to an active conformation with the electron transfer parameters $\Delta H_0 = 60\ \mathrm{kJ\ mol^{-1}}$ and $\Delta S_0 = 0.22$ eu.

Another aspect of the difficulties that arise in a quantitative interpretation of experimental data on electron transfer in proteins, especially at low temperatures, is associated with the fact that under certain conditions the molecular-dynamic state of the medium may have an effect not only on the frequency factors of equations of type (5.29) (the Marcus equation), but also on the apparent values of the reorganization energy (ΔF_r) and free energy (ΔF_0). In particular, such a situation occurs with the formation of conformationally and solvationally non-equilibrium states, which react before the corresponding equilibrium is set up [833]. It has been shown [844] for the model electron transfer reaction between donor and acceptor centres that the rate of the process is determined to a considerable extent by the ratio of the characteristic relaxation times of the solvent molecules and secondary processes in a pair of separated charges. Another example that illustrates the role of molecular dynamics of the protein medium is the long-distance electron transfer from ferroheme to the triplet-excited Zn-porphyrin in a hybrid homoglobin (Fig. 10.6).

The proximity of the active sites to the surface of metalloenzymes and the low reorganization energy are evidently responsible for the possibility of a direct electron transfer between laccase, hydrogenase, cytochrome b_2 and the electrodes observed experimentally [845].

Thus, the experimental data and theoretical considerations indicate that the electron transfer between protein macromolecules is characterized by the following optimal combination of parameters that provides a rapid and specific occurrence of the redox process: the distance between the redox centres 0.8 – 1.5 nm, a low free energy of reorganization of the medium and a relatively high value of the lifetime of the specific complex [815, 744]. The active centres of redox enzymes and carriers are buried to a certain extent in the protein matrix, which lowers the rate of non-specific processes. Enthalpy-entropy compensation phenomena for redox reactions in proteins (Fig. 11.6) provide evidence that conformation and solvatation transitions accompany the electron transfer.

It should be noted that cluster systems have certain advantages as electron carriers over polynuclear systems (Fig. 11.7) due to the specific features of their

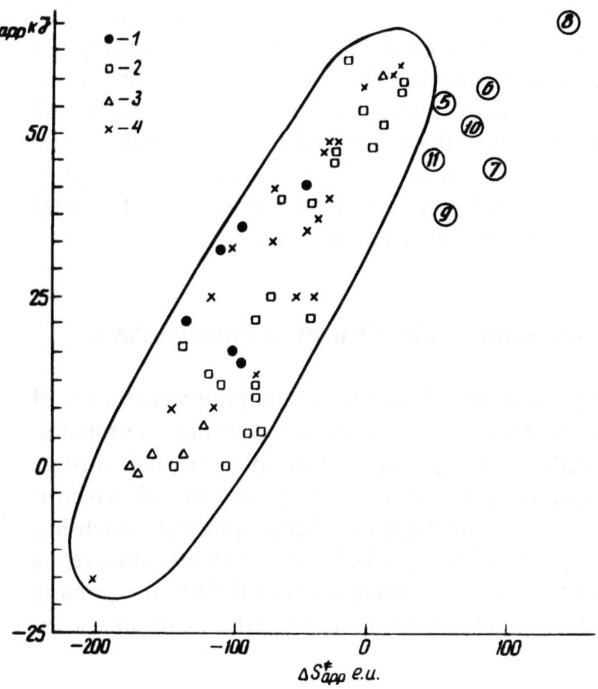

Fig. 11.6. Dependence of E_{app} on $\Delta S^{\neq}_{app}$ of electron transfer involving biological carriers. Constructed in accordance with the data of Ref. [820]. *1 – 4* Low molecular oxidizing agents with the following reducing agents: *1* cytochrome *c*; *2* HiPIP; *3* rubredoxin; *4* copper-containing carriers. *5 – 11* Protein reducing and oxidizing agents: *5* cytochrome *c*-azuring; *6* cytochrome c_{551}-azurin; *7* azurin-cytochrome c_{551}; *9* azurin-cytochrome c_{551}; *11* cytochrome *c*-cytochrome c_{551}; *1* cytochrome *c*-cytochrome c_{551} (*6* and *9*, *7* and *9 – 11* differ in ionic strength of the solution)

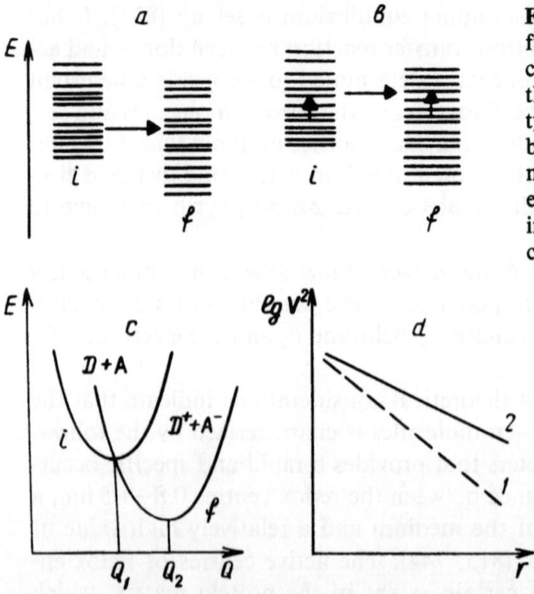

Fig. 11.7a–d. The specific features of the functioning of cluster systems as electron carriers: **a** overlap of the energy levels in the initial (*i*) and the final (*f*) state; **b** ability to by-pass quantum-mechanically forbidden routes; **c** minimal change of the nuclear configuration; **d** ability to undergo extensive electron-density delocalization in clusters (*1*) as compared with mononuclear systems (*2*)

structure (Section 5.6) [805]: (1) the tendency towards the overlap of the energy levels; (2) the possibility of by-passing the quantum-mechanically forbidden routes; (3) the absence of considerable nuclear displacements in redox processes; (4) a weaker damping of the wave functions in space as a result of the use of more strongly delocalized low-lying, unfilled orbitals.

In conclusion, we note that the available experimental data on the electron transfer in proteins support in general outline the well-known Vallee-Williams concept of the entatic state of redox active centres, i.e. of the state intermediate between the equilibrium oxidized and reduced states [471]. A typical example is the entatic structure of the ligand environment of the so-called blue copper in corresponding enzymes and carriers, which is intermediate between the tetrahedral and octahedral states (Chap. 8). Probably, the concept of the entatic state should also be extended to the above described cases of cluster, heme and chlorophyll centres, when the ligand, conformational and solvation environments of a metal atom are closely related in the oxidized and reduced states.

11.1.5 Electron Transfer in Photoseparation of Charges in Photosynthesis

The conversion of solar energy to chemical energy in the photoseparation of charges in the reaction centres of photosynthetic bacteria and plant chromatophores (Sect. 9.2.1) proceeds with a quantum yield close to unity. This process leads to the formation of reactive particles, such as the cation radical, which is a strong oxidizing agent with $E_0 = 0.7-0.9$ eV, and reduced quinone, which is a sufficiently strong reducing agent with $E_0 \approx 0$ eV. The process involves a photochemical electron transfer from a singlet-excited donor (D^*) to the primary acceptor (A_1) via two intermediate acceptors (A_1') and (A_1'') (Fig. 9.4) and is terminated with the emergence of an electron from the primary photosynthesizing cell to the secondary acceptor (A_2). The elementary acts of electron transfer occur at very high velocities (Fig. 9.4). The reverse electron transfer $A_1^- \rightarrow D^+$ occurs relatively slowly in spite of the favourable high energy of this process ($\Delta F_0 \approx -50.0\ \mathrm{kJ\ mol^{-1}}$).

The author of this monograph suggested in 1974 that the rapid electron transfer in the reaction centres in the forward direction and the slow transfer in the reverse direction may be accounted for by the tunnelling mechanism of the photoseparation of charges [846]. It was assumed that the donor and acceptor centres are at an optimal distance from each other and are separated by a non-conducting medium. The conceptions of the tunnelling mechanism of electron transfer in biological systems were originally worked out in the classical works of Chance and De Vault [842] for the electron-transfer reaction between oxidized chlorophyll and reduced cytochrome *c* in photosynthetic bacteria.

Let us explain the specific features of the tunnelling photoseparation of charges by using a simplified scheme involving a donor and two acceptors:

$$DA_1A_2 \xrightarrow{h\nu} D^*A_1A_2 \xrightarrow{k_1} D^+A_1^-A_2 \xrightarrow{k_2} D^+A_1A_2^-$$

$$D^+A_1^-A_2 \xrightarrow{k_{-1}} DA_1A_2$$

Under steady-state conditions the apparent rate constant of electron transfer to the acceptor A_2 is given by:

$$k_{app} = k_1k_2/(k_{-1}+k_2)\ . \qquad (11.18)$$

Here, k_1 and k_{-1} depend exponentially on the distance between the centres D and A_1: $k_1 = \exp(-a_1r)$ and $k_{-1} = \exp(-a_2r)$.

A necessary condition for obtaining a high quantum yield in the system is the fulfillment of the inequality $k_2 > k_{-1}$, i.e. the electron transfer to the secondary acceptor must proceed at a higher rate than the useless process of recombination of the photoseparated pair. This can be achieved by the location of the centres D and A_1 at an optimal distance r_{op} at which k_{-1} is already sufficiently low for $k_{app} \approx k_1$ and the value of k_1 is still sufficiently high as compared with the rate constant of the release from the primary cell ($k_1 \gtrsim k_2$). In other words, the separation of the donor and acceptor centres in space slows down both the forward and reverse reaction, but it nevertheless contributes to an increase in the quantum yield of the process. Here, the rate of the forward electron transfer may remain sufficiently high. Since the forward electron transfer involves the spatially extended excited orbitals of the donor, the wave functions of the donor should be expected to be damed in space more slowly than the corresponding wave functions for the ground state in the reverse transfer ($a_1 < a_2$). When the centres move farther frome each other, the rate of the process in the reverse direction will decrease much more rapidly than the rate of the forward transfer.

The theoretical conceptions described above are confirmed by experimentation [675, 676]. As can be seen from Fig. 9.4, the donor and acceptor groups in the reaction centres are separated by non-conducting barriers 0.4–0.8 nm wide. The foward electron transfer is effected by way of several tunnelling jumps. For the reverse electron transfer to occur the passage through a rather wide non-conducting barrier is required.

Of interest for the understanding of the nature of the physical factors that influence the rate constants of the reverse recombination in the pair $D^+A_I^-$ ($k_{A^- \to D^+}$) is the data of Ref. [838], the authors of which measured the rate of this process as a function of temperature and the standard redox potential of $A_I(E_0^0)$. The primary acceptors (A_I) used were the derivative of 1,4-benzoquinone, 1,4-naphthaquinone and 1,9-anthraquinone, which replace the natural quinone in the reaction centre. The values of E_0^0 varied from 0.11 to 0.94 eV. As shown by experiments, at low temperatures (5 – 100 K) the recombination rate is temperature-independent and depends relatively slightly (to within the factor 4) on E_0^0. In the opinion of the authors, such behaviour is due to the multi-phonon nature of the relaxation process. At T > 100 K the value of $k_{A^- \to D^+}$ increases with rise of temperature. At room temperature, as the exothermicity of the process increases (a decrease in ΔF_0) the rate constant for the recombination increases and then levels off. The authors of the work cited described the observed behaviour within the framework of the Marcus theory with $\Delta F_r^s = 0.375$ eV, $\Delta F_r^i = 0.200$ eV, $h\omega = 0.200$ eV and $V = 4$ to 8×10^{-8} eV. The point of interest is that to the value of the resonance integral just given with the distance between the edges of $(Bchl)_2$ and A_I^- [675, 676] being equal to r = 2.4 – 2.8 nm there corresponds the value of $\beta = 0.6$ nm [Eq. (11.14)]. This value is appreciably lower than the value of β for the known model and protein non-conducting systems, for which $\beta = 12 – 28$ nm (see Sect. 11.1.3). Presumably, this process does not occur via a direct overlap of the A_I^- and D^+ orbitals, but indirectly with the participation of the orbitals of the conducting portions of the intermediate acceptors $Bchl_1$ and Bph. The total length of the non-conducting gap between the edges of the pigments (~ 1.5 nm) is found to be sufficient for the tunnelling of an electron in the millisecond range.

An alternative explanation of the experimentally observed temperature dependence of $k_{A^-D^+}$ is that there are two conformational states: an "active" state and an "inactive" state with respect to the recombination. The fraction of the active state increases with increasing temperature [816, 841]. It should be noted, however, that the experimentally observed dependence of $K_{A^-D^+}$ on ΔF_0 [838] is an argument in favour of the validity of the Marcus theory.

A controversy has recently started [226, 667, 680] concerning the direct participation of the bacteriochlorophyll molecule in a direct electron transfer of the anion radical $Bchl_1$. The formation of this radical has been experimentally proved [680] and has been rejected by others [226]. Note that in any case, with the fixed distance between $(Bchl)_2$ and Bph equal to about 1.4 nm, the transfer with $k \sim 10^{12} s^{-1}$ seems to be improbable without direct (via the formation of an intermediate anion radical) or indirect (via the corresponding orbital overlap, which increases V) participation of a conjugated system of the bacteriochlorophyll molecule.

From a theoretical viewpoint, the absence of a strong dependence of the rate of electron transfer from the primary (Q_A) to the secondary (Q_B) acceptor in the presence of the metal ion is not clear [839]. The replacement of the iron atom by other metals, including the non-paramagnetic zinc ion, changes only very slightly (not more than three times) the electron transfer rate. The occurrence of a strong exchange between Fe^{2+} and the anion radicals Q_A^- and Q_B^- should have increased the electron transfer rate.

It is interesting to note that for electron-transfer and spin-exchange processes between the reaction-centre components close values of k_{tr} and k_{ex} are observed, which are reflected in the high values of the coefficient $\beta = k_{tr}/k_{ex}$. For example, $\beta_{A_I''-A_I} = 10$ and $\beta_{A_I''-D^+} = 0.1$ [675]. This means that the electron transfer between these components is not accompanied by a significant reorganization of the medium and that the vibrational relaxation reactions are highly efficient.

The reorganization of the medium introduces a substantial contribution to the steps of the electron-transfer process between the primary and the secondary acceptor. As has been shown in Section 10.2, this process takes place only in the case of the formation of the dynamic state of the water-protein matrix of the reaction centres with a correlation frequency of $\nu_c \approx 10^8\ s^{-1}$ and amplitudes of $A \gtrsim 0.03$ nm. In order to account quantitatively for this phenomenon, the following model may be considered.

The orientational relaxation of the surrounding electric dipoles stabilizes the charges after the electron transfer, lowering the free energy of the process (ΔF_0). If the relaxation is very slow, $\nu_c^{-1} \approx 0$, the electron-transfer process may be energetically unfavourable and the reaction will not proceed. In the other extreme case with $\nu_c^{-1} \approx \infty$, the surrounding nuclear system adjusts itself easily to the new electronic state and the reaction is energetically favourable. At a certain intermediate value of ν_c a region of the dependence is observed where the rate of the process depends on the frequency of motion of the terms, which is determined by the intensity of the molecular motion in the medium.

A decrease in the efficiency of the recombination of the pair $D^+A_I^-$ with an increase in temperature (Sect. 10.2.1) is presumably associated with a certain increase in the width of the tunnelling barrier brought about by a slow ($\nu_c \approx 10^2$ s^{-1}) temperature-dependent conformational transition.

The inferences based on the study of biological systems concerning the optimal spatial separation of the donor and acceptor centres and the necessity of a certain dynamic state of the protein matrix have been tested on model pigment-protein complexes [847]. Information on model reactions of photoseparation of charges can be found elsewhere [848–851]. For example, some authors [848] have succeeded in obtaining high quantum yields in photoseparation in solutions. Moreover, systems have been created, which catalyze simultaneously the photoevolution of molecular hydrogen [848].

11.2 Multi-Electron Mechanisms of Redox Reactions

11.2.1 General Principles

An important feature of polynuclear enzymes is their ability to involve inert molecules, such as N_2, O_2 and H_2O, into inner-sphere chemical conversions under mild conditions. As has been shown [24, 254, 539, 597, 598, 852], this ability is due to the conversion of these molecules, into the derivatives N_2H_4, H_2O_2

and O_2 respectively, without the formation of transition states, such as N_2H_2, HO_2^- or $^{\bullet}OH$ respectively. The formation of these transition states is energetically unfavourable, as has been shown by thermodynamic estimations. Since the products of the conversion of N_2H_4, H_2O_2 and O_2 differ from the reactant molecules by the presence of four and two electrons and by the absence of four electrons, the corresponding mechanisms of the conversions have been provisionally termed four- and two-electron mechanisms.

There are a considerable number of reactions in which the products contain two electrons more than the starting compound, and the consecutive multi-step electron-transfer process proves to be energetically unfavourable. In such cases, it is presumed that the two-electron process occurs in one two-electron step. A typical example of a two-electron process is the hydride transfer, when two electrons are transported together with a proton. Another possible two-electron mechanism involves the direct transport of two electrons. Such a transport alters sharply the electrostatic state of the system and obviously requires a substantial rearrangement of the nuclear configuration of the environment.

To illustrate, let us estimate the synchronization factor (a_{syn} and the rate constant (k_{tr}) for a two-electron transfer between two neutral particles with the formation of doubly charged ions, an anion and a cation. If the complexes are assumed to have an octahedral structure, then for the process to be described it is necessary that the elementary step involve the change of the configuration of at least 14 particles (two nuclei of metals) and 12 ligands moving as a single entity. According to formula (5.55), at an activation energy of 40–60 kJ mol^{-1} this leads to the values of $a_{syn} = 10^{-11}$ to 10^{-12} and to very low rates of reaction ($k_{tr} = 10^{-5}$ to $10^{-9}\,s^{-1}$). The rearrangement of the ligand environment can be reduced by a synchronous addition or abstraction of protons, but in such a case one may also expect low values of a_{syn}.

The above estimate points to the low probability of the two-electron processes considered above for reactions with a non-zero activation energy. The probability of two-electron processes, however, increases sharply if they take place in the coordination sphere of a metal, where the reverse transfer of electrons may occur which compensates for a change in charge. The following two-electron mechanism of the reduction of a substrate (S) in a binuclear complex may be visualized:

$$[2M^{+n}+S] \rightleftharpoons [M^{+n+\delta}\overset{\longrightarrow}{\leftarrow}S\overset{\leftarrow}{\longrightarrow}M^{+n+\delta}] \underset{}{\overset{2H^+}{\rightleftharpoons}} [2M^{+n+1}+SH_2]\ .$$

Here, the longer arrow indicates the direction of the preferred electron transfer from the metal to the substrate, and the shorter arrow indicates the direction of the reverse transfer.

Let us consider the possible four-electron mechanisms. Evidently, the direct transport of four electrons from the filled orbitals of a metal to the substrate in a binuclear complex is absolutely ruled out for energetic and relaxation reasons that have been analyzed for two-electron mechanisms. A simultaneous transfer of two hydrides is also unlikely. The only practically possible variant of the four-electron mechanism is the conversion in the coordination sphere of a polynuclear transition metal.

According to the literature data [74, 100, 539, 543], the interaction in transition-metal complexes with substrates occurs via multi-orbital binding: for example, via the transfer of four electrons from the orbitals of a metal to the anti-bonding (π^*-orbitals of the subtrate (Fig. 5.10) and the simultaneous transfer from the bonding orbitals of the substrate to the vacant d-orbitals of the metal atom (reduction) or in the reverse direction (oxidation). Such an electron transport may cause significant changes in the electronic structure of the substrate without substantially changing the total energy of the system and may play the role of the decisive step of the reduction or oxidation process.

According to the general theoretical concepts, the rate constant of the process $k \sim |V|^2$, where V is the resonance integral, which depends on the degree of the total overlap of the molecular orbitals of the reactants in the initial and final states. The multi-electron nature of the process does not evidently impose any new, additional restrictions on its velocity. This is confirmed by spin-exchange data. For example, the experimental data on the exchange triplet-triplet energy transfer and spin exchange in radical pairs, which may be regarded as a model of multi-electron processes, correspond approximately to the theoretical curves for a one-electron transfer (Fig. 11.3). Thus, the multi-electron nature of the chemical conversion must not significantly affect the dependence of k_{tr} on the distance: at short distances (0.2–0.3 nm) the oscillations of electrons occur very rapidly ($\tau \approx 10^{-13}$ to 10^{-15} s).

The determining factor in the conversion taking place inside the complex must be the adjustment of the nuclear environment (both vibrational and orientational) to the new electronic configuration. It is also essential that the transfer not be accompanied by a sharp change in the charge in any part of the system and by a change in the spin state or orbital symmetry.

In the variant of the four-electron process of N_2 reduction in the binuclear complex under consideration, the valency of each metal atom changes by two units, which, as we have seen above, must lead to the rearrangement of the ligand environment and to a decrease in the synchronization factor ($\alpha_{syn} = 10^{-11}$ to 10^{-15}). The probability of the elementary act sharply increases if these changes are prevented by a simultaneous transport of an additional number of electrons from the nearest electron-donating centres.

Thus, the actual route of the chemical reaction in the coordination sphere of the metal complex occurs by way of an optimal combination of the energy, entropy and synchronization factors. In certain cases, this combined effect leads to the necessity of limited synchronous multi-electron mechanisms. An external chemical sign of such a mechanism is the release of the corresponding reaction product from the complex, e.g. hydrogen peroxide in the reduction of O_2 (the two-electron mechanism) or N_2H_4 in the reduction of N_2 (the four-electron mechanism). It should be noted here that the chemical sign is valid only in those cases where the release of the product from the complex is close to an equilibrium process. In other words, the reagents that cause the release of the product, e.g. protons in the hydrolysis of complexes, must not upset significantly the redox equilibrium inside the complexes. It should be emphasized that enzymatic catalytic processes are, as a rule, characterized by a uniform distribution of the energy over the steps and by the equilibrium nature of certain steps, so that the release of the corresponding "multi-electron" intermediate product is a reliable sign of the multi-electron nature of the step.

Below, we will consider the multi-electron reduction processes of molecular oxygen, oxidation and hydroxylation of substrates, reduction of molecular nitrogen to ammonia and oxidation of water to molecular oxygen. Until recently, chemists have not been able to accomplish the last two reactions under mild conditions.

11.2.2 Two-Electron Reduction of Dioxygen

The data on the structure and chemical mechanism of action of polynuclear metalloenzymes and oxidases, which reduce O_2 to H_2O_2 or H_2O have been considered in Chapter 6 through 9. Here, we will be concerned with the possible elementary mechanisms of the decisive steps of these processes.

Reactions Catalyzed by Xanthine Oxidase and Related Enzymes. The basic thermodynamic data allow us to estimate the energetics of various mechanisms of the oxidation of one of the substrates of xanthine oxidase and aldehyde oxidase, acetic acetaldehyde, to acetic acid by molecular oxygen with the formation of H_2O_2. In principle, a one- or two-electron step of aldehyde oxidation and O_2 reduction may occur. Below, the equations of the corresponding reactions are given. The values of ΔH_0 or ΔF_0 of the processes expressed in kJ mol^{-1} are enclosed in parentheses have been calculated from handbook data [256].

$$CH_3C(=O)H + O_2 \rightleftharpoons CH_3\dot{C}{=}O + HO_2^{\bullet} \quad (\Delta H_0 = 155) \qquad (IV)$$

$$CH_3C(=O)H + 2O_2 + H_2O \rightleftharpoons CH_3C(=O)O^- + 2O_2^- + 3H^+ \quad (\Delta F_0 = -30) \qquad (V)$$

$$CH_3C(=O)H + O_2 + H_2O = CH_3C(=O)O^- + H_2O_2 + H^+ \quad (\Delta F_0 = -140) \qquad (VI)$$

As can be seen from mechanism IV, the postulated one-electron act of aldehyde oxidation with the formation of the acyl radical is completely excluded for energy reasons. The positive values of ΔH^0 for these reactions by far exceed the permissible value (+40 to +50 kJ mol^{-1}). It is therefore not surprising that all attempts to record the substrate radicals upon action of xanthine oxidase and aldehyde oxidase have failed.

Mechanism V is energetically possible in the xanthine-oxidase reaction. The generation of superoxide radicals has been observed experimentally (Sect. 7.1.1). The kinetic data permit one to presume the two-electron attack of $Mo^{6+}O$ on the hydride atom of the aldehyde or xanthine (Sect. 7.1.2).

Even more energetically favourable is scheme VI, which involves as two-electron mechanism of the reduction reaction. The two-electron reduction of O_2 by two reduced flavin groups may, however, be impeded by spatial and synchronization factors (Sect. 5.5.5). On the other hand, the products of the one-electron reduction (O_2^- radicals) may be disproportional in the active site of the enzyme. In any case, the liberation of O_2^- into solution, which is harmful to the living cell, is less intensive.

The experimental data on the redox potentials of the xanthine oxidase components agree well with the energy profile of mechanism VI (Sect. 7.1.1). The value of E_0 for the flavin components is quite sufficient for the reduction of O_2 by the one- or two-electron mechanism. Another noteworthy point is the smooth energy profile of the electron transfer through xanthine oxidase. The values of E_0 for the various components differ by no more than 100 mV.

On the whole, the process of substrates oxidation in the xanthine-oxidase reaction can be described by the following scheme:

$$[2 \rightarrow 1 \rightarrow 1 \rightarrow 1 \rightarrow 1(2)]\ .$$

Reactions Catalyzed by Polynuclear Copper-Containing Enzymes. The calculation of the energy profile for enzymes of this class can be made for three reactions of ascorbate (ascorbate oxidase) oxidation and divalent iron (ceruloplasmin), and oxidation of cytochrome *c* (cytochrome *c* oxidase), for which there are data not only on the redox potentials of the possible intermediate products, but also on the reaction-leading centres on the enzyme.

For the oxidation of the deprotonated form of ascorbic acid (DH^-) via the half-reduced form $D^{\bullet}$ we may formulate the following schemes [24] (the values of ΔF_0 calculated from the data of [853] at pH 6 and 25 °C are given in parentheses in kJ mol^{-1}):

$$DH^- + O_2 = D^{\bullet} + O_2^- + H^+ \qquad (73) \qquad (VII)$$

$$2DH^- + O_2 = 2D^{\bullet} + H_2O_2 \qquad (7) \qquad (VIII)$$

$$4DH^- + O_2 = 4D^{\bullet} + 2H_2O \qquad (-163) \qquad (IX)$$

As follows from the values of ΔF_0, step VII, which leads to the formation of a superoxide radical, is energetically unfavourable. There are no thermodynamic constraints for the occurrence of mechanisms VIII and IX, which involve the formation of the radical form $DH^{\bullet}$. Indeed, the formation of substrate radicals has been observed in many cases of reactions catalyzed by ascorbate oxidase and related enzymes (Sect. 8.1.2).

On the other hand, the potentials of the primary acceptor Cu^{+2}(I), the so-called blue copper, are about +500 mV, which is quite sufficient for the one-electron oxidation of ascorbate, but is not enough for the one-electron reduction of O_2 (at pH 7 for $DH_2/DH^{\bullet}$, the value of $E_0 = 310$ mV; and for O_2^-/O_2, the value of E_0 is equal to -330 mV [853]).

An analogous analysis can be performed for the oxidation of Fe^{2+} and ferrocytochrome *c*. There is, however, no need to describe here the details of such an analysis for these reactions. It should only be noted that the potentials of the primary acceptors of ceruloplasmin and cytochrome *c* oxidase, the role of which is also played by the blue form of copper Cu_a are 500 and 300 mV, respectively, which enable them to oxidize Fe^{2+} and cytochrome *c* (Fe^{2+}). The potentials of the primary donors, binuclear complexes of 2 Cu(III) an cytochrome a_3 and Cu_{a_3}, are approximately equal to 350 mV, which makes the possibility of producing superoxide as an intermediate doubful.

Thus, since molecular oxygen is a relatively weak one-electron oxidant, thermodynamic considerations lead to the conclusion that the rapid occurrence of one-electron processes in copper-containing oxidases in the O_2 reduction steps is impossible.

In order to choose between the other possible mechnisms (VIII and IX), it is expedient to compare the following schemes (for simplicity, the copper ligands are omitted in the schemes; the arrows indicate the direction of nuclear motion):

Two-electron mechanisms:

$$[Cu^{1+} \rightarrow \leftarrow O{=}O \rightarrow \leftarrow Cu^{1+}] \longrightarrow [Cu^{2+}\ \ O^{-}{-}O^{-}\ \ Cu^{2+}] \qquad \text{(VIIIa)}$$

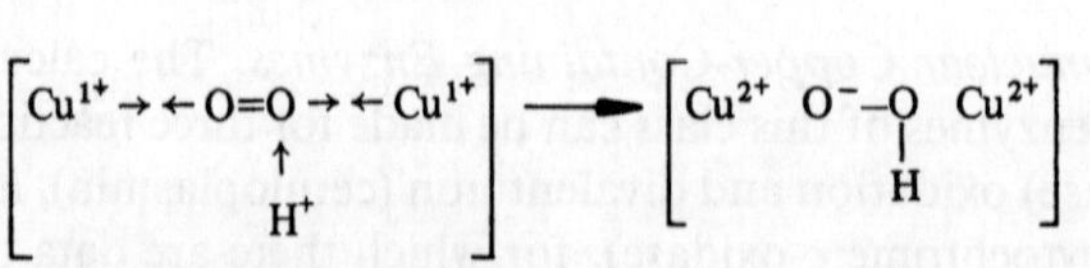

(VIIIb)

Four-electron mechanisms:

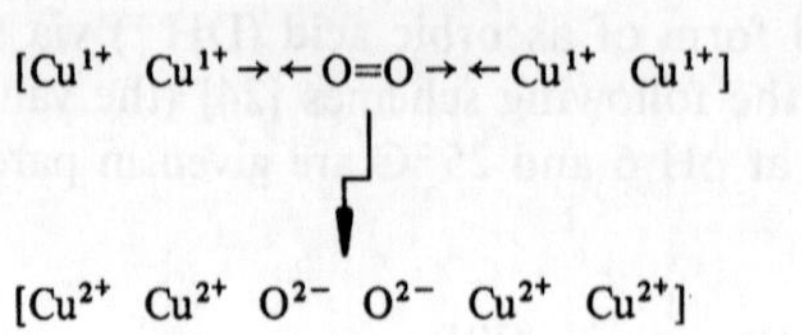

(IXa)

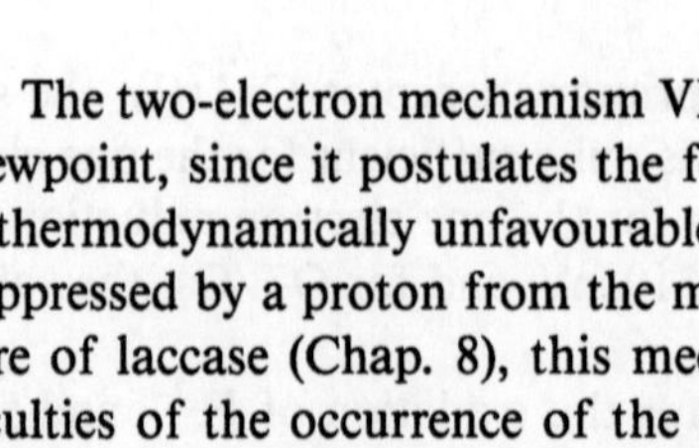

(IXb)

The two-electron mechanism VIIIa seems to be improbable from the energetic viewpoint, since it postulates the formation of a doubly charged O_2^{2-} ion which is thermodynamically unfavourable. In mechanism VIIIb one of these charges is suppressed by a proton from the medium. On the basis of the data on the structure of laccase (Chap. 8), this mechanism seems to be most probable. The difficulties of the occurrence of the four-electron reduction involving four copper atoms or the formation of two Cu^{3+} ions in a single elementary act arise either because of the energy complications associated with the formation of a doubly negative oxygen atom in the coordination sphere of copper (mechanism IXa) or due to the very low probability of the process because of the necessary synchronous transfer of two protons from water or the rearrangement of the ligand environment of copper (mechanism IXb). Analogous considerations allow one to prefer the two-electron reduction of O_2 in the active site of cytochrome *c* oxidase.

Thus, the oxidizing part of copper-containing oxidases acts by the one-electron mechanism and the reducing part by the two-electron mechanism and the process as a whole occurs according to the following scheme:

$$[1(2)\rightarrow 1\rightarrow\cdots\rightarrow 2]\ .$$

11.2.3 Oxidation and Hydroxylation Reactions Involving Heme Enzymes

In Chapter 6 we noted that there is a strong analogy in structure and mechanism of action between catalase and peroxidase. The two enzymes first form a complex with hydrogen peroxide [Fe^{3+} H_2O_2], which is rapidly converted into reactive compound I, formally presumed to have the state [$Fe^{3+}O$]. This state corresponds to two-electron oxidation. On the basis of the known energetics of the process $H_2O_2 = H_2O + O$ ($\Delta H_0 = 150$ kJ mol^{-1}), one can estimate the enthalpy of the reaction of compound I formation: $\Delta H_0 = 150 - D(Fe^{3+}O) + D(Fe^{3+}\cdots H_2O_2)$, where $D(Fe^{3+}O)$ and $D(Fe^{3+}\cdots H_2O_2)$ are the energies of dissociation of the corresponding bonds. Since in iron oxides the energy of the Fe–O bond usually lies in the range 240–360 kJ mol^{-1} [854] and the value of $D(Fe^{3+}\cdots H_2O_2)$ is not high, it follows that ΔH_0 has an essentially negative value and the formation of compound I is an energetically favourable process. It may be presumed that the rapid occurrence of this process is favoured by the presence of a distal histidine and arginine residue [404] and the elementary act of formation of compound I is assisted by an imidazole group and proceeds according to the following scheme (arginine is omitted):

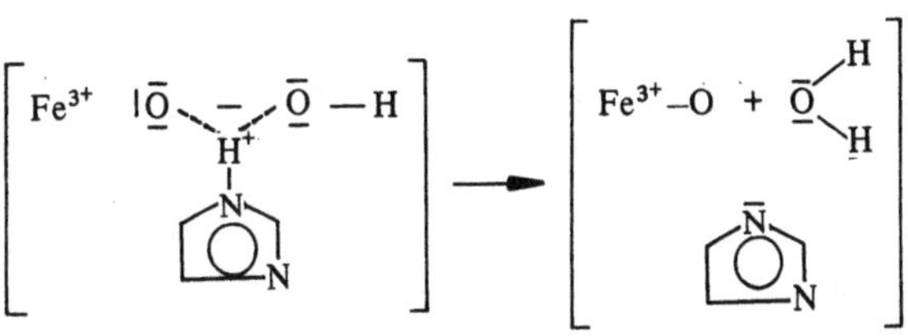

The elementary act involves a small number of nuclei and the value of the synchronization factor (a_{syn}, Sect. 5.5.5) must be sufficiently high. The only factor which slows down the reaction is the change of the spin state upon transition from the initial high-spin state to the low-spin state in compound I. The transmission coefficient of such a transition is sufficiently high ($\varkappa \approx 10^{-3}$; Sect. 11.1.1). This process may, however, be limited by a conformational transition, which accompanies the change of the spin state. The duration of such transitions in hemoproteins is $10^{-3} - 10^{-5}$ s [727].

In compound I one oxidizing equivalent is localized on the iron atom and the other in the form of the cation radical of the porphyrin ring or the nearest aromatic group (the tryptophan group in peroxidase). This is an indication that the formation of $Fe^{3+}O$ is followed by the abstraction of an electron from the aromatic group. The redox potentials of compounds I and II of peroxidase, which are formed upon its reduction ($E_0^{I} \approx E_0^{II} \approx 0.95$ eV) are sufficient for the one-electron oxidation of compounds, such as phenols and aromatic amines. Such potentials are, however, insufficient for the oxidation of H_2O_2 and other compounds with a relatively strong bond of the hydrogen atom or electron. Indeed, it can be shown [24] that the following relationship between the E_0 (in eV) of the oxidant, which abstracts hydrogen in compound RH, and the energy of dissociation of the R–H bond (in kJ mol^{-1}) is valid:

$$D_{RH} \lesssim 230 + 96 E_0 \ . \tag{11.19}$$

From relation (11.19) it follows that compounds I and II of peroxidase can oxidize substances, whose bond energy is $\lesssim 320$ kJ mol^{-1}. In hydrogen peroxide, $D_{RH} = 390$ kJ mol^{-1}, which requires the value of $E_0 \gtrsim 1.5$ eV.

The above estimate shows that compound I in catalase must have stronger oxidizing properties than compound I in peroxidase and the value of E^{I} must be equal to 1.3–1.4 eV. It is possible that these requirements are responsible for the presence of a tyrosine residue in the coordination sphere of the heme iron of catalase, in distinction to the histidine residue in peroxidase. A two-electron mechanism, in which the hydride and proton are synchronously split off from the substrate, is also possible.

Thus, the oxidation of donors of the DH_2 type by hydrogen peroxide in catalases and peroxidases may be described using the notation adopted in Section 1.3 as [2←1(2)]. The H_2O_2 molecule oxidizes the heme by the two-electron mechanism, this being followed by two one-electron steps of oxidation of the substrate. On the whole, the active site acts as a device which switches the reaction from the two-electron to the one-electron mechanism.

We will now consider the possible elementary mechanisms of the hydroxylation of substrates in the active site of cytochrome P-450. As has been shown in Section 6.1.5, the complexation of O_2 with heme and reduction lead to the formation of the state $[Fe^{2+}O_2^-]$, in which the oxygen adopts an active form capable of hydroxylating the substrates, including saturated hydrocarbons. These reactions can, in principle, occur by the following mechanisms.

1. The four-electron mechanism of the insertion of an "oxenoid" across the C–H bond of the hydrocarbon [434]:

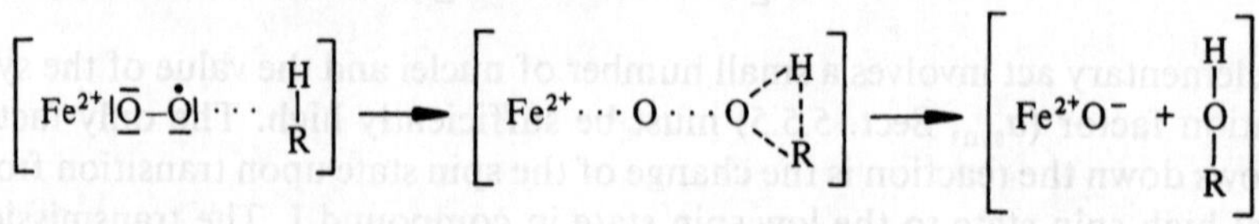

This mechanism is evidently energetically favourable, since it is accompanied by the formation of three bonds, two of which, O–H and C–O, are extremely strong, which undoubtedly then compensates more for the rupture of the C–H bonds and the weak O–O bond. The transition state, however, involves the formation of a three-membered ring with oxygen, whose formation is accompanied by a strain with an energy of 120 kJ mol^{-1} [283] and, probably, by a change in the spin state of iron due to the formation of a covalent bond with the oxygen atom. Both these factors can lower the rate of the step. This mechanism is in contradiction to experimental data, which point to the presence of a considerable isotope effect upon replacement of H by D in the hydrocarbon and to a change in the configuration at the asymmetric carbon atom during the course of the reaction (Sect. 6.1.3).

2. The two-electron mechanism of compound I formation followed by a one-electron abstraction of the hydrogen atom (by analogy with the peroxidase reaction or with the insertion reaction):

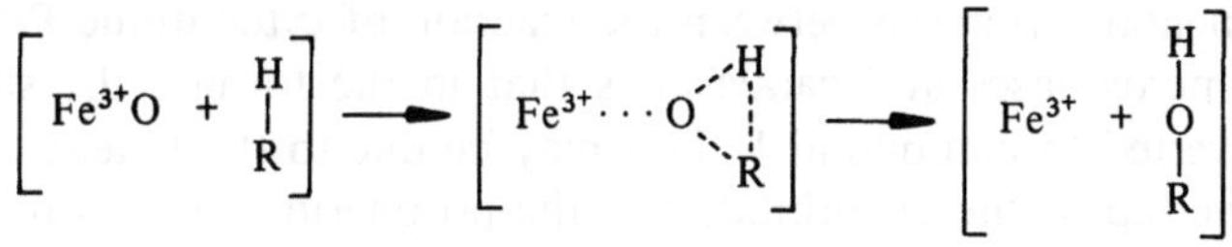

or

$$[Fe^{3+}O + HR] \longrightarrow [Fe^{3+}OH + R^{\bullet}] \longrightarrow [Fe^{3+} + ROH] .$$

A structure of the type $Fe^{3+}O$ may appear as a result of the oxidation $[Fe^{2+}O_2^-] \xrightarrow{2H^+} [Fe^{3+}O+H_2O]$ or the reaction with the carboxyl group (Sect. 6.1.5). The insertion reaction, though being thermodynamically favourable, may nevertheless fail to proceed for the reasons indicated above for the reaction of insertion of oxenoid and steric constraints. According to Eq. (11.19), for the reaction to occur by the one-electron mechanism of hydrogen abstraction with the consumption of an energy of about 400 kJ mol^{-1}, this being characteristic of saturated hydrocarbons, it is necessary that the compound $[Fe^{3+}O]$ have the potential of $E_0 \gtrsim 1.8$ eV. Since the potential of the known structure of compound I of peroxidase is 0.9 eV lower, the suggestion that compound I be included in reactions of cytochrome P-450 is in contradiction to thermodynamic requirements. It may be presumed that in contrast to peroxidase, in cytochrome P-450 both oxidizing equivalents are localized in the $Fe^{3+}O$ group and that the redox potential of the state $[Fe^{3+}O]$ is significantly higher than that of compound I. In such cases, however, it is difficult to explain why with such a high redox potential of the $Fe^{3+}O$ group, the porphyrin ring, whose potential is estimated as $E_0 = 0.8-0.9$ eV, is not oxidized. Note also that as has been shown [52], in camphor cytochrome P-450 in the distal region of the active site, there are no polar groups that could have contributed to the formation of compound I.

3. The masked radical mechanism involves the reaction of the superoxide radical in the coordination sphere, which is the rate-limiting step [24, 429]:

$$[Fe^{2+}|\bar{\underline{O}}-\bar{\underline{O}}|+HR] \rightarrow [Fe^{2+}|\bar{\underline{O}}-\bar{\underline{O}}-H+R^{\cdot}] ,$$

followed by fast radical steps which also take place in the coordination sphere:

$$[Fe^{2+}|\bar{\underline{O}}-OH+R^{\cdot}] \rightarrow [Fe^{2+}O^- + R-OH] .$$

This reaction is to a certain extent analogous to the reaction of hydrogen atom abstraction by the peroxide or hydroperoxide radicals from the hydrocarbons:

$$HO_2^{\cdot}+H_3C-R_1 \rightarrow HO_2H+ {}^{\cdot}CH_2R_1 .$$

The heat effect of such a reaction in a non-polar solvent for linear hydrocarbons (e.g. decane) is equal to about 16 kJ mol^{-1}. Substitution of this value into the Polanyi-Semenov formula yields the value of the activation energy $E \approx 60$ kJ

mol^{-1} at which the reaction can occur at measurable velocities at room temperatures. In solutions this reaction is practically not observed because of the very effective competing reaction of disproportion of the $HO_2^{\bullet}$ radicals. The presence of peroxide radicals in the coordination sphere of the metal without introducing a free valence into the coordinate bond may, however, prevent their decay, their ability to take part in the reaction of hydrogen being retained or even enhanced.

Thus, the fundamental difference between the reaction of cytochrome P-450 and the reaction of peroxidases and catalases is that in the former, the state $[Fe^{2+}O_2^{\dot{-}}]$ is not converted to compound I. This may be due to the kinetic factors, e.g. the weak overlap of the O_2 orbitals and the porphyrin ring as a result of a spatial separation and also to the absence of a distal protonated histidine group, which is capable of contributing to the formation of compound I. Mechanism 3 accounts for a number of known experimental factors, such as the presence of an isotope effect upon deuteration of substrates, the change of the configuration of substrates, the absence of an isotope exchange of oxygen in complexes with the oxygen of water, the ability of unsaturated compounds to undergo epoxidation, etc. Note also that in cytochrome P-450, no ionogenic groups that could contribute to the formation of compound I have been detected.

4. The mechanism, involving the elementary reaction act, starts with a partial abstraction of the hydrogen atom from the CH group by the fragment $Fe^{3+}O$ and ends with the insertion of an oxygen atom across the C–H bond [721].

At present, there are no direct arguments in favour of or aginst the mechanisms of O_2 activation and the hydroxylation of substrates considered above. The problem of the mechanisms of these processes awaits a fundamental solution.

11.2.4 The Four-Electron Mechanisms of Dinitrogen Reduction and Water Oxidation

Reductive Fixation of N_2. The Nitrogenase Reaction. Since the time of Daniel Rutherford, who discovered molecular nitrogen about 200 years ago, this colourless and odourless gas has served as an example of an inert substance. With the development of the physical chemistry of nitrogen the chemical inertness of this gas has been proved repeatedly. In particular, from the works of Pauling [257] it follows that the energy of the first bond being broken is very high (560 kJ mol^{-1}) and that the detachment energy of an electron and its excitation energy to the higher electron orbitals is also very high (1550 and 600 kJ mol^{-1} respectively).

In 1968 Likhtenshtein and Shilov advanced the supposition that the enzyme nitrogenase by-passes the above mentioned energy difficulties by way of realizing a reaction mechanism that provides the rupture of two bonds in molecular nitrogen with a simultaneous compensation due to the formation of four new bonds with the metal atoms [254]. This mechanism was provisionally termed the four-electron mechanism. Since the energy of the second bond being broken is much lower than that of the first (Sect. 4.4), the reduction potential per bond in the four-electron mechanism involving the formation of a hydrazine derivative is much lower than in the case of the one- or two-electron mechanism, in which the derivatives of $^{\bullet}N_2H$ and diimide N_2H_2 are formed. A more detailed analysis has shown that such an optimal energy profile is possible at which the structure of the compound formed in the decisive step will be intermediate between the hydrazine and the diimide structure, being closer, however, to the hydrazine structure [254, 884].

On the basis of the analysis carried out, we can draw two important conclusions. For the catalytic reduction of nitrogen to be accomplished under mild conditions, it is necessary that a polynuclear (at least, binuclear) complex capable of donating three of four electrons to the nitrogen molecule takes part in the reaction. An example or such a complex could be a binuclear complex in which both metal atoms would have the values of the redox potential of a two-electron transfer in the range of -500 to -700 mV. The second conclusion is that the first intermediate product of fixation under mild conditions must be a hydrazine derivative.

The two conclusions have been confirmed by studies on model systems, in which use is made of the catalyst of a binuclear complex of trivalent molybdenum on titanium hydroxide or compounds of divalent vanadium in an alkaline medium [100, 786]. These systems operate at low temperatures and pressures of nitrogen, converting molecular nitrogen to hydrazine or ammonia, depending on the conditions. Homogeneous fixation systems were later found which include solutions of a vanadium complex of pyrocatechol or other polyphenols; systems were also found, in which the reduction is effected by means of sodium amalgam or electrochemically by means of a solid electrode [885]. The efficiency of the latter system was sharply increased after a lipid layer was applied to the amalgam. This system was found to be capable of effecting the fixation of molecular nitrogen at room temperatures [542].

As seen in Fig. 11.8, the formation of the free products N_2H and N_2H_2 in the nitrogenase reaction is energetically unfavourable even when account is taken of the additional consumption of the energy released in the hydrolysis of ATP. Stabilization of these products in the coordination sphere of the metal in the ac-

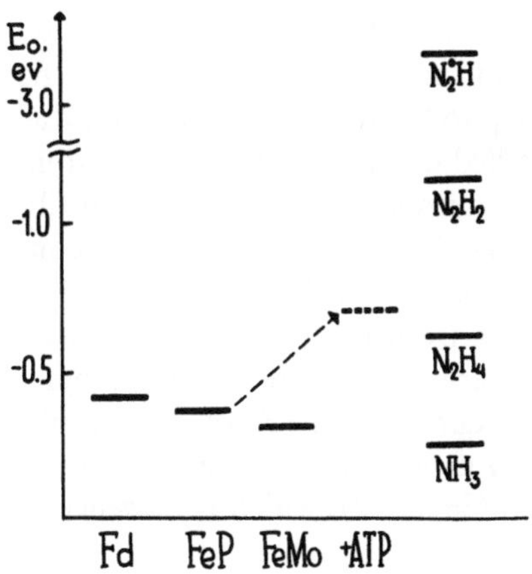

Fig. 11.8. The energy profile of a nitrogenase reaction. E_0 is the standard redox potential of the reactants, intermediates and products of the reaction; *Fd* = ferredoxin; *FeP* = Fe protein; *FeMo* = FeMo protein. The *arrow* indicates the increase of the reduction potential upon ATP hydrolysis

tive site would require the formation of bonds with an energy of about 160 kJ mol^{-1} and this is evidently impossible without an additional electron transfer from the transition-metal atoms to nitrogen (for more details, see Shestakov's calculation [24]).

On the other hand, the hydrolysis of ATP can, in princinple, provide an increase in the reduction potential of the system up to −600 to −700 mV, which is quite sufficient for the formation of the hydrazine derivative (or the derivative of diimide stabilized by the transfer of electrons from the metal, which is basically the same thing).

Thus, the energy considerations are in favour of the four-electron reduction of N_2 on nitrogenase. The elementary act of reduction must likely occurs on a binuclear complex of trivalent molybdenum (Sect. 7.3.3). It may be noted that in the xanthine-oxidase system the value of E_0 for the two-electron transfer, $Mo^{6+}/Mo^{4+} = -355$ mV is clearly insufficient for the formation of the hydrazine derivative. This is an indication that in nitrogenase super-reduced by ATP the valence of molybdenum is equal to three or is even lower.

In nitrogenase and in model systems one can visualize a simultaneous, six-fold change in the valency of two octahedral complexes of molybdenum with molecular nitrogen, with the formation, e.g. of metal nitrides. On the basis of general considerations and the results of X-ray diffraction analysis of molybdenum compounds [855], in this case one can expect a substantial change in the ligand environment of each metal atom. Altogether, at least 14 nuclei must undergo rearrangement, not to mention the atoms of the medium (the protein or solvent). In order to describe the motion of nuclei, which provides this rearrangement, generally 36 normal coordinates are required; the value of α_{syn} will be equal to 10^{-12} and 10^{-18} respectively, for the activation energies 40 and 80 kJ mol^{-1} [calculated according to Eqs. (5.55) and (5.56) and Fig. 6.7]. To these values of α_{syn} there correspond the pre-exponential factors (k_0) approximately equal to 10^1 and 10^{-5} s^{-1}, which are much lower than the experimentally observed values: for the rate-limiting step of the nitrogenase reaction $k_0 \approx 10^{13}$ s^{-1} and $E_a \approx 60$ kJ mol^{-1} (Sect. 7.3.2).

Thus, the energetically favourable six-electron mechanism cannot evidently be effected because of the constraints imposed on nuclear synchronization. The most optimal mechanism is presumably the four-electron mechanism of N_2 reduction, which involves two molecules of the Fe-Mo cofactor similar to the one shown in Fig. 11.9. According to this mechanism, the N_2 molecule in the binuclear complex is converted in the decisive step to the hydrazine derivative due to a single displacement of electrons from the filled d-orbitals of the metal atom to the anti-bonding orbitals of nitrogen and of the nitrogen electrons to the unfilled orbitals of the metal. The presence in the immediate vicinity of the iron-containing cluster FeMoCo allows the rearrangement of the nuclear environment to be reduced to a minimum upon electron transfer from molybdenum to N_2. At present, the possibility cannot be excluded that the second metal in direct contact with N_2 is an iron atom and that the reduction occurs on one FeMoCo molecule.

The hydrazine derivative has been detected by means of the nitrogenase reaction and in model reactions of nitrogen fixation (Sect. 7.3), which is evidence in favour of the four-electron mechanism of the reduction reaction.

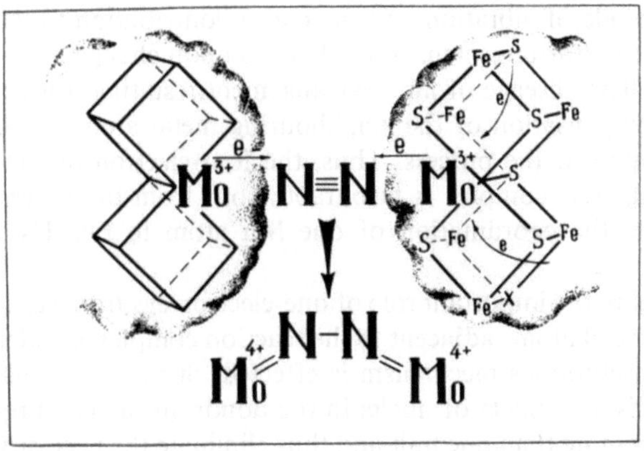

Fig. 11.9. Suggested optimal mechanism of N_2 reduction at the active site of nitrogenase

Water Oxidation. In biological photosynthesis, the decomposition of water with evolution of oxygen occurs under the action of a relatively mild oxidant: the cation radical of chlorophyll which is the product of a one-electron photo-oxidation with $E_0 = 0.7-0.9$ eV (Sect. 9.2.1). The potentials of the oxidation of water by the one-, two- and four electron mechanisms are equal to 2.7 eV (the hydroxyl radical), 1.36 eV (hydrogen peroxide) and 0.81 eV (molecular oxygen) respectively. Enclosed in parentheses are the products evolved at the most endothermic step. It is evident that the smoothest thermodynamic profile of the reaction is provided by the four-electron mechanism [786], which can be realized in biological systems in tetranuclear or binuclear manganese-containing complexes (see Sect. 9.2.2). The presence of such O_2 complexes in biological systems of O_2 evolution and in model reactions has been demonstrated experimentally (Sect. 9.2.2).

The evolution of oxygen from water may be described as a sequence of elementary steps: four one-electron steps of oxidation of the manganese complex by the chorophyll cation radical and one four-electron step of O_2 evolution.

For example, the following simplified scheme may be visualized:

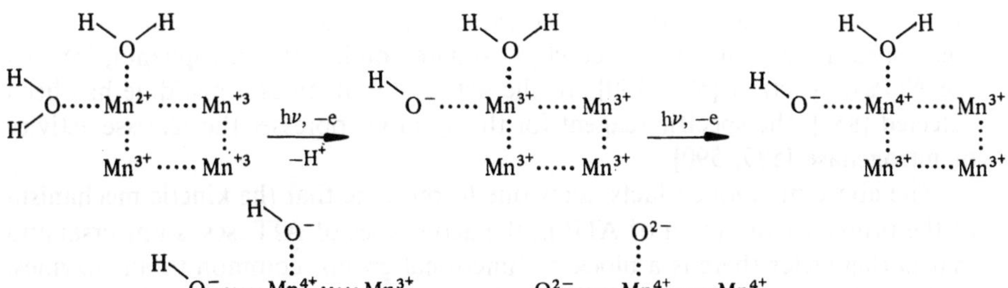

Each one-electron step is accompanied by the elimination of a proton, which contributes to the preservation of the total charge of the complex and considerably simplifies the last, decisive step of the process. As can be seen from the above scheme, the motion of the central nuclei at the step of oxygen formation corresponds to the normal mode of vibration and the major contribution to the total value must come from the rearrangement of the ligands upon change of the Mn^{+n} valency by two units. The absence of any rearrangement resulting, for example, from the electron compensation of the neighbouring metal atoms must sharply increase the probability of the process. Thus, the four-electron mechanism in a four-nuclear manganese complex is favourable both from the energy and kinetic viewpoint. Here, the coordination of one Mn atom to two H_2O molecules is preferred.

In conclusion, we emphasize the important role of one-electron electronic tunnelling transitions from the metal atoms adjacent to the reaction complexes under consideration, in which a snychronous mechanism is effected. Such transitions, which do not require large displacements of nuclei in the donor group, can prevent the change of valency by more than one unit and thus eliminate the necessity of the rearrangement of ligands in the acceptor group. Tunnelling proton transfers can also contribute to an increase in the probability of synchronization in processes involving protons.

11.3 The Chemical Mechanism of the ATPase Reaction in Energy-Conversion Systems

At present, much attention is devoted to enzymes that utilize the energy of ATP hydrolysis for the realization of energy-rich mechanical (myosin), transport (Na^+, K^+-ATPase, Ca^{2+}-ATPase) and chemical (nitrogenase) processes and for the creation of an electrochemical gradient in membranes (H^+-ATPase) [645, 857, 866, 870]. Despite the differences in the structure, functions and reaction conditions, these enzymes have some common properties.

In particular, all ATPases make use of Mg^{2+} ATP as the substrate or form of complex with these compounds, catalyze the direct exchange of oxygen between free phosphate and water and cleave the phosphate bond at the terminal phosphorus atom. Acetyl phosphate intermediates have been detected in cation-transport enzymes and H^+ ATPase [857, 858]. In the case of nitrogenase and myosin indirect data are available, indicating the formation of such an intermediate: a specific reagent for the acetyl-phosphate bond, p-thionitrophenol, inhibits the ATPase reaction [589, 859]. In the active site of myosin histidine has been detected [859]; the specific reagent for this group suppresses the ATPase activity of nitrogenase [589, 590].

The above mentioned facts allow one to presume that the kinetic mechanism of the primary conversion of ATP in the active sites of ATPases is universal and that in these sites there is a block of functional groups common to all enzymes.

It has been suggested [864] that the first result of ATP hydrolysis in the active site of enzymes may be the forced protonation of one of the functional groups of the active site X followed by the formation of the protonated energy-rich intermediate $\sim XH^+$. The energy of $\sim XH^+$ is then utilized for the performance of chemical, mechanical of electrical work. Such a mechanism will be efficient if this intermediate is shielded from the water environment during a time long enough for the performance of work and if a system of proton transport into the internal portions of the protein globule exists, this system being coupled with the system that directly cleaves the phosphate bond of ATP.

Taking into acocunt the data on the active-site structure of ATPases, one may presume that the ATPase portion of enzymes includes a charge-transfer (charge-relay) block, which is partly embedded in the protein matrix, as shown in Fig. 7.7, and a X group, which is capable of accepting a proton.

In the initial state, one of the protons is localized on the carboxyl group and the other on the imidazole residue and the X group is deprotonated. Under the action of ATP the carboxyl group is phosphorylated and the proton is knocked out of the carboxyl group into the histidine, which turn leads to the transfer of the second proton to X with the formation of the energy-rich state $\sim XH^+$.

The equilibrium deprotonation of intermediates of the XH^+ type is followed by the formation of the corresponding states, which are in non-equilibrium with respect to the given conditions and, hence, are capable of performing the corresponding work: the super-reduced state of X^- (X is an iron-sulphur cluster Fe_nS_n) in the case of nitrogenase, the charge state of the membrane in the H^+ ATPase system, the conformationally strained state in myosin. In cation-transport enzymes, the forced protonation of X probably contributes to the electrostatic knocking-out of the cation M^{n+} from a complex with protein groups into the non-equilibrium complex, which triggers a further displacement of the cations in the system.

It should be noted that there are data indicating that in oxidative phosphorylation and photophosphorylation systems the electron transport is coupled with the transfer of a proton through the membrane [637, 644, 860–863, 883]. The reaction considered above may be induced by the elementary step of the process.

The conceptions described above, which are based on chemical concepts, allow one to treat, from a unified standpoint, a large amount of data on the chemical conversions in ATPases in redox and other processes and may be used as the working hypotheses in planning further experiments. Other approaches to the problem have been described in the literature [637–645, 727]. It should be noted that the above described chemical concept is an extension of the conception of the proton local chemical potential proposed as early as 1961 by Williams [645].

11.4 Conclusions

Metal enzymes and carriers catalyze a large number of redox biological processes. They participate in the conversion of various substrates (both organic and inorganic), thus performing the same functions as do other enzymes; they provide high selectivity and specificity, high rates of the chemical steps of addition or detachment of electrons, protons or other groups, and provide fine regulation of the enzymatic process. In metal catalysis, new phenomena arise. The main and most surprising phenomenon is that chemically inert substances such as N_2, H_2O and O_2, while entering the active sites, become extremely reactive and begin to the converted at rates equal to or exceeding 1 mol substrate per mole enzyme per second.

Modern chemical physics provides an armamentarium of theoretical and experimental methods for the study of metalloenzyme catalysis; the joint application of these methods makes it possible to obtain deeper insights into the structure and mechanism of action of biological catalysts. Quantum chemistry and statistical physics form the foundation of the physical methods of investigation of the enzyme structure and lie, together with chemical kinetics, at the basis of the theoretical conceptions of chemical reactions in the active sites of enzymes. The experience that has been accumulated shows that the combined use of modern physico-chemical methods makes it possible to solve practically any structural and kinetic problems. Apart from X-ray diffraction methods, efficient methods are the following selective methods of studying the structure of the active sites of metalloenzymes: EXAFS, microwave and gamma resonance spectroscopy, optical methods, X-ray absorption and X-ray photoelectron spectroscopy, methods based on the scattering of X-rays, electron and neutron radiation. Substantial information on the relative arrangement of metal atoms and other portions of the active sites and their localization on the enzyme macromolecule can be obtained with the aid of new methods making use of spin, luminescent, Mössbauer, and electron-dense (electron-scattering) labels. These methods, together with other physical techniques (NMR, dielectric and mechanical losses, scattering of neutrons and gamma quanta, etc.), enable one to study the molecular-dynamic properties of enzymes.

The spin-label techniques coupled with other experimental and theoretical approaches can contribute to the establishment of the tertiary and quaternary structures of enzymes from the known primary amino-acid sequence without using direct X-ray studies of single-crystal preparations. The theoretical approaches developed at present allow one to predict the secondary structure of proteins from the known primary structure [877, 878]. It may be expected that in the near future it will be possible to select theorectically a certain number of the most probable tertiary structures. The choice between the probable structures can be made by measuring the distances between spin, luminescent, Mössbauer and electron-scattering (electron-dense) labels introduced into certain portions of the primary structure. Analysis of the relaxational characteristics and NMR spectra of the paramagnetic nuclei of substrates and enzymes can provide additional information on the distances between spatially separated protein groups.

The application of modern kinetic methods enables one to establish the sequence of the individual steps of the enzymatic process and their nature and also to measure the rate constants of the steps over a wide range of their values.

In spite of the large variety of macromolecular structures, compositions and structures of the active sites and catalyzed reactions, it is possible at present to distinguish the common features in the structure and elementary mechanisms of the functioning of metalloenzymes. For example, iron-containing enzymes may use, as the elementary units, the heme group and mononuclear, binuclear, trinuclear and tetranuclear iron-sulphur centres. All these centres, with the exception of special cases of the interaction of the heme group with O_2 and H_2O_2, perform the functions of one-electron carriers. An important new property of tetranuclear systems is the ability to adopt any of the four redox states. The copper atoms in oxidases and carriers enter into the composition of three elementary complexes: two mononuclear electron carriers, Cu(I) and Cu(II), and one binuclear carrier, 2Cu(III), which have different ligands and different redox properties. The Cu(I) complex directly detaches electrons from the substrate being oxidized, the 2Cu(III) probably binds O_2 and, together with Cu(II), exerts a two-electron attack on oxygen. In cytochrome *c* oxidase such an attack is effected with the aid of the Cu(II)-heme complex.

In molybdenum-containing enzymes, oxidases and reductases, the molybdenum atom also enters into the composition of a certain complex, which has a similar structure in different enzymes. The ligands of the complex are sulphur and oxygen atoms. The function of this complex is to effect a two-electron hydride oxidation of substrates, which involves the transition $Mo^{6+} \rightarrow Mo^{4+}$ and is followed by the transfer of electrons along the chain of one-electron carriers (xanthine and aldehyde oxidase) or into the system of two- or one-electron reduction of nitrates and sulphates. In nitrogenase, the molybdenum atom is surrounded by sulphur and iron atoms in the second coordination sphere. There is every ground for believing that the union of two such elementary complexes into a binuclear centre in nitrogenase and their "super-reduction" due to ATP hydrolysis makes it possible to realize the four-electron reduction of molecular nitrogen to the hydrazine derivative.

The combination of the elementary structures considered above and the adjustment of their chemical properties through a local electrostatic or conformational effect to the optimal value (for a given process) provide the proper functioning of a large number of enzymes under the most favourable conditions. The evolution of the amino-acid composition of the protein matrix has contributed to the appearance of a large variety of their macromolecular structures and surface properties, which is necessary for the realization of their specific interactions with other biological structures.

The most important prerequisite for the appearance of catalytic properties of the active sites of polynuclear enzymes is the intramolecular dynamics of the protein-matrix structure. The required changes in the conformation of electronic transitions, the dynamic adjustment of the enzyme shape to the structure of the reaction complex at various stages, the allosteric interactions, etc., cannot be realized on a completely rigid substrate. The universal, local mobility of protein macromolecules in the nanosecond range, recorded by means of a number of physico-chemical methods, provides rapid and reversible conformational changes. There is every ground for believing that among the entire spectrum of local motions a special role is played by reversible displacements of relatively large blocks of the protein matrix that bear catalytic and contact groups. Such displacements are not accompanied by significant entropy changes and provide the strucutral entity of the active site of enzymes with their local mobility being retained.

Of special interest is the problem polynuclear enzymes and carriers[2]. Analysis of the specific energy and kinetic features of the conversion of chemically inert N_2, H_2O and O_2 molecules under mild conditions leads to the conclusion that the requisite step of the process is a multi-electron synchronous attack

[2] A general conception on the special role of these systems in catalysis was introduced in 1960 by Purmal [889]. For recent developments of the problem, see for example reference [890].

of the catalyst on the substrate: the four-electron reduction of N_2, the two-electron reduction of O_2 and the four-electron oxidation of H_2O. Since in all these cases use is made of one-electron external reductants and oxidants, polynuclear enzymes play the role of specific devices that switch the mechanism of the reaction from the one-electron mechanism to a multi-electron one.

Convenient devices for the realization of multi-electron elementary acts are cluster systems that contain several closely spaced and interacting transition-metal atoms. The transport and storage functions can here be performed by one-electron carriers, which are also of the cluster nature.

Another, qualitatively new property of enzyme systems that contain several transition-metal atoms is the ability to transport electrons over large distances (up to 5.0–6.0 nm). This ability appears when metal-containing carriers are inserted into the chain in approximate conformity with their redox potentials. The experimental data and theoretical considerations allow one to presume that the transport occurs, as a rule, by a tunnelling mechanism; the elementary act of electron transport here extends to a distance of at least 0.8–1.0 nm.

The experimental data and theoretical considerations indicate that the electron transfer between protein macromolecules is characterized by the following optimal set of parameters which provide the rapid and specific occurrence of the redox process: the distance between the redox centres (0.8–1.2); an insignificant free energy of the reorganization of the medium; and a relatively long lifetime of the specific complex. The active centres of redox enzymes and carriers are buried to some extent in the protein macromolecule.

The most important bioenergetic process of transformation of the sunlight energy into the chemical energy of ATP and reduced compounds is accomplished in the reaction centres of specific photoenzymes. There is every ground for believing that a strong analogy exists between the structures of the reaction centres in bacteria, algae and plants. In these cases, the primary act of sunlight storage, the photoseparation of charges, is effected by a system of chlorophyll, pheophytin and quinone molecules separated by an optimal distance (0.5–0.8 nm) from one another. Such an arrangement allows one to account for the rapid electron transfer and the slow process of reverse recombination.

A quantum-mechanical analysis of the electronic structure of transition-metal clusters shows that the main feature of these systems is the presence of a large number of electron levels separated from one another by small energy gaps [24, 830]. A further important feature of cluster systems is the sharing of the nuclear core of the complex, which responds weakly to a change in the electronic state of the cluster. These two qualitatively new properties allow cluster systems to perform the functions of the elements of an optimal multi-centre catalyst, namely: (1) to provide a smoothing of the energy profile of the reaction due to the "optimal adjustment" of the redox potential; (2) to effect a multi-centre spatial interaction with the substrate; (3) to realize multi-electron synchronous elementary acts; (4) to avoid quantum-mechanically forbidden reaction paths by way of the thermal excitation of electrons to the low-lying orbitals, the transitions from which are allowed; (5) to preserve the nuclear configuration upon transfer of one or two electrons due to the high electron density of the cluster; (6) to contribute to the formation of more flattened potential curves due to the effect of the low-lying, unfilled orbitals on them (the vibronic effect of the second kind; [304]); this property of clusters must lead to a decrease in the activation energy of chemical conversions in which they participate; (7) to provide a higher probability of fitting between the energy levels and an increase in the delocalization of electron density in a tunnelling electron transfer.

The wide range of phenomena associated with the energy transformation of ATP hydrolysis into mechanical, chemical, electrical and other forms of energy are accounted for within the framework of the hypothesis that suggests a formation step of a non-equilibrium, protonated, energy-rich transition state [546, 864, 869]. The electron transfers in biological systems are, as a rule, accompanied by protonation of the reduced forms and deprotonation of the oxidized forms.

It may be expected that the detailed study of redox enzymes and the development and industrial application of model chemical catalysts will in the near future make an important contribution to the theory and practice of chemical catalysis and, furthermore, will contribute to the creation of a theoretical foundation for molecular biology, biochemistry, pharmacology, medicine, agriculture and other related sciences.

References

1. Bertrand MG (1894) Chimie Industielle. Sur le latex de l'arbre a'laque. C R Acad Sci 118:1215–1218
2. Keilin, D, Hartree EF (1938) Cytochrome *a* and cytochrome oxidase. Nature 141:870–871
3. Mason HS (1955) Comparative biochemistry of the phenolase complex. In: Nord FF (ed) Advances in Enzymology. Interscience Publishers, New York London 16:105–184
4. Brodie BB, Axelrod J, Cooper JR, Gaudette Z, La Du BN, Mitoma C, Udenfriend S (1955) Detoxication of drugs and other foreign compounds by liver microsomes. Science 121:603–604
5. Omura T, Sato R (1964) Carbon monoxide-binding pigment of liver microsomes. I Evidence for its hemoprotein nature. J Biol Chem 239:2370–2378
6. Van der Hoeven TA, Coon MJ (1974) Preparation and properties of partially purified cytochrome P-450 and reduced nicotinamide adenine dinucleotide phosphate-cytochrome P-450 reductase from rabbit liver microsomes. J Biol Chem 249:6302–6310
7. Imai Y, Sato R (1974) A gel-electrophoretically homogeneous preparation of cytochrome P-450 from liver microsomes of phenobarbital-pretreatment rabbits. Biochem Biophys Res Commun 60:8–14
8. Ball EG (1939) Xanthine oxidase: purification and properties. J Biol Chem 128:51–67
9. Avis PG, Bergel F, Bray RC, Shooter KV (1954) A crystalline material with xanthine oxidase activity. Nature 173:1230–1231
10. Beinert H, Sands RH (1960) Study of succinic and reduced diphosphopyridine nucleotide dehydrogenase preparations by paramagnetic resonance spectroscopy. Biochem Biophys Res Commun 3:41–47
11. Mortenson LE, Valentine RC, Carnahan JE (1962) An electron transport factor from Clostridium pasteurianum. Biochem Biophys Res Commun 7:448–452
12. Lovenberg W, Sobel E (1965) Rubredoxin: a new electron transfer protein from Clostridium pasteurianum. Proc Natl Acad Sci USA 54:194–199
13. Boussingault J (1838) Recherches sur la vegetation entreprises dans le but d'examiner si les plantes fixent dans leur organisme l'azote qui est a l'atet gaseux dans l'atmosphere. CR Acad Sci 6:102–112
14. Vinogradsky SN (1895) Concerning the assimilation of atmospheric free nitrogen by microbes. Arkh Biol Nauk 3:293–315 (in Russian)
15. Vinogradsky SN (1952) Microbiology of soil. Izv Akad Nauk SSSR, Moscow (in Russian)
16. Bakh AN, Ermolieva ZV, Stepanian MP (1934) Fixation of atmospheric nitrogen at an ordinary temperature and pressure by means of enzymes extracted from nitrogeneous bacteria. Bokl Akad Nauk SSSR 1:22–24 (in Russian)
17. Bulen WA, LeComte JR (1966) The nitrogenase system from Azotobacter: two-enzyme requirement for N_2 reduction, ATP-dependent H_2 evolution and ATP hydrolysis. Proc Natl Acad Sci USA 56:979–986
18. Mortenson LE, Morris JA, Jeng DY (1967) Purification, metal composition, and properties of molybdoferredoxin and azoferredoxin, two of the components of the nitrogen-fixing system of Clostridium pasteurianum. Biochim Biophys Acta 141:516–538
19. Bortels H (1936) Molybden als Katalysator bei der biologischen Stickstoffbindung. Arch Mikrobiol 1:333–342

20. Villee C (1962) Biology. 4th edn. Saunders, Philadelphia London
21. Xavier AV (ed) (1986) Frontiers in bioorganic chemistry. VCH, Weinheim
22. Harrison PM (ed) (1982) Metalloproteins, vol 1. Macmillan, London
23. Brill AS (1977) Transition metals in biochemistry. Springer, Berlin Heidelberg New York
24. Likhtenshtein GI (1979) Polynuclear redox metalloenzymes. Nauka, Moscow (in Russian)
25. Addison AW, Cullen WR, Dolphin D, James BB (eds) (1977) Biological aspects of inorganic chemistry. Wiley Interscience, New York
26. Enzyme nomenclature (1972) Recommendations of the International Union of Biochemistry on the nomenclature and classification of enzymes and also on enzyme units and notation for the kinetics of enzymatic reactions. Translated from English (with additions adopted up to 1975). Edited by AE Braunshtein. Moscow, 1979
27. Enzyme nomenclature (1972) Recommendations of the International Union of Pure and Applied Chemistry and the International Union of Biochemistry. Biochim Biophys Acta (1976) 426:1–45
28. Hall DO, Rao KK, Cammack R (1975) The iron-sulphur proteins: structure, function and evolution of a ubiquitous group of proteins. Sci Progr 62:285–317
29. Kent TA, Moura J, Moura IIG, Lipscomb ID, Ituynh BH, Le Gall I, Xavier AV, Munck E (1982) Conversion of 3Fe-3S into 4Fe-4S clusters in a Desulfovibrio gigas ferredoxin and isotopic labeling of iron-sulfur cluster subsites. FEBS Lett 138:55–58
30. Stout CD, Ghosh D, Pattabhi V, Robbins AH (1980) Iron-sulphur clusters in Azotobacter ferredoxin at 2.5 Å resolution. J Biol Chem 255:1797–1800
31. Moura J, Bruschi M, Le Gall I, Moura IIG, Xavier AV (1977) Isolation and characterization of desulforedoxin, a new type of non-heme iron protein from Desulfovibrio gigas. Biochem Biophys Res Commun 75:1037–1044
32. Krüger HJ, Der-Vartanian DV, Peck HD, Le Gall J (1984) Desulfovibrio vulgaris hydrogenase: a nonheme iron enzyme lacking nickel that exhibits anomalous EPR and Mössbauer spectra. Proc Natl Acad Sci USA 81:3728–3732
33. Chen JS, Blanchard DK (1978) Isolation and properties of a unidirectional H_2-oxidizing hydrogenase from strictly anaerobic N_2-fixing bacterium Clostridium pasteurianum W5. Biochem Biophys Res Commun 84:1144–1150
34. Hagen WR, Berkel-Arts A van, Krüse-Wolters KM, Voordouw G, Veeger C (1986) The iron-sulfur composition of the active site of hydrogenase from Desulfovibrio vulgaris (Hildenbourough) deduced from its subunit structure and total iron-sulfur content. FEBS Lett 203:59–63
35. Reinhammar BRM (1972) Oxidation-reduction potentials of the electron acceptors in laccases and stellacyanin. Biochim Biophys Acta 275:245–259
36. Palmer G (1975) Iron-sulphur proteins. In: Boyer PD (ed) The Enzymes, 3rd edn. vol XII, Oxidation-reduction, Part B. Academic Press, New York San Francisco, pp 1–56
37. Hatefi J, Stiggall DL (1976) Metal-containing flavoprotein dehydrogenases. In: Boyer PD (ed) The Enzyme, 3rd edn, vol 13. Academic Press, New York, San Francisco, pp 175–297
38. Nelson CA, Handler P (1968) Preparation of bovine xanthine oxidase and subunit structures of some iron flavoproteins. J Biol Chem 243:5368–5373
39. Johnson MK, Morningstar JE, Cecchini G, Ackrell BA (1985) Detection of a tetranuclear iron-sulfur center in fumarate reductase from E. coli by EPR spectroscopy. Biochem Biophys Res Commun 131:756–762
40. Crichton RR (1973) Annotation: The biochemistry of ferritin. Brit J Haematol 24:677–680
41. Fletcher J, Huehns ER (1968) Function of transferrin. Nature 218:1211–1214
42. Sanders-Loehr J (1985) Oxygen binding of the binuclear center of hemerythrin. Rev Port Quim 27:41–42
43. Wilson DF, Erecinska M (1978) Cytochrome oxidase. In: Dolphin DC (ed) The Porphyrins, vol VII, Biochemistry, part B. Academic Press, New York, pp 2–70
44. Mathews FS, Levine M, Argos P (1972) Three-dimensional Fourier synthesis of calf liver cytochrome b_5 at 2–8 Å resolution. J Mol Biol 64:449–464
45. Takano T, Dickerson RF (1980) Redox conformation changes in refined tuna cytochrome c. Proc Natl Acad Sci USA 77:6371–6375
46. Schonbaum GR, Chance B (1976) Catalase. In: Boyer PD, Lardy H, Myrback K (eds) The Enzymes. 3rd edn. vol 13. Academic Press, New York, pp 363–408

47. Paul KG (1963) Peroxidases. In: Boyer PD, Lardy H, Myrback K (eds) The Enzymes. 2nd edn, vol 8. Academic Press, New York London, pp 227–283
48. Poulos TL, Freer ST, Alden RA, Xuong NH, Edwards SL, Hamlin RC, Kraut J (1978) Crystallographic determination of the heme orientation and location of the cyanide binding site in yeast cytochrome c peroxidase. J Biol Chem 253:3730–3735
49. Coval ML, Taurog A (1967) Purification and iodinating activity of hog thyroid peroxidase. J Biol Chem 242:5510–5523
50. Sutherland J, Greenwood C, Peterson J, Thomson AJ (1986) An investigation of the ligand-binding properties of Pseudomonas aeruginosa nitrite reductase. Biochem J 233:893–898
51. Mitoma C, Udenfriend S (1962) Aryl-4-hydroxylase. In: Colowick SP, Kaplan NO (eds) Methods in enzymology, vol 5. Academic Press, New York London, pp 816–819
52. Poulos TL, Finzel BC, Gunsalus IC, Wagner GC, Kraut J (1985) The 2.6 Å crystal structure of Pseudomonas putida cytochrome P-450. J Biol Chem 200:16122–16130
53. McKenna EJ, Coon MJ (1970) Enzymatic ω-oxidation (IV. Purification and properties of the ω-hydroxylase of Pseudomonas oleovorans). J Biol Chem 245:3882–3889
54. Zuidweg MHJ (1968) Hydroxylation of Reichstein's compound S with cell-free preparations from Curvularia lunata. Biochim Biophys Acta 152:144–202
55. Lino AR, Moura JJ, Xavier AV, Moura I, Fauque G, Le Gall J (1985) Characterization of two low-spin bacterial siroheme proteins. Rev Port Quim 27:215–216
56. Eady P, Smith B (1979) Physico-chemical properties of nitrogenase. In: Hardy RWF, Bottomley F (eds) A treatise on dinitrogen fixation. Wiley Interscience, New York, pp 352–427
57. Syrtsova LA, Pisarskaya TN, Nazarova II, Uzenskaya AM, Likhtenshtein GI (1977) Coupling of ATPase and substrate-binding sites in the nitrogenase Azotobacter vinelandii. Bioorg Khim 3:1251–1260 (in Russian)
58. Losada M (1976) Metalloenzymes of the nitrate-reducing system. J Mol Catal 1:245–264
59. Andres RY (1976) Aldehyde oxidase and xanthine dehydrogenase from wild-type Drosophila melanogaster and immunologically gross-reacting material from ma-1 mutants. Eur J Biochem 62:591–600
60. Bray RC (1977) Molybdenum-containing enzymes. J Less Common Metals 54:527–536
61. Tanaka M, Haniu M, Yasunobu KT (1977) Amino acid sequence of the heme *a* subunit of bovine heart cytochrome *c* oxidase and sequence homology with hemoglobin. Biochem Biophys Res Commun 76:1014–1019
62. Malmström BG, Andreasson LE, Reinhammar B (1975) Copper-containing oxidases and superoxide dismutase. In: Boyer PD (ed) The Enyzmes. 3rd edn, vol 12. Academic Press, New York San Francisco, pp 507–579
63. Ullrich V, Duppel W (1975) Iron- and copper-containing monooxygenases. In: Boyer PD (ed) The Enzymes. 3rd edn, vol 12. Academic Press, New York San Francisco, pp 253–297
64. Yamada H, Kumagai H, Kawasaki H, Matsui H, Koichi O (1967) Crystallization and properties of diamine oxidase from pig kidney. Biochem Biophys Res Commun 29:723–727
65. Blaschko H (1963) Amine oxidase. In: Boyer PD, Lardy H, Myrback K (eds) The Enzymes. 2nd edn, vol 8. Academic Press, New York London, pp 337–351
66. Kendrew JC (1963) Myoglobin and protein structure. Biofizika 8:273–283 (in Russian)
67. Richardson JS, Thomas KA, Rubin BH, Richardson DC (1975) Crystal structure of bovine CuZn superoxide dismutase at 3 Å resolution: chain tracing and metal ligands. Proc Natl Acad Sci 72:1349–1353
68. Powers L, Chance B, Ching Y, Angiolillo P (1981) Structure features and the reaction mechanism of cytochrome *c* oxidase. Iron and copper X-ray absorption fine structure. Biophys J 34:465–498
69. Teixeira M, Moura I, Fauque G, Czechowski M, Berlier X, Lespinat PA, Le Gall J, Xavier AV, Moura JJ (1986) Redox properties activity studies on nickel-containing hydrogenase isolated from a halophilic sulphate reducer Desulfovibrio salexigens. Biochimie 68:75–84
70. Trautwein AX (1985) New aspects of Mössbauer spectroscopy. Rev Port Quim 27:24–25
71. Vainshten BK (ed) (1979) Modern crystallography, vol 2. Nauka, Moscow (in Russian)
72. Schenborn BP (ed) (1975) Neutron scattering for the analysis of biological structure. Report of symposium held in June 2–6, 1975. Brookhaven National Laboratory, New York
73. Alpert Y (1980) Tentative use of NSE in biological systems. In: Mezei F (ed) Neutron spin-echo. Springer, Berlin Heidelberg New York, pp 87–93

74. Muilenberg GE (ed) (1979) Handbook of X-ray photoelectron spectroscopy. Perkin-Elmer, Eden Prarie
75. Davis LE, McDonald NC, Palmberg PW, Riach GE, Weber RE (1976) Physical Electronics. Perkin-Elmer, Eden Prarie
76. Griffith JS (1972) On the general theory of magnetic susceptibilities of polynuclear transition-metal complexes. Systems with two or three spins. In: Memmerich P (ed) Structure and bonding, vol 10. Springer, Berlin Heidelberg New York, pp 87–126
77. Dickson DP (1984) Application to biological systems. In: Mössbauer spectroscopy. Appl Inorg Chem, New York, pp 339–389
78. Goldansky VI (1963) Mössbauer effect and its applications in chemistry. Izv Akad Nauk SSSR, Moscow (in Russian)
79. Suzdalev IP (1979) Dynamic effects in gamma-resonance spectroscopy. Atomizdat, Moscow (in Russian)
80. Afanasiev AM, Kagan YuM (1968) Stabilization of the hyperfine structure of the Mössbauer line in paramagnetics in a low external magnetic field. Zh Eksp Teor Fiz Pis'ma 8:620–624
81. Czanderna AW (1975) Methods of surface analysis. Elsevier, Amsterdam
82. Hercules DM (1976) Electron spectroscopy: X-ray and electron excitation. Anal Chem 48:294R–314R
83. Andrews PT, Johnson ChE, Wallbauk B, Cammack R, Hall DO, Rao KK (1975) X-ray photoelectron spectra of iron-sulphur proteins. Biochem J 149:471–474
84. Kramer LN, Klein MP (1971) Application of extended Huckel theory to X-ray photoelectron spectra of transition metal complexes. Correlations between electron binding energy and calculated atomic charge in iron and sulphur compounds. Chem Phys Lett 8:183–186
85. Borodko YuG, Vetchinkin SI, Zimont SL, Ivleva IN, Shulga YuM (1976) Nature of satellites in X-ray photoelectron spectra (XPS) of paramagnetic cobalt (II) compounds. Chem Phys Lett 42:264–267
86. Hodgson KO (1978) The molybdenum site in nitrogenase: structural elucidation by X-ray absorption spectroscopy. In: Newton WE, Orme-Johnson WH (eds) Nitrogen fixation, vol 1. Free living systems and chemical models. University Park Press, Baltimore, pp 261–282
87. Teo BK, Lee PA, Simons AL, Eisenberger P, Kincaid BM (1977) EXAFS: Approximation, parametrization and chemical transferability of amplitude functions. J Am Chem Soc 99:3854–3856
88. Ashley CA, Doniach S (1975) Theory of extended X-ray absorption edge fine structure (EXAFS). Phys Rev B11:1279–1288
89. Jaffe HH, Orchin M (1962) Theory of ultraviolet spectroscopy. Wiley Interscience, New York
90. Drago RS (1977) Physical methods in chemistry. Saunders, Philadelphia London
91. Djerassi C (1960) Optical rotatory dispersion. McGraw Hill, New York Toronto
92. Volkenstein MV (1979) Electron-conformational interactions in biopolymers. Pure Appl Chem 51:801–829
93. Mineyev AP, Sharonov YuA, Sharonova NA, Figlovsky VA (1979) Orbit-orbit interaction in the NO-Fe(I) hemoproteins. Low temperature MCD. Int J Quant Chem 16:883–895
94. Stephens PJ (1974) Magnetic circular dichroism. Annu Rev Phys Chem 25:201–232
95. Barnes AJ, Orville-Thomas WJ (eds) (1977) Vibrational spectroscopy. Modern trends, Elsevier, Amsterdam
96. Al'tshuler SA, Kozyrev BM (1972) Electron paramagnetic resonance of compounds of intermediate-group elements. Nauka, Moscow (in Russian)
97. Beinert H (1985) Electron-paramagnetic resonance spectroscopy: past, present, future. Biochem Soc Trans 13:542–547
98. Palmer H (1985) The electron paramagnetic resonance of metalloproteins. Biochem Soc Trans 13:548–560
99. Likhtenshtein GI (1976) Spin-labeling in molecular biology. Wiley Interscience, New York Toronto
100. Abragam A, Bleaney B (1970) Electron paramagnetic resonance of transition ions. Clarendon, Oxford
101. Lebedev YaS, Muromtsev VV (1972) EPR and relaxation of stabilized radicals. Khimiya, Moscow (in Russian)

102. Salikhov KM, Semenov AG, Tsvetkov YuD (1976) Electron spin-echo and its application. Nauka, Moscow (in Russian)
103. Mims WB (1974) Measurement of the linear electric field effect in EPR using the spin echo method. Rev Sci Intrum 45:1583–1591
104. Munck E, Rhodes H, Orme-Johnson WH, Davis LC, Brill WJ, Shah VK (1975) Nitrogenase. VIII Mössbauer and EPR spectroscopy of the Mo-Fe protein component from Azotobacter vinelandii OP. Biochim Biophys Acta 400:32–53
105. Mims WB, Peisach J, Davis JL (1977) Nuclear modulation of the electron spin echo envelope in glassy materials. J Chem Phys 66:5536–5550
106. Peisach J, Mims WB (1978) The linear electric field effect in stellacyanin, azurin and in some simple model compounds. Eur J Biochem 84:207–214
107. Kulikov AV (1976) Determination of the distance between label spins and paramagnetic centre spins in spin-labeled proteins from the saturation curve parameters of the EPR spectrum of the label at 77 K. Mol Biol (Mosc) 10:132–141
108. Kulikov AV, Likhtenshtein GI (1977) The use of spin relaxation phenomena in the investigation of the structure of model and biological systems by the methods of spin label. Adv Mol Relaxation Interact Processes 10:47–50
109. Orbach R (1961) Spin-lattice relaxation in rare-earth salts. Proc R Soc (Lond) A 264:458–484
110. Burbaev DSh, Levanidze AV (1979) Evaluation of the exchange integral of antiferromagnetic interaction between Fe atoms at the active sites of ferredoxins. Biofizika 24:392–394 (in Russian)
111. Benderscky VA, Blumenfeld LA, Stunzhas PA, Sokolov EA (1968) Double electron-electron resonance of triplet excitons in ion-radical salts. Nature 220:365–367
112. Hyde JS, Chien CW, Freed JH (1968) Electron-electron double resonance of free radicals in solution. J Chem Phys 48:4211–4226
113. Hyde JS (1975) Saturation-transfer spectroscopy. In: Hirs CHW, Timasheff SN (eds) Methods in enzymology, vol 49G, No 19. Academic Press, New York, pp 480–511
114. Grinberg OYa, Dubinsky AA, Shuvalov VF, Oransky LG, Kurochkin VI, Lebedev YaS (1976) Submillimeter EPR spectroscopy of free radicals. Dokl Akad Nauk SSSR 230:884–887 (in Russian)
115. Mims WB, Peisach J (1981) Electron spin echo spectroscopy and the study of metalloproteins. In: Berliner L, Reuben J (eds) Biological magnetic resonance, vol VIII. Plenum Press, New York, pp 213–263
116. Cohm M, Reuben J (1971) Paramagnetic probes in magnetic resonance studies of phosphoryl transfer enzymes. Account Chem Res 4:214–222
117. Jardetzky O, Roberts GC (1981) NMR in molecular biology. Academic Press, New York
118. Bertini I, Luchinat C (1985) NMR of paramagnetic molecules in biological systems. Benjamin, Cumming
119. Lamer CN, Holm R, Horrocks WD (eds) (1974) NMR in paramagnetic molecules. Academic Press, New York
120. Dwek RA (1977) NMR in biology. Academic Press, New York
121. Nowak T, Mildwan AS (1972) Nuclear magnetic resonance studies of selectively hindered internal motion of substrate analogs at the active site of pyruvate kinase. Biochemistry 11:2813–2818
122. Wien RW, Morrisett JD, McConnell HM (1972) Spin-label-induced nuclear relaxation distance between bound saccharides, histidine-15 and tryptophan-123 on lysozyme in solution. Biochemistry 11:3707–3716
123. Syrtsova LA, Pisarskaya TN, Nazarova II, Uzenskaya AM, Likhtenshtein GI (1977) Coupling of ATPase and substrate-binding sites of the nitrogenase of Azotobacter vinelandii. Bioorg Khim 3:1251–1260 (in Russian)
124. Solomon J, Blombergen N (1956) Nuclear magnetic interactions in the HF molecule. J Chem Phys 25:261–266
125. Likhtenshtein GI, Kulikov AV, Kotelnikov AI, Levchenko LA (1986) Methods of physical labels – a combined approach to the study of microstructure and dynamics in biological systems. J Biochem Biophys Methods 12:1–28
126. Kuznetsov AN (1974) Spin probe technique. Nauka, Moscow (in Russian)

127. Watts A (1985) Spin-labels in biochemistry. Biochem Soc Trans 13:588–592
128. McConnell HM, McFarland BG (1970) Physics and chemistry of spin labels. Quart Rev Biophys 3:91–136
129. Berliner L (ed) (1976) Spin labeling. Theory and application, vol 1. Academic Press, New York
130. Berliner L (ed) Spin labeling. Theory and application, vol 2. Academic Press, New York
131. Weber C, Teal FG (1959) Electronic energy transfer in haem proteins. Discuss Faraday Soc 27:134–141
132. Stryer L (1968) Fluorescence spectroscopy of proteins. Science 162:526–533
133. Likhtenshtein GI, Pivovarov AP, Smolina N (1968) Study of structural transitions in serum albumins by the luminescent active label technique. Mol Biol (Mosc) 2:291–302 (in Russian)
134. Alfimova EYa, Likhtenshtein GI (1979) Investigation of structure and conformational changes of proteins by inductive-resonance energy transfer method. Adv Mol Relaxation Interact Processes 14:47–89
135. Levchenko LA, Raevsky AV, Sadkov AP, Likhtenshtein GI (1973) On the localization of an iron-containing active site on the nitrogenase macromolecule. Dokl Akad Nauk SSSR 211:238–240 (in Russian)
136. Likhtenshtein GI, Levchenko LA, Sadkov AP, Raevsky AV, Pivovarova TS, Gvozdev RI (1973) Study of the active site structure of nitrogenase by electron microscopy. Dokl Akad Nauk SSSR 213:1442–1443 (in Russian)
137. Strothkamp KG, Lehman J, Lippard SJ (1978) Tetrakis (acetoxymercuri) methane: A polymetallic reagent for labeling sulfur in nucleic acids. Proc Natl Acad Sci USA 75:1181–1184
138. Marakushev SA, Levchenko LA, Likhtenshtein GI, Kaplun AP, Mekler VM, Raevskii AV, Shvets VI (1986) A new mercarbide electron dense label for the hydrocarbon moiety of lipid bilayer. Biol Membrany 3:423–427 (in Russian)
139. Levchenko LA, Likhtenshtein GI, Raevsky AV, Salitra IS (1979) Application of mercarbide labels is a new approach in electron microscopy of proteins and membranes. Usp Biokhimii 20:257–274 (in Russian)
140. Feigin LA, Gonchar NA, Likhtenshtein GI, L'vov YuM, Marakushev SA (1978) Determination of the distance between heavy atom labels in the histidinedecarboxylase molecule by means of small-angle X-ray scattering. Sov Phys Crystallogr 23:420–424
141. Vainstein BK, L'vov YuM, Feigen LA, Gvozdev RI, Marakushev SA, Likhtenshtein GI (1980) Determination of the distance between heavy atom labels in hemoglobin by means of small-angle X-ray scattering. FEBS Lett 16:107–110
142. Al'terman MA, Marakushev SA, Levchenko LA, Raevsky AV, Devichinsky VM, Bachmanova GI, Likhtenshtein GI, Archakov AI (1981) Study of the distribution of cytochrome P-450 in rat microsomes. Biokhimiya 46:222–229 (in Russian)
143. Frolov EN, Mokrushin AD, Likhtenshtein GI, Trukhtanov VA, Goldansky VI (1973) Investigation of the dynamic structure of proteins using gamma-resonance labels. Dokl Akad Nauk SSSR 212:165–168 (in Russian)
144. Frolov EN, Belonogova OV, Likhtenshtein GI (1977) Investigation of the mobility of spin and Mössbauer labels bound to macromolecules. In: Burshtein EA (ed) Equilibrium dynamics of the native structure of protein. NTSBI Akad Nauk SSSR, Pushchino, pp 99–142 (in Russian)
145. Likhtenshtein GI (1976) The water-protein interaction and dynamic structure of proteins. In: Alfsen A, Bertreaud AJ (eds) L'eau les st systems biologique. Editions CNRS, Paris, pp 45–52
146. Belonogova OV, Frolov EN, Krasnopol'skaya SA, Atanasov BP, Gins VK, Mukhin EN, Levina AA, Andreeva AP, Likhtenshtein GI, Goldansky VI (1978) Study of the effect of the degree of hydration on the mobility of Mössbauer atoms at the active sites of metalloenzymes and metal carriers. Dokl Akad Nauk SSSR 241:219–222 (in Russian)
147. Stone TJ, Buchman T, Nordio PL, McConnell HM (1965) Spin-labeled biomolecules. Proc Natl Acad Sci USA 54:1010–1017
148. Rosantsev EG (1970) Free nitroxyl radicals (translated by Blazzard BJ). Plenum Press, New York

149. Keana JFW (1979) New aspects of nitroxide chemistry. In: Berliner L (ed) Spin-labeling. Theory and application, vol 2. Academic Press, New York, pp 115–172
150. Volodarskii LB, Kurikova GA, Kobrin VS, Sagdeev RZ, Molin YuN (1971) Synthesis and spectral studies of 2,2,5,5-tetramethyl-3-imidazoline-1-oxil-4 carboxylic acid and its amide, salt and Cu(II) complex. Izv Sib Otd Akad Nauk SSSR, Ser Khim Nauk, pp 101–103
151. Ullman EF, Osiecki YN, Boocock DGB, Darcy R (1972) Studies of stable free radicals. J Am Chem Soc 94:7049–7059
152. Gaffney BJ (1976) Chemistry of spin labeling. In: Berliner L (ed) Spin-labeling. Theory and application, vol 1. Academic Press, New York, pp 207–263
153. Zhdanov RI (1981) Paramagnetic models of biologically active compounds. Nauka, Moscow (in Russian)
154. Neiman MB, Rosantsev EG, Mamedova YuG (1962) Free radical reactions involving no unpaired electron. Nature 196:472–474
155. McConnell HM (1956) Effect of anisotropic hyperfine interactions on paramagnetic relaxation in liquids. J Chem Phys 25:709–711
156. Kivelson D (1960) Theory of ESP linewidths of free radicals. J Chem Phys 33:1094–1106
157. Freed JH (1976) Theory of ESR spectra of nitroxide radicals. In: Berliner L (ed) Spin-labeling. Theory and application, vol 1. Academic Press, New York, pp 64–155
158. Kuznetsov AN, Wasserman AM, Volkov AYu, Korst NN (1971) Determination of rotational correlation time of nitric acid radicals in viscous medium. Chem Phys Lett 12:103–106
159. Shimshick EJ, McConnell HM (1972) Rotational correlation time of spine-labeled α-chymotrypsin. Biochem Biophys Res Commun 46:321–327
160. Hyde JS, Dalton LR (1976) In: Berliner L (ed) Spin-labeling. Theory and application, vol 1. Academic Press, New York, pp 2–71
161. Livshitz VA, Kuznetsov AN (1980) Study of hyperfine superslow rotations of spin-labeled proteins using the effects of microwave saturation effects in EPR spectra. Mol Biol (Moscow) 14:182–189
162. Grinberg OYa, Dubinskii AA, Lebedev YaS (1983) ESP of free radicals in the 2-mm range. Usp Khim 52:1490–1513 (in Russian)
163. Anziferova LI, Lubashevskaya EB (1986) Atlas of EPR spectra of nitroxide radicals in the 2-mm range. Nauka, Chernogolovka (in Russian)
164. Griffith ON, Jost PC (1976) In: Berliner L (ed) Spin Labeling. Theory and application, vol 1. Academic Press, New York, pp 488–569
165. Grinberg OYa, Dubinsky AA, Levedev YaS (1972) Kinetics of the annealing of radical pairs in solids. Kinetika Kataliz 13:660–665 (in Russian)
166. Likhtenshtein GI, Bogatyrenko VR, Kulikov AV, Khideg K, Khankovskaya GO, Lukoyanov NV, Kotelnikov AI, Tanaseichuk BS (1980) Study of superslow motions in solid solutions using physical probes. Dokl Akad Nauk SSSR 253:481–484 (in Russian)
167. Safronov SI, Mstislavsky VI, Safronova YuI, Muromtsev VI (1969) A method of determination of relaxation times from signal saturation curves under rapid passage conditions. Zavodskaya Laboratoriya 35:1463–1466 (in Russian)
168. Likhtenshtein GI (1968) Determination of the topography of protein groups using specific paramagnetic labels. Mol Biol (Mosc) 2:234–240 (in Russian)
169. Likhtenshtein GI, Bobodzhanov PKh (1968) Investigation of the structure and local conformational changes of proteins and enzymes using double paramagnetic labels. Biofizika 13:757–764 (in Russian)
170. Parmon VN, Kokorin AI, Zhidomirov GM (1980) Stable biradicals. Nauka, Moscow (in Russian)
171. Kulikov AV, Cherepanova ES, Bogatyrenko VR (1981) Determination of the closest-approach distance between a radical and a paramagnetic ion. Teor Eksp Khim 17:788–797 (in Russian)
172. Taylor JC, Leigh JS, Cohn M (1969) Magnetic resonance studies of spin-labeled creatine kinase system and interaction of two paramagnetic probes. Proc Natl Acad Sci USA 64:219–226
173. Likhtenshtein GI, Grebenshchikov YuB, Bobodzhanov PKh, Kokhanov YuV (1970) Investigation of the microstructure of proteins by the paramagnetic label-paramagnetic probe technique. Mol Biol (Mosc) 4:682–691 (in Russian)

174. Zamaraev KI, Molin YuN, Salikhov KM (1981) Spin Exchange. Theory and physicochemical application. Springer, Berlin Heidelberg New York
175. Kulikov AV, Yudanova EI, Likhtenshtein GI (1983) Investigation of the spin exchange of nitroxide radicals using the continuous EPR spectrum saturation technique. Zhur Fiz Khim 56:2982–2987 (in Russian)
176. Grebenshchikov YuB, Likhtenshtein GI, Ivanov VP, Rozantsev EG (1972) Investigation of electrostatic probes in a protein using the paramagnetic probe technique. Mol Biol (Mosc) 6:498–506 (in Russian)
177. Grebenshchikov YuB, Ponomarev GV, Evstigneeva RP, Likhtenshtein GI (1972) Spin relaxation of 2,2,6,6-tetramethyl-4-oxopiperidine-1-oxil radical upon interaction in solutions with porphyrin complexes of VO(II), Fe(III), Ni(II), Cu(II) and hemoglobin. Biofizika 17:910–913 (in Russian)
178. Hyde JS, Swartz HM, Antholine WE (1976) The spin probe-spin label methods. In: Berliner L (ed) Spin labeling. Theory and application, vol 2. Academic Press, New York, pp 72–113
179. Burshtein EA (1976) Luminescence of protein chromophores. In: Vladimirov YuA (ed) Advances in science and technology. Biophysics, vol 6. VINITI, Moscow (in Russian)
180. Vladimirov YuA, Dobretsov GE (1980) Fluorescent probes for the investigation of biological membranes. Nauka, Moscow (in Russian)
181. Stryer L (1978) Fluorescence energy transfer as a spectroscopic ruler. Annu Rev Biochem 47:819–846
182. Kotelnikov AI, Kuznetsov SN, Fogel VR, Likhtenshtein GI (1979) Investigation of the microstructure of biological systems by the triplet label technique. Mol Biol (Mosc) 13:152–159 (in Russian)
183. Barltrop JA, Coyle JD (1975) Excited states in organic chemistry. Wiley, London New York
184. Mazurenko YuT, Bakhshiev NG (1970) The effect of orientational dipolar relaxation on the spectral, time and polarization characteristics of the luminescence of solutions. Optika Spektrosk 28:905–913 (in Russian)
185. Vavilov SI (1945) On the photoluminescence of solutions. Izv Akad Nauk SSSR Ser Fiz 9:283–304 (in Russian)
186. Dexter DL (1953) A theory of sensitized luminescence in solids. J Chem Phys 21:836–850
187. Forster Th (1948) Zwischenmolekulare Energiewanderung und Fluoreszenz. Ann Phys 2:55–75
188. Galanin MD (1951) Quenching of the fluorescence of solutions by absorbing substances. Zhur Eksp Teor Fiz 21:126–132 (in Russian)
189. Dale RE, Eisinger J (1976) Intramolecular energy transfer and molecular conformation. Proc Natl Acad Sci USA 73:271–278
190. Ermolaev VL (1963) Energy transfer in organic systems with participation of the triplet state. Solid solutions and crystals. Usp Fiz Nauk 80:3–41 (in Russian)
191. Ermolaev VL, Bodunov EN, Sveshnikova EB, Shakhverdov TA (1977) Radiationless transfer of electronic excitation energy. Nauka, Leningrad (in Russian)
192. Takikolov AS, Kuzmin VA (1975) Quenching of the triplet states of aromatic hydrocarbons by stable nitric acid radicals. Dokl Akad Nauk SSSR 223:403–406 (in Russian)
193. Yang NC, Loeschen R, Mitchell D (1967) On the mechanism of the Paterno-Buchi reaction. J Am Chem Soc 89:5465–5466
194. Mekler VM, Kotelnikov AI, Likhtenshtein GI, Berkovich MA (1982) Application of luminescent probes to the investigation of model and biological membranes. Biofizika 27:641–645 (in Russian)
195. Shin IV, Fogel VR, Kotelnikov AI, Likhtenshtein GI (1982) Study of the localization of chromophores in a codein globule using the method of phosphorescence quenching by nitroxide radicals. Biofizika 27:5–9 (in Russian)
196. Kotelnikov AI, Fogel VR, Likhtenshtein GI, Postnikova GB, Shlapnikova EA (1981) The use of the exchange deactivation of triplet excited states for the investigation of the structure and electronic conductivity of proteins. Mol Biol (Mosc) 15:281–289 (in Russian)
197. Mekler VM, Kotelnikov AI, Likhtenshtein GI (1983) Application of probes emitting annihilation delayed fluorescence for investigation of model and biological membranes. Biofizika 28:503–504 (in Russian)

198. Mekler VM, Likhtenshtein GI (1986) Study of dynamic contacts of macromolecules according to kinetics of isomerization of fluorescent labels sensitized by the dye in a triplet state. Biofizika 31:568–571 (in Russian)
199. Ulanov BP, Malysheva LP, Moshkovsky YuSh (1967) On the determination of the base sequence in nucleic acids. Biofizika 12:326–330 (in Russian)
200. Beer M (1964) Problems and progress in submolecules. Microsc Soc 83:153–160
201. Grdenic D, Kamenar B, Korpar-Colig B, Sikirca M, Jovanovski G (1974) Tetrakis (trifluoroacetoxymercuri) methane and Tetrakis (acetoxymercuri) methane as the reaction products of Hofmann's base with the corresponding acid: X-ray crystallographic evidence. J Chem Soc Chem Commun 16:646–647
202. Matesson DS, Castle RB, Larson GL (1970) Tetrakis (acetoxymercuri) methane. J Am Chem Soc 92:231–232
203. Kalamkarov GR, Raevsky AV, Salitra IS, Pivovarova TS, Ostrovsky MA, Levchenko LA, Likhtenshtein GI (1977) On the localization of the visual pigment rhodopsin in the photoreceptor membrane. Bioorg Khim 3:775–779 (in Russian)
204. Kratky O, Pilz J (1972) Recent advances and applications of diffuse X-ray small-angle scattering on biopolymers in dilute solutions. Quart Rev Biophys 5:481–537
205. Fedorov BA (1976) Physical methods in molecular biology. In: Volkenshtein MV (ed) Advances in science and technology. Molecular Biology, vol 8, part I, pp 6–69 (in Russian)
206. Vainshtein BK, Safonov NI, Feigin LA (1970) The X-ray method of determination of the distances between heavy atoms in macromolecules in solution and its application to the study of derivatives of gramicidin C. Dokl Akad Nauk SSSR 190:574–579 (in Russian)
207. Keller H, Kundig W (1975) Mössbauer studies of Brownian motion. Solid State Commun 16:253–256
208. Parak F, Frolov EN, Mössbauer RL, Goldanskii VI (1981) Dynamics of metmyoglobin crystals investigated by nuclear-gamma resonance absorption. J Mol Biol 145:825–833
209. Berg AI, Noks PP, Kononenko AA, Frolov EN, Khrymova IN, Rubin AB, Likhtenshtein GI, Goldansky VI, Parak F, Bukl M, Mössbauer RL (1979) Conformational control of functional activity in photosynthetic membranes of purple bacteria. Mol Biol (Mosc) 13:81–89 (in Russian)
210. Parak F, Frolov EN, Kononenko AA, Mössbauer RL, Goldanskii VI, Rubin AV (1980) Evidence for a correlation between the photoinduced electron transfer and dynamic properties of the chromatophore membranes from Rhodospirillum rubrum. FEBS Lett 117:368–372
211. Keller H, Debruner PG (1980) Evidence for Conformational and diffusional mean square displacements in frozen aqueous solution of oxymyoglobin. Phys Rev Lett 45:68–71
212. Parak F, Knapp EW, Kucheida D (1982) Protein dynamics. Mössbauer spectroscopy on deoxymyoglobin crystals. J Mol Biol 161:177–194
213. Likhtenshtein GI (1979) Protein dynamics studied by spin-labeling, Mössbauer spectroscopy and NMR. In: Losche A (ed) Special colloque ampere on dynamical processes in molecular systems studied by rf spectroscopy. Karl Marx Universität, Leipzig, pp 100–107
214. Hammet LP (1970) Physical Organic Chemistry. McGraw-Hill, New York
215. Franco R, Gavalda MT, Canela EI (1986) A computer program for enzyme kinetics that combined modern discrimination, parameter refinement and sequential experimental design. Biochem J 238:855–862
216. Jencks WP (1969) Catalysis in chemistry and enzymology. McGraw-Hill, New York
217. Fersht AR (1977) Enzyme structure and mechanism. Freeman, San Francisco
218. Jordan PS (1979) Chemical kinetics and transport. Plenum Press, New York
219. Dixon M, Webb EC (1979) Enzymes. 3rd edn. Longman, London
220. Hilinski EF, Rentzepis PM (1983) Biological applications of picosecond spectroscopy. Nature 302:481–487
221. Klesov AA, Berezin IV (1980) Enzyme catalysis. Moscow University Press, Moscow (in Russian)
222. Purich DL (ed) (1980) Enzyme kinetics and mechanism. In: Methods in enzymology, vol 64, part B. Academic Press, New York, 418 pp
223. Volkenstein MV (1977) Molecular biophysics. Academic Press, New York

224. Koshland D, Nemethy G, Filmer D (1966) Comparison of experimental binding data and theoretical models in protein containing subunits. Biochemistry 5:365–375
225. Monod J, Wyman J, Changeux JP (1965) On the nature of allosteric transitions: a plausible model. J Mol Biol 12:88–118
226. Martin JL, Breton J, Hoff AJ, Migus A, Antonetti A (1986) Femtosecond spectroscopy of electron transfer in the reaction center of the photosynthetic bacterium Rhodopseudomonas spheroides R-26. Proc Natl Acad Sci USA 83:957–961
227. Likhtenshtein GI, Sukhorukov BI (1964) On the kinetics of cooperative processes. Zh Fiz Khim 38:747–751 (in Russian)
228. Sukhorukov BI, Likhtenshtein GI (1965) On the kinetics and mechanism of denaturation of biopolymers. Biofizika 10:935–943 (in Russian)
229. Likhtenshtein GI (1969) Water contribution to protein stability. In: Kajushin LP (ed) Water in biological systems. Consultants Bureau, New York
230. Likhtenshtein GI (1970) On the nature of conformational compensation effect in liquid-phase reactions. II. Kinetic models of solvation processes. Zh Fiz Khim 44:1913–1918 (in Russian)
231. Likhtenshtein GI, Troshkina TV (1968) Kinetic features of cooperative processes in proteins. Mol Biol (Mosc) 2:654–663 (in Russian)
232. Kurganov BI (1978) Allosteric enzymes. Nauka, Moscow (in Russian)
233. Caldin EF (1964) Fast reactions in solutions. Blackwell, Oxford
234. Hammes GG (ed) (1974) Investigation of rates and mechanisms of reactions. Wiley, New York
235. Porter JM, Mukle B, de Maria AJ, Chenn NU (1971) Continuous dye laser (Polarized light from white dwarf). Astra monitors pollution (200-MeV linac/Argonne neutrinos). Phys Today 24:19–26
236. Klevanik AV, Kryukov PG, Matveets YuA, Semchishin VA, Shuvalov VA (1980) Measurement with subpicosecond time resolution of the rates of energy and electron transfer at the physical steps of photosynthesis. Zh Eksp Teor Fiz Pis'ma 32:107–111 (in Russian)
237. Rochley MG, Windsor NW, Cogdell R, Parson W (1975) Picosecond detection of an intermediate in photochemical reactions of bacterial photosynthesis. Proc Natl Acad Sci USA 72:2251–2254
238. Ataullakhanov FI (1972) Application of the method of periodic influence on the kinetics of enzymatic reactions. Biofizika 17:647–654 (in Russian)
239. Trestman GA, Likhtenshtein GI (1979) Study of the photosensitive systems using the pulse "kinetic resonance" method. I. Theory of the method. Biofizika 27:779 (in Russian)
240. Trestman GA, Likhtenshtein GI, Frolov EN, Kochetkov VV (1979) Study of the photosensitive systems using the pulse "kinetic resonance" technique. 2. Apparatus. Applications. Biofizika 27:779 (in Russian)
241. Ingold CK (1969) Structure and mechanism in organic chemistry, 2nd edn. Cornell University Press, Ithaca London
242. Collins CJ, Bowman NS (eds) (1970) Isotope effect in chemical reactions. Van Nostrand, Reinhold, New York
243. Bell RP (1980) The tunnel effect in chemistry. Chapman and Hall, London
244. Melander L, Saunders WH (1980) Reaction rates of isotopic molecules. Wiley Interscience, New York
245. Brunschwig BS, Ehrenson S, Sutin N (1986) Solvent reorganization in optical and thermal electron transfer processes. J Phys Chem 90:3657–3668
246. Salikhov K, Molin Yu, Sagdeev R, Buchachenko A (1982) Spin polarization and magnetic effects in chemical reactions. Elsevier, Amsterdam
247. Nagakura S, Hayasy H (1986) Magnetic field and chemical reactions. Saiensu 1:80–90
248. Buchachenko AL, Galimov EM, Ershov VV (1976) Enrichment of isotopes induced by magnetic interactions in chemical reactions. Dokl Akad Nauk SSSR 228:379–382 (in Russian)
249. Bargon J, Fisher H, Johnoson U (1967) Kernresonanzemissionslinien während rascher Radikalreaktionen. Z Naturforsch 22a, 5:1551–1555
250. Werner HJ, Schulten K, Weller A (1978) Electron transfer and spin exchange contributing to the magnetic field dependence of the primary photochemical reaction of bacterial photosynthesis. Biochim Biophys Acta 502:255–268

251. Elfimov EI, Voznjak VM, Kazantsev AP (1980) The effect of the magnetic field on the fluorescence yield of the cells of Rhodopseudomonas sphaeroides, wild strain and R-26. Temperature dependence. Biofizika 25:498–502 (in Russian)
252. Buchachenko AL, Berdinsky VN (1981) Radio emission in chemical reactions. Vestnik Akad Nauk SSSR 1:91–98 (in Russian)
253. Buchachenko AL (1974) Chemical nuclear and electron polarization. Nauka, Moscow (in Russian)
254. Likhtenshtein GI, Shilov AE (1970) On the thermodynamic fixation of molecular nitrogen. Zh Fiz Khim 44:849–856 (in Russian)
255. Semonov NN (1958) Some problems in chemical kinetics and reactivity (translated by M Boundart). Princeton University Press, Princeton
256. Kondratiev VN (ed) (1962) The dissociation energy of chemical bonds. Ionization potentials and electron affinity. Izd Akad Nauk SSSR, Moscow (in Russian)
257. Pauling L (1962) Die Natur der chemischen Bindung. Verlag Chemie, Weinheim
258. Eyring H (1935) The activated complex in chemical reactions. J Chem Phys 3:107–115
259. Karplus M, McCammon JA (1986) Dynamics of protein structure. Sci Am 251:4–12
260. Marcus RA, Sutin N (1985) Electron transfer in chemistry and biology. Biochim Biophys Acta 811:265–322
261. Levich VG, Dogonadze RR, German E, Kuznetsov AN, Kharakats YuI (1970) Theory of homogeneous reactions involving proton transfer. Electrochem Acta 15:353–368
262. Bjorn GS, Bjorn LO (1984) Thermal energy dissipation in proteins. Photochem Photobiol 39:875–880
263. Blumenfeld LA, Chernavskii DS (1973) Tunneling of electrons in biological processes. J Teor Biol 39:1–10
264. Chernavskaya NM, Chernavsky SN (1977) Tunneling transport of electrons in photosynthesis. Moscow University Press, Moscow (in Russian)
265. Hopfield JJ (1974) Electron transfer between biological molecules by thermally activated tunneling. Proc Natl Acad Sci USA 71:3640–3644
266. Jortner J (1976) Temperature dependent activation energy for electron transfer between biological molecules. J Chem Phys 64:4860–4867
267. Pearson RG (1977) Symmetry rules for chemical reactions. Orbital topology and elementary processes. Wiley, New York
268. Woodward RB, Hoffmann R (1970) The conservation of orbital symmetry. Verlag Chemie, Weinheim
269. Bersuker IB, Stavrov S (1981) Vibronic effects in geometry and stereochemistry of metalloporphyrins and hemoproteins. J Chem Phys 54:331–340
270. Samson R, Deutch JM (1978) Diffusion-controlled reaction rate to a buried active site. J Chem Phys 68:285–290
271. Burshtein AI (1980) Long-range theory of transport transfer reactions in condensed media. Khim Vysok Energ 14:291–306 (in Russian)
272. Wiegel FW, Heringa JR (1985) Diffusion coefficient of a protein in a fluid membrane: numerical calculations. Can J Phys 63:44–45
273. Longworth LG (1952) Diffusion measurements at 1 °C of aqueous solutions of amino acids, peptides and sugars. J Am Chem Soc 74:4155–4159
274. Razi-Nagvi K (1974) Diffusion-controlled reactions in two-dimensional fluids: discussion of measurement of lateral diffusion of lipids in biological membranes. Chem Phys Lett 28:280–285
275. Roszlow HS, Jaeger JC (1959) Conduction of heat in solids. Clarendon, Oxford
276. Burshtein AI (1978) Molecular-kinetic aspects of chemical physics of the condensed state. Usp Khim 47:212–234 (in Russian)
277. Temkin SJ, Jakobson BJ (1984) Diffusion controlled reaction of chemically anisotropic molecules. J Chem Phys 88:2679–2685
278. Wherland S, Gray HB (1976) Metalloprotein electron transfer reactions: Analysis of reactivity of horse heart cytochrome c with inorganic complexes. Proc Natl Acad Sci USA 73:2950–2954
279. Postnikova GB, Shlyapnikova EA, Volkenshtein MV, Atanasov BP (1981) Study of electron transfer in hemoproteins. IV. Dependence of the rate of reduction of ferricytochrome

c by oxyhemoglobin on the ionic strength of solution. Mol Biol (Mosc) 15:526–537 (in Russian)
280. Staninets VI, Shilov EA (1971) Addition reactions involving the intramolecular formation of the ring. Usp Khim 40:491–512 (in Russian)
281. Storm DR, Koshland DE (1970) A source for the special catalytic power of enzymes: Orbital steering. Proc Natl Acad Sci USA 66:445–452
282. Likhtenshtein GI (1977) Kinetic analysis of orientation effects in model chemical systems. Zh Fiz Khim 41:842–846 (in Russian)
283. Benson SW (1968) Thermochemical kinetics. Methods for the estimation of thermochemical data and rate parameters. Wiley, New York
284. Kondratiev VN, Nikitin EE (1974) Kinetics and mechanism of gas-phase reactions. Nauka, Moscow (in Russian)
285. Landau L (1932) Zur Theorie der Energieübertragung. Phys Z, Sowjetunion 2:46–51
286. Zener C (1933) Dissociation of excited molecules by external perturbations. Proc R Soc A140:660–668
287. Likhtenshtein GI (1970) On the nature of the compensation effect in liquid-phase reactions. I. Solvation phenomena in liquids. Zh Fiz Khim 44:1908–1912 (in Russian)
288. Rudakov ES (1967) The effect of the solvent on the rate of chemical reactions in solution. Inst Kataliza Sibirsk Otd Akad Nauk SSSR, Novosibirsk (in Russian)
289. Gamow GA (1926) Zur Quantentheorie des Atoms. Z Phys 51:204–212
290. Alexandrov IV, Khairutdinov RF, Zamaraev KI (1977) Electron tunneling in solid-phase redox reactions. Chem Phys 32:123–141
291. Grigorov LN, Chernavsky DS (1972) Quantum-mechanical model of electron transfer from cytochrome to chlorophyll in photosynthesis. Biofizika 17:195–202 (in Russian)
292. Banks E, Jortner I (1980) Activationless electron transfer processes in biological systems. FEBS Lett 109:117–120
293. Dogonadze RR, Kuznetsov AM, Ulstrup J (1977) Conformational dynamics in biological electron and atom transfer reactions. J Theor Biol 64:239–263
294. De Vault D (1984) Quantum mechanical tunneling in biological systems. Cambridge University Press, Cambridge
295. Bendersky VA, Ovchinnikow AA (1980) In: Kolotyrkin YaM (ed) Physical chemistry. Modern Present-day problems. Khimiya, Moscow (in Russian)
296. Benderskii VA, Goldanskii VI, Ovchinnikov AA (1980) Effect of molecular motion at low-temperature other anomalously fast reactions in the solid phase. Chem Phys 73:492–496
297. Ivanov GK, Kozhushner MA (1978) Theory of interimpurity tunneling in solids. Fiz Tverdogo Tela 20:9–16 (in Russian)
298. Likhtenshtein GI, Avilova TV (1973) Kinetic features of biological catalysis and the dynamic structure of enzymes. Usp Sovrem Biol 75:26–45 (in Russian)
299. Likhtenshtein GI (1977) On the principles of "optimal motion" in elementary acts of chemical and biological processes. I. Estimation of the synchronization factor for model processes. Kinetika Kataliz 28:878–882 (in Russian)
300. Likhtenshtein GI, Mullokandov EA (1977) On the principles of "optimal motion" in elementary acts of chemical and biological processes. II. Mechanism of the recution of molecular nitrogen and liberation of molecular oxygen from water. Kinetika Kataliz 28:883–888 (in Russian)
301. Likhtenshtein GI (1977) On the principles of "optimal motion" in elementary acts of chemical and biological processes. III. Analysis of probabilistic mechanisms of some chemical and biochemical processes. Kinetika Kataliz 28:1255–1260 (in Russian)
302. Bordwell FG (1970) Are nucleophilic bimolecular concerted reactions involving four or more bonds a myth? Accounts Chem Res 3:281–290
303. Aleksandrov IV (1976) Dependence of the rate of concerted chemical reactions on the number of degrees of freedom involved in the transition. Teor Eksp Khim 12:299–306 (in Russian)
304. Bersuker IB (1978) Mutual vibronic influence of weakly coordinated molecular systems in chemical reactions and catalysis. Chem Phys 31:85–93
305. Vol'pin ME, Moiseev II, Shilov AE (1980) Metal-complex catalysis in chemistry and chemical industry. Zh Vses Khim Ova Im DI Mendeleeva 25:515–523 (in Russian)

306. Henrici-Olive G, Olive S (1977) Coordination and catalysis. Verlag Chemie, Weinheim
307. Shestakov AF (1982) Structural and catalytic features of cluster compounds of transition metals. In: Likhtenshtein GI (ed) Redox enzymes and their models. Part II. Nauka, Moscow, pp 25–49 (in Russian)
308. Likhtenshtein GI (1978) Structure and mechanism of action of nitrogenase active center. In: Newton WE, Orme-Johnson WE (eds) Nitrogen fixation, vol 1. Free living systems and chemical models. Univ Park Press, Baltimore, pp 195–210
309. Hammes GG (1968) Relaxation spectrometry of biological systems. Accounts Chem Res 1:321–353
310. Douzou P (1977) Cryobiochemistry. An introduction. Academic Press, London New York
311. Norton JR (1977) Metal clusters in catalysis. In: Tsutsui M, Ugo R (eds) Fundamental research in homogeneous catalysis. Plenum Press, New York, pp 99–140
312. Karpeisky MYA, Yakovlev GI, Antonov VK (1980) Does an enzyme reduce the activation energy of the elementary act of a chemical reaction? Bioorg Khim 6:645–654 (in Russian)
313. Likhtenshtein GI (1972) The dynamic structure of proteins and enzyme catalysis. Stud Biophys 33:185–193
314. Alexandrov IV, Goldanski VI (1984) Dependence of the rate constant of an elementary chemical process on the viscosity of the medium. Khim Fiz 3:185–198 (in Russian)
315. Yang CX, Johnson KH, Holm RH, Norman JC (1975) Theoretical model for the 4Fe active sites in oxidized ferredoxin and reduced "high-potential" proteins. J Am Chem Soc 97:6596–6598
316. Shaitan KV, Rubin AB, Chernavskii DS (1985) Kinetics of 40 conformational transitions in proteins. Biofizika 30:31–35 (in Russian)
317. Dunford HB, Dolphin D, Raymond KN, Sieker L (eds) (1982) The biological chemistry of iron. Reidel, Dordrecht
318. Ingram LL, Meyer DL (1985) Biochemistry of dioxygen. Plenum Press, New York
319. Eaton WA, Hanson LK, Stephens PJ, Sutherland JC, Dunn JBR (1978) Optical spectra of oxy- and deoxyhemoglobin. Extended Huckel calculation. J Am Chem Soc 100:4991–5003
320. Lehninger AL (1972) Biochemistry. The molecular basis of cell structure and function. Worth, New York
321. Granic S (1942) Ferritin 1. Physical and chemical properties of horse spleen ferritin. J Biol Chem 146:451–461
322. Klotz JM, Klotz TA (1955) Oxygen-carrying proteins. Science 121:477–478
323. Mortenson LE, Valentine RC, Carnahan JE (1962) An electron transport factor from Clostridium pasteurianum. Biochem Biophys Res Communs 7:448–452
324. Tagava K, Arnon DJ (1962) Ferredoxins as electron carriers in photosynthesis and in biological production and consumption of hydrogen gas. Nature 195:537–542
325. Bunn HF, Forget BG (1986) Hemoglobin. Molecular, genetic, and clinical aspects. Saunders, Philadelphia
326. Antonini E, Brunori M (1971) Hemoglobin and myoglobin in their reactions with ligands. North-Holland, Amsterdam
327. Perutz MF (1979) Regulation of oxygen affinity of hemoglobin: Influence of structure of the globine on the heme iron. Annu Rev Biochem 48:327–387
328. Baldwin J (1980) Structure and cooperativity of hemoglobin. Trends Biochem Sci 5:224–229
329. Kendrew JC, Dickerson RE, Strandberg BE, Hart RG, Davies DR, Phillips DC, Shore VC (1960) Structure of myoglobin. A three-dimensional Fourier synthesis at 2 Å resolution. Nature 185:422–427
330. Takano T (1977) Structure of myoglobin refined at 2.0 Å resolution. I. Crystallographic refinement of metmyoglobin from sperm whale. J Mol Biol 110:537–568
331. Phillips SE (1978) Structure of oxymyoglobin. Nature 273:247–248
332. Perutz M (1972) Nature of haem-haem interactions. Nature 237:495–497
333. Arutyunyan EG, Kuranova IP, Vainshtein BK, Staigeman VR (1980) X-ray diffraction study of leghemoglobin. VI. Structure of ferrileghemoglobin acetate at 2.0 Å resolution. Kristallografia 25:80–103 (in Russian)
334. Traylor TG (1981) Synthetic model compounds for hemoprotein. Accounts Chem Phys 14:102–109

335. Shaanan B (1982) The iron-oxygen bond in human oxyhaemoglobin. Nature 296:683–684
336. Tovros BS, Vitko DJ, Drago RS (1975) Nature of the bond O_2 in a series of cobalt dioxygen adducts. J Am Chem Soc 98:5144–5152
337. Christahl M, Raap A, Gersonde K (1981) pH dependece of oxy and deoxy cobalt-substituted leghemoglobin from soybean. Biophys Struct Mech 7:171–186
338. Simbok K, Korszun LR, Stucky G, Moffat K, McLendon G (1986) EXAFS studies of Zn_2Fe_2 hybrid hemoglobins: absence of heme band length changes in half-ligated species. Biochemistry 25:3773–3778
339. Chance MR, Parkhurst LJ, Powers LS, Chance B (1986) Movement of Fe with respect to heme plane in the RT transition of carp hemoglobin. EXAFS study. J Biol Chem 261:5689–5693
340. La Mar GN, Budd DL, Sick H, Gersonde K (1978) Acid Bohr effects in myoglobin characterized by proton NMR shifts and oxygen binding studies. Biochim Biophys Acta 537:270–283
341. Eisenberger P, Shulman RG, Kincaid BM, Breen GS, Ogawa S (1978) Extended X-ray absorption fine structure determination of iron-nitrogen distances in haemoglobin. Nature 274:30–34
342. Perutz ME, Hasnain SS, Dake PJ, Sessler JL, Hahn JE (1982) Stereochemistry of iron in deoxyhaemoglobin. Nature 295:535–538
343. Hanson JC, Schoenborn BP (1981) Real space refinement of neutron diffraction data from sperm whale carbon monooxymyoglobin. J Mol Biol 153:117–146
344. Dilworth MJ, Appleby CA (1979) Leghemoglobins and hemoproteins from Rhizobium. In: Hardy RWF, Bottomley F (eds) A treatise on dinitrogen fixation. Wiley Interscience, New York
345. Abaturov LV, Tinoria KSh, Varshavski YuM, Yakobashvily NN (1977) Effect of ligand and heme on conformational stability (intermolecular conformational mobility) of hemoglobin as revealed by hydrogen exchange. FEBS Lett 77:103–106
346. Lesk AM, Janin J, Wodak S, Chothia C (1986) Hemoglobin: the surface buried between $\alpha_1\beta_1$ and $\alpha_2\beta_2$ dimers in the deoxy and oxy structure. J Mol Biol 188:267–270
347. Bell JA, Korszun R, Moffat K (1981) Structure of imidazole methemoglobin. J Mol Biol 147:325–335
348. Jones WC, Rothgeb TM, Gurd FRN (1976) Nuclear magnetic resonance studies of sperm whale myoglobin specifically enriched with ^{13}C in the methionine methyl group. J Biol Chem 251:7452–7460
349. Wilson DF, Evecinska M (1979) Cytochrome *c* oxidase. In: Dolphin D (ed) The porphyrins. VII. Biochemistry, part B. Academic Press, New York, pp 2–70
350. Palmer G, Babcock GT, Vickery LE (1976) A model for cytochrome *c* oxidase. Proc Natl Acad Sci USA 73:2206–2210
351. Lang G, Lippard SJ, Rosen S (1974) Mössbauer spectroscopy studies of beff heart cytochrome *c* oxidase. Biochim Biophys Acta 336:6–14
352. Martin JL, Migus A, Poyart C, Lacerpentier Y, Astier R, Antonetti A (1983) Femtosecond photolysis of CO-ligation protoheme and hemoprotein. Proc Natl Acad Sci USA 80:173–177
353. Karlsson B, Andreasson LE (1981) The identity of a new Cu(II) ESP signal in cytochrome *c* oxidase. Biochim Biophys Acta 635:73–80
354. Gersonde K, Kerr E, Yu N, Parish DW, Smith KM (1986) Resonance Raman investigation of CO-ligated monomeric insect hemoglobins. J Biol Chem 261:8678–8685
355. Smith KM, Sick H, Petershans A, Gersonde K (1986) Acid Bohr effect of a monomeric hemoglobin from Dicrocoelium denditicum. Mechanism of the allosteric conformation transition. Eur J Biol 155:231–237
356. Jagow G, Sebald W (1960) B-type cytochromes. In: Snell EE (ed) Annual Review of Biochemistry, vol 49. Annual Reviews, Palo Alto California, pp 281–314
357. Mathews FS, Levine M, Argos P (1972) Three-dimensional Fourier synthesis of calf liver cytochrome b_5 at 2.8 Å resolution. J Mol Biol 64:449–464
358. Rasmussen UF, Rasmussen UN, Jorgensen BB (1982) Three functionally different cytochrome b redox centers in pigeon heart mitochondria. Biochem J 201:311–320
359. Mathews FS, Bethge PU, Czerwinski EW (1979) The Structure of cytochrome b from Escherichia coli at 2.5 Å resolution. J Biol Chem 254:1699–1706

360. Dickerson RJ, Timkovich R (1975) Cytochrome c. In: Boyer PD (ed) The enzymes, vol 11. Academic Press, New York, pp 397–448
361. Sulemme FR (1977) Structure and function of cytochrome c. Annu Rev Biochem 46:299–329
362. Takano T, Dickerson RE (1981) Redox conformation changes in refined tuna cytochrome *c*. J Mol Biol 159:95–116
363. Weber PC, Howord A, Xuong NH, Salemme FR (1981) Crystallographic structure of Rhodospirillum molischianum ferricytochrome *c* at 2.5 Å resolution. J Mol Biol 153:399–424
364. Bruschi M (1981) The primary structure of tetrahaem cytochrome c_3 from Desulfovibrio desulfuricans (strains Norway 4). Description of a new class of low-potential cytochrome *c*. Biochim Biophys Acta 677:219–226
365. Shin-ichi K, Yasuhiro A (1986) Purification of a hexaheme cytochrome c_{552} Escherichia coli K 12 and its properties as a nitrite reductase. Eur J Biochem 154:457–463
366. Yonetani T (1970) Cytochrome *c* peroxidase. Adv Enzymol 33:309–335
367. Yonetani T, Yamanoto H (1973) Optical and electron paramagnetic resonance properties of the nitric oxide compounds of cytochrome *c* peroxidase and horseradish peroxidase. In: King TE, Mason HS, Morrison M (eds) Oxidases and related systems, vol 1. University Park Press, Maryland, pp 279–298
368. Dunford HB, Stillman JS (1976) On the function and mechanism of action of peroxidases. Coord Chem Rev 19:187–251
369. Poulos TL, Freer ST, Alden RA, Xuong H, Edwards SL, Hamlin RC, Kraut J (1978) Crystallographic determination of the heme orientation and location of the cyanide binding site in yeast cytochrome *c* peroxidase. J Biol Chem 253:3730–3735
370. Wang M-YuK, Hoffman BM, Hollenberg PF (1977) Cobalt-substituted horseradish peroxidase. J Biol Chem 252:6268–6275
371. Mauk MR, Girotti AW (1974) The protoporphyrin-apoperoxidase complex photooxidation studies. Biochemistry 13:1757–1763
372. Nozawa T, Koboyashi N, Hatano M (1976) Magnetic circular dichroism studies on horseradish peroxidase. Biochim Biophys Acta 427:652–662
373. Rakshit G, Spiro ThG (1974) Resonance Raman spectra of horseradish peroxidase: Evidence for anomalous heme structure. Biochemistry 13:5317–5323
374. Williams RJP, Whright PE, Mazia G, Richard JR (1975) Proton magnetic resonance studies of peroxidases from turnip and horseradish. Biochim Biophys Acta 412:127–147
375. Blumberg WE, Peisach J, Wittenberg BA, Wittenberg JB (1968) The electronic structure of protoheme proteins. I. An electron paramagnetic resonance study of horseradish peroxidase and its derivatives. J Biol Chem 243:1854–1862
376. Ugarova NN, Roznikova GD, Berezin IV (1979) Chemical modification of the ε-amino groups of lysine residues in horseradish peroxidase and its effect on the catalytic properties and thermostability of the enzyme. Biochim Biophys Acta 570:31–42
377. Azakura T, Leigh JS, Drott HR, Yonetani T, Chance B (1971) Structural measurements in hemoproteins: Use of spin-labeled protoheme as a probe of heme environment. Proc Natl Acad Sci USA 68:861–865
378. Morrison M, Schonbaum GK (1976) Peroxidase catalyzed halogenation. Annu Rev Biochem 45:861–888
379. Vuk-Pavlovic C, Siderer Y (1977) Probing axial ligands in ferric haemproteins: An ESP study of myoglobin and horseradish peroxidase in $H_2^{17}O$. Biochem Biophys Res Commun 79:885–899
380. Chance BC (1949) The properties of the enzyme-substrate compounds of horse-radish and lacto-peroxidase. Science 109:204–208
381. Makino R, Uno T, Nishimura Y, Iizuka T, Tsuboi M, Ishimura Y (1986) Coordination structures and reactivities of compound II in iron and manganese horseradish peroxidase. J Biol Chem 261:8376–8383
382. Schonbaum GR, Lo S (1972) Interaction of peroxidases with aromatic peracids and alkyl peroxides. J Biol Chem 247: 3353–3360
383. Lang, G, Spartalian K, Yonetani T (1976) Mössbauer spectroscopic study of compounds of cytochrome *c* peroxidase. Biochim Biophys Acta 451:250–258

384. Rakshit G, Spiro TG (1976) Resonance Raman evidence for Fe(IV) in compound II of horseradish peroxidase. Biochem Biophys Res Commun 71:803–808
385. Roberts JE, Hoffman BM, Rutter R, Hager LP (1981) Electron-nuclear double resonance of horseradish peroxidase compound I. Biol Chem 256:2118
386. Roberts JE, Hoffman BM, Rutter R, Hager LP (1981) ^{17}O ENDOR of horseradish peroxidase compound I. J Am Chem Soc 103:7654–7656
387. Browlett WR, Stillman MJ (1981) Evidence for heme π-cation radical species in compound I of horseradish peroxidase and catalase. Biochim Biophys Acta 660:1–7
388. Felton RH, Dolphin D, Fajer J (1971) Iron (IV) porphyrins. J Am Chem Soc 93:6332–6334
389. Hanson LK, Chang ChK, Davis MS, Fajer J (1981) Electron pathways in catalase and peroxidase enzyme catalysis. Metal and macrocycle oxidations of iron porphyrins and chlorins. J Am Chem Soc 103:663–670
390. Hayashi Y, Yamazaki T (1979) The oxidation-reduction potentials of compound I/compound II and compound III ferric couples of horseradish peroxidases A and C. J Biol Chem 254:9101–9106
391. Job D, Ralston I, Dunford HB (1978) Horseradish peroxidase XXXII. pH Dependence of the oxidation of L(−)-tyrosine by compound I. Can J Biochem 56:702–707
392. Ugarova NN, Lebedeva OV, Berezin IV (1981) Horseradish peroxidase catalysis. I. Steady-state kinetics of peroxidase-catalyzed individual and cooxidation of potassium ferrocyanide and o-dianisidin by hydrogen peroxide. J Mol Cat 13:215–225
393. Yonetani T (1976) Peroxidases. In: Boyer PD (ed) The enzymes, vol 8. Academic Press, New York, pp 345–393
394. Sakurado J, Takahashi S, Hosoya T (1986) NMR studies on the spatial relationship of aromatic donor molecules to the heme iron of horseradish peroxidase. J Biol Chem 261:9657–9662
395. Goff HM, Behere DV, Gonzales-Vergara E (1985) The heme environment of peroxidases: ^{15}N NMK of bound $^{15}CN^-$. Chem Rev Port 27:217–218
396. Dawson JH, Sono M, Eble ES, Hager LP (1985) Spectroscopic properties of chloroperoxidase compounds II and III-possible structural models for analogous cytochrome P-450 derivative. Chem Rev Port 27:205–206
397. Dunford HB, Lambier AM, Bohne C, Boader W (1985) Steady state and burst kinetic studies on peroxidases. Chem Rev Port 27:125–126
398. Yamada H, Itoh N, Izumi Y (1985) Chloroperoxidase catalyzed halogenization of trans-cinnamic acid. J Biol Chem 260:11962–11969
399. Van Wart HE, Zimmer J (1985) Resonance Raman evidence in the activation of dioxygen in horseradish oxyperoxidase. J Biol Chem 260:8372–8377
400. Elenkova EV, Sono M, Dawson JH, English AM (1985) Magnetic circular dichroism studies of cytochrome c peroxidase and its ligand complexes. Rev Port Quim 27:221–222
401. Hewson WD, Hager LP (1979) Peroxidase, catalase, chloroperoxidase. In: Dolphin D (ed) The porphyrins. VII. Biochemistry, part B. Academic Press, New York, pp 229–339
402. Schonbaum GR, Chance B (1976) Catalases. In: Boyer PD (ed) The enzymes, vol 13. Academic Press, New York, pp 363–408
403. Murthy MRN, Reid TJ, Sicignano A, Tanako N, Rossman MG (1981) Structure of beef liver catalase. J Mol Biol 152:465–499
404. Fita I, Rossman MG (1985) The active center of catalase. J Mol Biol 183:21–37
405. Fridovich J, Kano J (1983) Isolation and characterization of the pseudocatalase of Lactobacillus plantarum. A new manganese-containing enzyme. J Biol Chem 258:6015–6019
406. Vainshtein BK, Melik-Adamyan VP, Barynin VV, Vagin AA, Grebenko AI (1981) Structure of the catalase of Penicillium vitale. Kristalografia 26:1003–1016 (in Russian)
407. Archakov AI (1980) Intermembrane electron transfer. In: Sculachev VP (ed) Biological review of physico-chemical aspects. Scientific Review, vol 1, Moscow, pp 43–81
408. Coon MJ, White RE, Blake RC (1981) Oxygen activation by cytochrome P-450. In: King TE (ed) Oxidases and related systems, vol 1. University Park Press, Baltimore, pp 93–118
409. Griffin BW, Peterson JA, Estabrook RW (1979) Cyt-P450 Biophysical properties and catalytic function. In: Dolphin D (ed) The porphyrins. II. Biochemistry, part B. Academic Press, New York, pp 378–398

410. Lyakhovich VV, Tsyrlov IB (1978) Structural aspects of monooxidases. Nauka, Moscow (in Russian)
411. Guengerich PP, Macdonald TL (1984) Chemical mechanisms of catalysis by cytochrome P-450: a unified view. Acc Chem Res 17:9–16
412. Akhrem AA, Adamovich TB, Lapko VN, Sherman SA, Usanov SA, Shkumatov VM, Chashcin BL (1985) A model for molecular organization of the cholesterol side chain cleavage cytochrome P-450 from adrenal cortex mitochondria. In: Cytochrome P-450. Biochem Biophys and Induction. Proc 5th Int Conf Budapest. 21–24 Aug. Budapest, pp 155–158
413. Peisach J, Blumberg WE (1970) EPR study of the high and low spin forms of cytochrome P-450 in liner and in liver microsomes from a methylcholanthrene-treated rabbit. Proc Natl Acad Sci USA 67:172–179
414. Dolphin D, James BR, Welborn HC (1981) Oxygenation and carbonylation of a reduced P-450$_{cam}$ enzyme and derivatives with mesodeutero-, dibromodeutero and diacetyl deuteroheme. J Mol Cat 7:21 201–208
415. Budyka MF, Khenkin AM, Shteinman AA (1981) The mercaptide anion as a trans-ligand in the oxycytochrome P-450. Biochem Biophys Res Commun 101:5 615–622
416. Champion PM, Stallard BR, Wagner GC, Gunsalus IC (1982) Resonance Raman detection of a Fe-S bond in cytochrome P-450. In: Cytochrome P-450. Structure and function. National Symposium, Minsk, p 7
417. Gelb MH, Toscano WA, Sligar SG (1982) Preparation and chemical properties of manganese substituted cytochrome P-450. In: Cytochrome P-450. Structure and function. National Symposium, Minsk, p 24
418. Sakurai H, Yoshimura T (1982) The active site structure as thiolate-hemin-hydroxyl group coordination mode. In: Cytochrome P-450. An approach from model investigation. Inorg Chem Acta 66:225–227
419. Ruckpaul K, Rein H, Ballon DP, Coon MJ (1980) Analysis of interaction among purified components of the liner microsomal cytochrome P-450 containing monooxygenase system by second derivative spectroscopy. Biochim Biophys Acta 626:41–56
420. Hahn JE, Hodgson KO, Andersson LA, Dawson JH (1982) Endogenous cysteine ligation in ferric and ferrous cytochrome P-450. Direct evidence from X-ray absorption spectroscopy. J Biol Chem 257:10934–10941
421. Ruckpaul K, Rein H (1984) Cytochrome P-450. Akademie-Verlag, Berlin
422. Jung C (1985) Quantum chemical explanation of the "hyper" spectrum of the carbon monoxide complex of cytochrome P-450. Chem Phys Lett 113:589–596
423. Rohmer MM, Loew GH (1979) Electronic structure and properties of model oxy and carboxy ferrous cytochrome P-450: comparison of semiempirical and ab initio calculations. Int J Quant Chem Quant Biol Symp 6:93–104
424. Hansson LK (1979) Axial ligand effects on iron and manganese porphyrins: extended Hückel calculations of cytochrome P-450 analogs and O_2 binding to iron and manganese. Int J Quant Chem Quant Biol Symp 6:73–87
425. Dawson JH, Andersson LA, Davies IM, Hahn JE (1980) Spectroscopic investigation of the active site structure of cytochrome P-450. X-ray absorption spectroscopy. In: Gustafsson JA (ed) Biochemistry, biophysics and regulation of cytochrome P-400. Elsevier, Amsterdam, pp 565–573
426. Lo Brutto R, Scholes CP, Wagner GS, Gunsalus IC, Debrunner PG (1980) Electron nuclear double resonance of ferric cytochrome P-450$_{cam}$. J Am Chem Soc 102:1167–1174
427. Weiner LM (1986) Magnetic resonance study of the structure and function of cytochrome P-450. Crit Rev Biochem 20:139–200
428. Groh SE, Nagahisa A, Tan SL, Orme-Johnson WH (1983) Electron spin echo modulation demonstrates P-450$_{scc}$ complexation. J Am Chem Soc 105:7445–7450
429. Reichman LW, Belova VS, Borukaeva MR, Likhtenshtein GI (1971) On the mechanism of hydrocarbon interaction with microsomal hydroxylating system. Stud Biophys 28:43–47
430. Blum H, Leigh JS, Salerno JC, Ohnishi T (1978) The orientation of bovin adrenal cortex cytochrome P-450 in submitochondrial particle multilayers. Arch Biochem Biophys 187:153–162

431. Novak RF, Vatsis KP (1982) ^{1}H NMR relaxation rate studies on the interaction of acetanilide with purified isozymes of rabbit liver microsomal cytochrome P-450 and with cytochrome b_5. Mol Pharmacol 21:1982–1987
432. Metelitsa DI (1982) Activation of oxygen by enzyme systems. Nauka, Moscow (in Russian)
433. Rahimtula AD, O'Brien PJ (1974) Hydroperoxide catalyzed liver microsomal aromatic hydroxylation reactions involving cytochrome P-450. Biochem Biophys Res Commun 60:440–447
434. Hamilton GA, Giacin JR, Hellman ThT, Snook ME, Weller JW (1973) Oxenoid models for enzymic hydroxylation. Ann Acad Sci New York 212:4–32
435. Miva GT, Carland WA, Hodshon BJ, Lu AYH (1980) Kinetic isotope effects in cytochrome P-450-catalyzed oxidation reactions. J Biol Chem 255:5049–6054
436. Groves JT, McClusky GA, White RE, Coon MJ (1978) Aliphatic hydroxylation by highly purified liver microsomal cytochrome P-450. Evidence for a carbon radical intermediate. Biochem Biophys Res Commun 81:154–161
437. Gelb MH, Heimbrook DC, Malkonen P, Sligar SG (1982) Stereochemistry and deuterium isotope effects in camphor hydroxylation by the cytochrome P-450$_{cam}$ monooxygenase system. Biochemistry 21:370–377
438. Heimbrook DC, Sligar SG (1981) Multiple mechanisms of cytochrome P450-catalyzed substrate hydroxylation. Biochem Biophys Res Commun 99:530–535
439. Sligar SG, Kennedy KA, Pearsson DC (1980) Chemical mechanisms for cytochrome P-450 hydroxylation. Evidence for acylation of heme-bound dioxygen. Proc Natl Acad Sci USA 77:1240–1244
440. Hardman KD, Eylar EH, Ray DK, Banaszak LJ, Gurd FRN (1966) Isolation of sperm whale myoglobin by low temperature fractionation with ethanol and metallic ions. J Biol Chem 241:432–437
441. Vickery L, Nozawa T, Sauer K (1976) Magnetic circular dichroism studies of myoglobin complexes. Correlations with heme spin state and axial ligation. J Am Chem Soc 98:343–350
442. Kent TA, Spartalian K, Lang G, Yonetani T, Reed CA, Collman JP (1979) High magnetic field Mössbauer studies of deoxymyoglobin, deoxyhemoglobin and synthetic analogues. Biochim Biophys Acta 580:245–258
443. Maeda Y (1979) Mössbauer studies on the iron-ligand binding in hemoproteins and their related compounds. J Phys (France) 40, 3, Colloq 2:514–522
444. Banerjee R, Cassoly R (1969) Preparation and properties of the isolated and chains of human hemoglobin in the ferri state. Investigation of oxidation-reduction equilibria. J Mol Biol 42:337–349
445. Waterman MR (1978) Spectral characterization of human hemoglobin and its derivatives. In: Fleisher S (ed) Methods in Enzymology, vol 52. Academic Press, New York, pp 456–463
446. Treu JI, Hopfield JJ (1975) Magnetic circular dichroism in hemoglobin. J Chem Phys 63:613–623
447. Johnson CE (1975) Mössbauer spectroscopy in biology. In: Gonser U (ed) Mössbauer spectroscopy. Springer, Berlin Heidelberg New York
448. Babcock GT, Vickery LE, Palmer G (1976) Electronic state of heme in cytochrome oxidase I. Magnetic circular dichroism of the isolated enzyme and its derivatives. J Biol Chem 251:7907–7919
449. Babcock GT, Ondrias MR, Gobeli DA, Vanstellandt J, Leroi GE (1979) The optical properties of heme a: resonance Raman scattering with visible excitation. FEBS Lett 108:147–151
450. Aasa R, Albracht SPJ, Falk KE, Lanne B, Vangrand T (1976) EPR signals from cytochrome *c* oxidase. Biochim Biophys Acta 422:260–272
451. Lang G, Lippard SJ, Rosen SR (1974) Mössbauer spectroscopy studies of beef-heart cytochrome *c* oxidase. Biochim Biophys Acta 336:6–14
452. Babcock GT, Vickery LE, Palmer G (1978) The electronic state of heme in cytochrome oxidase II. Interactions and heme iron spin state behavior observed in reductive titrations. J Biol Chem 253:2400–2411

453. Sutherland JC, Klein MP (1972) Magnetic circular dichroism of cytochrome *c*. J Chem Phys 57:76–86
454. Brautigan DL, Feinberg BA, Hoffman BM, Margoliash E, Peisach J, Blumberg WE (1977) Multiple low-spin forms of the cytochrome *c* ferrihemochrome. EPR spectra of various eukaryotic and prokaryotic cytochromes *c*. J Biol Chem 252:574–582
455. Lang G, Herbert D, Yonetani T (1968) Mössbauer spectroscopy of cytochrome *c*. J Chem Phys 49:944–950
456. Hettinger TP, Harbury HA (1965) Guanidinated cytochrome *c*. Biochemistry 4: 2585–2589
457. Vickery LE, Salman AG, Sauer K (1975) Magnetic circular dichroism studies on microsomal aryl hydrocarbon hydroxylase: comparison with cytochrome b_5 and cytochrome P-450_{cam}. Biochim Biophys Acta 386:87–98
458. Shimizu T, Nozawa T, Hatano M, Imai Y, Sato R (1975) Magnetic circular dichroism studies of hepatic microsomal cytochrome P-450. Biochemistry 14:4172–4178
459. Chevion M, Peisach J, Blumberg WE (1977) Imidazole, the ligand trans to mercaptide in ferric cytochrome P-450. An EPR study of proteins and model compounds. J Biol Chem 252:3637–3645
460. Sharrock M, Debrunner PG, Schulz C, Lipscomb JD, Marshall Y, Gunsalus IC (1976) Cytochrome P-450_{cam} and its complexes. Mössbauer parameters of the heme iron. Biochim Biophys Acta 420:8–26
461. Waterman MR, Mason HS (1970) The redox potential of liver cytochrome P-450. Biochem Biophys Res Commun 39:450–454
462. Browett WR, Stillman MJ (1980) Magnetic circular dichroism studies on the electronic configuration of catalase compounds I and II. Biochim Biophys Acta 623:21–31
463. Maeda Y, Trautwein A, Gonser U, Yoshicka K, Kikuchi-Toru K, Homma T, Ogura Y (1973) Mössbauer effect in bacterial catalase. Biochim Biophys Acta 303:230–236
464. Rein H, Ristau O, Hackenberger F, Jung F (1968) Electronenspinresonanzuntersuchungen von Katalasekomplexen. Biochim Biophys Acta 167:538–546
465. McIntosh AR, Stillman MJ (1977) Electron-paramagnetic resonance studies on a photochemically produced species of horseradish peroxidase compound I. Biochem J 167:31–37
466. Moss TH, Ehrenberg A, Bearden AJ (1969) Mössbauer spectroscopic evidence for electronic configuration of iron in horseradish peroxidase and its peroxide derivatives. Biochemistry 8:4159–4162
467. Cammack R, Pastil DS, Fernandez VM (1985) EPR spectroscopy of iron-sulphur enzymes. Biochem Soc Trans 13:572–578
468. Spiro TG (ed) (1982) Iron-sulfur proteins, vol 4. Wiley, New York
469. Cammack R, Rao KK, Hall DO (1985) Ferredoxins: Structure and function of a ubiquitous group of protein. Physical Végét 23:649–658
470. Münck E (1985) Structure and magnetism of iron-sulfur clusters in proteins. Rev Port Quim 27:5–6
471. Williams RJ (1985) Metallo-enzyme catalysis: the entatic state. J Mol Cat 30:1–26
472. Veda T, Coon MJ (1972) Reduced disphosphopyridine nucleotide-rubredoxin reductase: properties and function as an electron carrier in hydroxylation. J Biol Chem 247:5010–5019
473. Jensen LH (1975) Rubredoxin. In: Lovenberg W (ed) Iron-sulfur proteins. Academic Press, New York, pp 11–45
474. Holm RH (1977) Identification of active sites in iron-sulfur proteins. In: Addison AW (ed) Biological aspects of inorganic chemistry. Wiley Interscience, New York, pp 71–111
475. Dunham WR, Palmer G, Sands RH, Bearden AJ (1971) On the structure of the iron-sulfur complex in two-iron ferredoxins. Biochim Biophys Acta 253:373–384
476. Gibson JF, Hall DD, Thornley JHM, Whatley FR (1966) The iron complex in spinach ferredoxin. Proc Natl Acad Sci USA 56:987–990
477. Mortenson LE, Gillum WO (1980) Nitrogenase. In: San Pietro A (ed) The methods in enzymology, vol 69, part C. Academic Press, New York, pp 779–792
478. Cline JF, Hoffman BM, Mims WB, La Haie E, Ballou D, Fee JA (1985) Evidence for N-coordination to Fe in the [2Fe.2S] clusters of Thermus Rieske protein and phthalate dioxygenase from Pseudomonas. J Biol Chem 260:3251–3254

479. Blum H, Adar F, Salerno JC, Leigh JS (1977) Exchange coupling in spinach ferredoxin determined by resonance Raman spectroscopy. Biochem Biophys Res Commun 77:650–653
480. Kurtz DM (1982) A new method for extrusion of iron-sulfur cores from active centres of proteins. Biochem Biophys Res Commun 104:437–443
481. Adam ET, Sieker LC, Jensen LH (1973) The structure of a bacterial ferredoxin. J Biol Chem 248:3987–3996
482. Carter CW, Kraut JJ, Freer ST, Xuong N, Alden RA (1973) Two-angstrom crystal structure of oxidized chromatium high potential iron protein. J Biol Chem 249:4212–4225
483. Carter CW, Kraut J, Freer ST, Alden RA, Sicker LC, Adam E, Jensen LH (1972) A comparison of Fe_4S_4 clusters in high-potential iron protein and in ferredoxin. Proc Natl Acad Sci USA 69:3526–3529
484. Mathews R, Charleton S, Sands RH, Palmer G (1974) On the nature of the spin coupling between the iron-sulfur clusters in eight-iron ferredoxins. J Biol Chem 249:4326–4328
485. Meyer J, Moulis JM, Lutz M (1985) Resonance Raman spectroscopy of iron-sulfur proteins. Rev Port Quim 27:26–27
486. Malkin R, Bearden AJ (1980) Bound iron-sulfur centres in photosynthetic membranes. In: San-Pietro A (ed) Methods in Enzymology, vol 69, part C. Academic Press, New York, pp 233–267
487. Prince RC, Crowder MS, Bearden AJ (1980) The orientation of the magnetic axes of the membrane bound iron-sulfur clusters of spinach chloroplasts. Biochim Biophys Acta 592:323–337
488. Chen JS, Mortenson LE (1974) Purification and properties of hydrogenase from Clostridium pasteurianum W5. Biochim Biophys Acta 371:283–298
489. Mayhew SG, O'Connor ME (1982) Structure and mechanism of bacterial hydrogenase. Trends Biochem Sci 7:18–27
490. Chen JS, Mortenson LE, Palmer G (1976) Hydrogenases. In: Yasonulu (ed) Iron and copper proteins. Plenum Press, New York, pp 68–97
491. Graf EG, Thauer RK (1981) Hydrogenase from Methanobacterium thermoautotrophicum, a niclel-containing enzyme. FEBS Lett 136:165–169
492. Scott RA, Czechowski M, der Vartanjan DV, le Gall J, Peck HD, Moura I (1985) Nickel X-ray absorption spectroscopy of Desulfovibrio gigas hydrogenase. Rev Port Quim 27:67–70
493. Albracht SPJ (1985) The use of EPR spectroscopy to establish the properties of nickel and the iron-sulphur cluster in hydrogenase from Chromatium vinosum. Biochem Soc Trans 13:582–585
494. Telser J, Emptage MH, Merkle H, Kennedy MC, Beinert H (1986) ^{17}O ENDOR characterization of substrate binding to the $[4Fe.4S]^{+1}$ cluster of reduced active aconitase. J Biol Chem 261:4840–4846
495. Kurtz DM, Holm RH, Ruzicka FJ, Beinert H, Coles ChJ, Singer ThP (1979) The high potential iron-sulfur cluster of aconitase is a binuclear iron-sulfur cluster. J Biol Chem 254:4967–4969
496. Albracht SPJ, Unden G, Kroger A (1981) Iron-sulfur clusters in fumarate reductase from vibriosuccenogenes. Biochim Biophys Acta 661:295–302
497. Vega JM, Cardenas J, Losada M (1980) Ferredoxin-nitrite reductase. In: San-Pietro A (ed) Methods in Enzymology, vol 69, part C. Academic Press, New York, pp 255–270
498. Aisen Ph, Listowsky J (1980) Iron transport and storage proteins. Annu Rev Biochem 49:357–393
499. Christner JA, Münck E, Kent TA, Janick PA, Salerno JC, Siegel LM (1984) Exchange coupling between siroheme and [4Fe.4S] cluster in E. coli sulfite reductase. Mössbauer studies and coupling models. J Am Chem Soc 106:6786–6791
500. Banyard SH, Stammers DK, Harrison PM (1978) Electron density map of apoferritin at 2–8 Å resolution. Nature 71:282–284
501. Stenkamp RE, Sieker LC, Jensen LH (1981) Structure of binuclear iron complex in metazidohaemerytin from Themiste dyscritum at 2.2 Å resolution. Nature 291:263–264
502. Aisen P (1985) The binuclear iron center of uteroferrin. Rev Port Quim 27:10–11

503. Likhtenshtein GI, Frolov EN, Nesnaiko NR, Levchenko LA, Sklyar YuS (1972) Investigation of a model iron-sulfur protein using spin and luminescent labels. Mol Biol (Mosc) 6:201–209 (in Russian)
504. Bray RC, George GH (1985) EPR studies using pre-steady state kinetics and substitution with stable isotopes on mechanism action of molybdoenzymes. Biochem Soc Trans 13:561–567
505. Bray RC (1975) Molybdenum iron-sulphur fravin hydroxylases and related enzymes. In: Boyer PD (ed) The Enzymes, vol 12. Academic Press, New York, pp 299–419
506. Gerge GN (1984) Coupling between Mo(IV) and Fe/S in xanthine oxidase. In: Flavins and Flavoproteins. Proc 8th Int Symp. Brighton, July 9–13, pp 325–330
507. Johnson JL, Hainline BE, Jones HP, Rajagopalan KV (1978) Molybdenum cofactor. In: Newton WE, Orme-Johnson WH (eds) Nitrogen fixation, vol 1. University Park Press, Baltimore, pp 249–260
508. Hodgson E, Fridovich I (1976) The mechanism of the activity dependent luminescence of xanthine oxidase. Arch Biochem Biophys 172:202–205
509. Watt AB (1972) Xanthine dehydrogenase and pteridine metabolism in Colias butterflies. J Biol Chem 247:1445–1451
510. Hawkes TR, Bray RC (1984) Studies by EPR spectroscopy of the environment of the metal in the molybdenum cofactor of molybdenum-containing enzymes. Biochem J 222:587–600
511. Bray RC, Engel PC, Mayhew SG (eds) (1985) Flavins and Flavoproteins. de Gruyter, Berlin
512. Newton WE, Gheller SF, Hedman B, Hodgson KO, Lough SM (1986) Extraction of thiomolybdates from the iron-molybdenum cofactor of nitrogenase. Comparison with synthetic FeMoS complex. Europ J Biochem 159:111–115
513. Yourno J, Kohno T, Roth JR (1970) Enzyme evolution: generation of a bifunctional enzyme by fusion of adjacent genes. Nature 228:820–824
514. Williams PIP, Wentworth RAD (1974) Molybdenum in Enzymes. J Less Common Metals 36:405–412
515. Ketchum PA, Cambier HY, Frazier WA, Madonsky ChH, Nason A (1970) In vitro assembly of Neurospora assimilatory nitrate reductase from protein subunits of a Neurospora mutant and the xanthine oxidizase or aldehyde oxidase systems of higher animals. Proc Natl Acad Sci USA 66:1016–1023
516. Bray RC, George GH (1985) EPR studies using pre-steady state kinetics and substitution with stable isotopes on mechanism action of molybdoenzymes. Biochem Soc Trans 13:561–567
517. L'vov NP, Alikulov ZA, Kildibekov NA, Kretovich VL (1981) Molybdenum cofactor of MO-containing enzymes. Izv Akad Nauk SSSR (Ser Biol) 2:214–235 (in Russian)
518. Johnson JL, Hainline BE, Rajagopalan KV (1980) Characterization of the molybdenum cofactor of sulfite oxidase, xanthine oxidase and nitrate reductase. J Biol Chem 255:1783–1786
519. Bordas J, Bray RC, Garner D, Gutteridge S, Hasnain HS (1980) X-ray absorption spectroscopy of xanthine oxidase. The molybdenum centers of the functional and the desulpho forms. Biochem J 191:499–508
520. Jalthouse JRG, Bray RC (1980) The nature of the sulphur atom liberated from xanthine oxidase by cyanide. Biochem J 191:265–268
521. Lowe DJ, Hide JS (1975) Electron-electron double resonance measurements on xanthine oxidase. Biochim Biophys Acta 377:205–210
522. Cammack R, Barber MI, Bray RC (1976) Oxidation-reduction potentials of molybdenum, flavin and iron-sulphur centers in milk xanthine oxidase. Biochem J 157:469–478
523. Mary H, McGartoll MA, Bray RC (1969) Reaction with iodoacetamide and the number of active centres in xanthine oxidase. Biochem J 114:443–444
524. Hille R, Hagen WR, Dunham WR (1985) Spectroscopic studies on the iron-sulfur centers of milk xanthine oxidase. J Biol Chem 260:10569–10575
525. Stevenson RC, Dunham WR, Sands RH, Singer TP, Beinert H (1986) Studies on the spin-spin interaction between flavin and iron-sulfur cluster in an iron-sulfur flavoprotein. Biochem Biophys Acta 869:81–88

526. Lowe DJ, Lynden-Bell RM, Bray RC (1972) Spin-spin interaction between molybdenum and one of the iron-sulphur systems of xanthine oxidase and its relevance to the enzyme mechanism. Biochem J 130:239–249
527. Moura JJG, Xavier AV, Bruschi M, le Gall J (1977) NMR characterization of three forms of ferredoxin from Desulphovibrio gigas, a sulphate reducer. Biochim Biophys Acta 459:278–289
528. Andres RY (1976) Aldehyde oxidase and xanthine dehydrogenase from wild-type Drosophila melanogaster and immunologically cross-reacting material from ma-1 mutants. Eur J Biochem 62:591–600
529. Fhaolain JN, Coughlan MP (1976) Turkey liver xanthine dehydrogenase. Relation between nicotinamide-adenine dinucleotide oxidoreductase activity and the content of functional enzyme. Biochem J 157:283–285
530. Heimer IM, Krashin S, Riklis E (1976) The use of affinity chromatography for the purification of nitrate reductase. FEBS Lett 62:30–32
531. Lewis NJ, Scazzoccio G (1977) The genetic control of molybdoflavoproteins in Aspergillus nidulans. A xanthine dehydrogenase half-molecule mutant strains of Aspergillus nidulans. Eur J Biochem 76:441–446
532. Aslam M, Oaves A (1975) Effect of glucose on the induction of nitrate reductase in corn roots. Plant Physiol 56:634–639
533. Guerro MG, Jetschmann K, Volker W (1977) The stereospecificity of nitrate reductase for hydrogen removal from reduced pyridine nucleotides. Biochim Biophys Acta 482:19–26
534. Cohon HJ, Fridovich I, Rajagopalan KV (1971) Hepatic sulfite oxidase. A functional role for molybdenum. J Biol Chem 246:374–382
535. Burns RC (1979) Molecular nitrogen reaction mechanism. In: Hardy RWF, Bottomley F, Burns RC (eds) A treatise on nitrogen fixation. Sect I and II. Wiley Interscience, New York, pp 433–453
536. Likhtenshtein GI, Gvozdev RI, Levchenko LA, Syrtsova LA (1978) Structure and mechanism of action of the nitrogenase active center. Izv Akad Nauk SSSR (serbiol) pp 1–28
537. Lowe DJ, Smith BE (1985) Nitrogenase. Biochem Soc Trans 13:579–589
538. Nikonova LA, Shilov AE (1977) Dinitrogen fixation in homogeneous protic media. In: Newton W, Postgate JR, Rodriguez-Barrueco G (eds) Recent developments in nitrogen fixation. Academic Press, New York, pp 41–52
539. Bersuker IB (1978) Mutual vibronic influence of weakly coordinated molecular systems in chemical reaction and catalysis. J Chem Phys 31:85–93
540. Mishustin EN (ed) (1983) Molecular mechanism of the nitrogen utilization by plants. Nauka, Moscow (in Russian)
541. Vol'pin ME, Shur VB (1970) Nitrogen fixation by transition-metal complexes in aprotic media. Organomet React 1:50–78
542. Didenko LP, Gavrilov AB, Shilova AK, Strelets VV, Tsarev VN, Shilov AE, Makhaev VD, Banerjee AK, L. Pocpišil (1986) Dinitrogen fixation. Fast amalgam and electrochemical catalytic N_2 reduction at ambient temperature and pressure. Nouv Chim 10:583–588
543. Shilov AE (1984) Dinitrogen fixation in model systems. In: Mechanisms of catalysis. Part II. Nauka, Novosibirsk, pp 135–155 (in Russian)
544. Whiting MJ, Dilworth MJ (1974) Legume root nodule nitrogenase. Purification, properties and studies on its genetic control. Biochim Biophys Acta 371:337–351
545. Bulen WA, le Compte JR (1972) Nitrogenase. In: San-Pietro A (ed) Methods in Enzymology, vol 24B. Academic Press, New York, pp 456–470
546. Syrtsova LA, Pysarskaya TN, Nazarova II, Uzenscaya AM, Likhtensthein GI (1977) Coupling of ATPase and substrate binding sites in Azotobacter vinelandii nitrogenase. Bioorgan Khim 3:1251–1260
547. Hageman RV, Burris RH (1978) Nitrogenase and nitrogenase reductase associate and dissociate with each catalytic cycle. Proc Natl Acad Sci USA 75:2699–2702
548. Watt GD, Burns A, Lough S (1980) Redox properties of oxidized MoFe protein. In: Newton WE, Orme-Johnson WH (eds) Nitrogen fixation, vol 1. University Park Press, Baltimore, pp 159–170

549. Lough S, Burns A, Watt CD (1983) Redox properties of the dinitrogenase. Biochemistry 20:4062–4070
550. Orme-Johnson WH, Münck E, Zimmerman R, Brill WJ, Shah VK, Rawling J, Henzel MT, Averill BA, Orme-Johnson NR (1978) On the metal center or nitrogenase. In: Singer TP, Ondarza RN (eds) Mechanism of oxidizing enzymes. Elsevier, New York, pp 85–107
551. Syrtsova LA, Levchenko LA, Frolov EN, Likhtenshtein GI, Pisarscaja TN, Vorob'ev LV, Gromoglasova VA (1971) Structure and function of the nitrogenase components from Azotobacter vinelandii. Mol Biol (Mosc) 5:726–734 (in Russian)
552. Dunham WR, Hagen WR, Broaksma A, Grande HJ, Haaker H (1985) Iron contents in the iron-protein of the dinitrogenase. Eur J Biochem 146:497–503
553. Tsuprun VL, Mitsova IZ, Blazhchuck IS, Gvozdev RI, Orlova EV, Kiselev NA (1985) Electron microscopy of the MoFe protein from Azotobacter vinelandii nitrogenase. Eur J Biochem 149:389–392
554. Sosfenov NI, Andrianov BI, Vagin AA, Strokopytov BV, Vainshtein BK, Shilov AE, Gvozdev RI, Likhtenshtein GI, Mizova IZ, Blashchuk IS (1986) X-ray study of the nitrogenase MoFe-protein from Azotobacter vinelandii. Dokl Akad Nauk SSSR 291:1123–1127
555. Meyer J, Zaccai G (1980) Neutron small-angle scattering of the MoFe Protein from Clostridium pasteurianum. In: Appleby CA, Bergersen FJ (eds) The IV International Symposium on Nitrogen Fixation. Programs and Abstracts. Canberra Reprographic Printers, Canberra, p 77
556. Hardy RWF, Knight E (1967) ATP-dependent reduction of azide and HCN by N_2-fixing enzymes of Azotobacter vinelandii and Clostridium pasteurianum. Biochim Biophys Acta 139:69–90
557. Linde VR, Alfimova EYa, Slepko GI, Uzenskaya AM, Likhtenshtein GI (1969) The kinetics and mechanism of enzymatic reduction of molecular nitrogen. Dokl Akad Nauk SSSR 185:636–638 (in Russian)
558. Hardy RWF (1979) Reducible substrates of nitrogenase. In: Hardy RWF, Bottomley F, Burns RC (eds) A treatise on dinitrogen fixation. Sects. I and II. Wiley Interscience, New York, pp 453–503
559. Burris RH (1979) Inhibition. In: Hardy RWF, Bottomley F, Burns RC (eds) A treatise on dinitrogen fixation. Sects. I and II. Wiley Interscience, New York, pp 504–536
560. Jensen BB, Burris RH (1985) Effect of high pN_2, pD_2 on NH_3 production, H_2 evolution, HD formation by nitrogenases. Biochemistry 24:1141–1147
561. Burgess BK, Wherland S, Newton WE, Stiefel EJ (1981) Nitrogenase reactivity: insight into the nitrogen fixing process through hydrogen-inhibition and HD forming reactions. Biochemistry 20:5140–5146
562. Kochetkov VV, Linde VR, Likhtenshtein GI (1979) Some kinetic peculiarities of nitrogenase reaction. Mol Biol (Mosc) 13:402–409 (in Russian)
563. Hageman RV, Burris RH (1980) Electron allocation to alternative substrate of A. vinelandii is controlled by the electron flux through dinitrogenase. Biochim Biophys Acta 591:63–75
564. Thornley RNF, Lowe DJ (1983) Mechanistic studies on nitrogenase from Klebsiella pneumoniae using the rapid quench technique. Isr J Bot 31:61–71
565. Rawling J, Shah VK, Chisnel JR, Brill WJ, Zimmerman R, Münck E, Orme-Johnson WH (1978) Novel metal clusters in the iron-molybdenum cofactor of nitrogenase. J Biol Chem 253:1001–1004
566. Likhtenshtein GI, Shilov AE (1970) Thermodynamic and kinetic peculiarities of dinitrogen reductive fixation. Zh Fiz Khim 44:849–856 (in Russian)
567. Likhtenshtein GI (1978) Nitrogenase und ihre Modelle. Wiss Fortschr 28:10–16
568. Simpson FB, Jihong L, Burris RH (1985) The effect of high pressures of H_2 on HD formation in the nitrogenase product lag period. In: 6th International Symposium on Nitrogen Fixation. Book of abstracts. Corvallis, Oregon, State University. Abstract poster 3–29
569. Howard KS, Holes BJ, Socolofsky MD (1985) In vivo interaction between N_2ase MoFe proteins and membrane in Azotobacter vinelandii and Rhodospirillum rubrum. Biochim Biophys Acta 812:575–585

570. Klugkist M, Haaker H, Wassink H, Veeger C (1985) The catalytic activity of nitrogenase in intact Azotobacter vinelandii cells. Eur J Biochem 146:509–515
571. Stephens PJ, McKenna CE, Smith BE, Nguyen HT, McKenna MC, Thomson AJ, Devlin F, Jones JB (1979) Circular dichroism and magnetic circular dichroism of nitrogenase proteins. Proc Natl Acad Sci USA 76:2585–2589
572. Lindahl PA, Orme-Johnson WH, Kent TA, Day EP, Münck E (1985) A [4Fe.4S] cluster with S = 3/2: studies of the nitrogenase in the presence of 0.4 M urea. Rev Port Quim 27:191–193
573. Orme-Johnson WH, Davis LC, Henzel MT, Averill BA, Orme-Johnson NR, Münck E, Zimmerman R (1977) Extrusion of Cp1 clusters into apoproteins of Bacillus polymyxa ferredoxin and adrenodoxin. In: Newton W, Postgate JR, Rodriguez-Barrueco C (eds) Recent developments in nitrogen fixation. Academic Press, New York, pp 131–178
574. Shah VK, Brill WJ (1977) Isolation of an iron-molybdenum cofactor from nitrogenase. Proc Natl Acad Sci USA 74:3249–3253
575. Shah VK, Ugalde RA, Imperial J, Brill WJ (1985) Inhibition of iron-molybdenum cofactor binding to component I of nitrogenase. J Biol Chem 260:3891–3894
576. Cramer SP, Eccles T, Kutzler F, Hodgson K, Mortenson LE (1976) Molybdenum X-ray absorption edge spectra: The chemical state of molybdenum in nitrogenase. J Am Chem Soc 98:1286–1287
577. Flank AM, Weininger M, Mortenson LE, Cramer SP (1986) Single-crystal EXAFS of nitrogenase. J Am Chem Soc 108:1049–1055
578. Teo BK, Antonio MR, Orme-Johnson WH, Nelson MJ, Groh SE, Lindahl PA, Kunzlarich SM, Averill BA (1982) Iron EXAFS of the iron-molybdenum cofactor of nitrogenase. J Am Chem Soc 104:4703–4705
579. Zumft WG (1974) Separation of low molecular weight components from the molybdenum-iron protein of Clostridial nitrogenase. Ber Dtsch Bot Ges 87:135–143
580. Syrtsova LA, Likhtenshtein GI, Popko EV, Druzhinin SYu (1982) On the composition of the active site of the nitrogen-fixing enzyme of nitrogenase. Kinetika Kataliz 23:249–250 (in Russian)
581. Burgess BK, Stiefel EI, Newton WE (1980) Oxidation-reduction properties and complexation reaction of the iron-molybdenum cofactor of nitrogenase. J Biol Chem 255:353–356
582. Newton WE (1981) Nitrogen fixation. In: Encyclopedia of chemical technology. 3rd edn, vol 15. Wiley, New York, pp 942–968
583. Syrtsova LA, Nazarova II, Nazarov WB (1972) NMR investigation of ATP binding by nitrogenase from Azotobacter vinelandii. Dokl Akad Nauk SSSR 206:367–369 (in Russian)
584. Gvozdev RI, Kotelnikov AI, Pivovarov AP, Sadkov AP, Kost AA (1975) Use of the affinity luminescent labeling for the study of nitrogenase structure. Bioorg Chem 1:1207–1214 (in Russian)
585. Levchenko LA, Sadkov AP, Raevski AV, Pivovarova TS, Likhtenshtein GI (1984) Localization of the FeMo cofactor in nitrogenase. Dokl Akad Nauk SSSR 277:1003–1005 (in Russian)
586. Druzhinin SYu, Fogel VR, Syrtsova LA, Kotelnikov AI, Likhtenshtein GI (1986) Location study of the cofactor-binding center in nitrogenase MoFe protein by the triplet probe method. Biofizika 31:16–20 (in Russian)
587. Venters RA, Nelson MJ, McLean PA, True E, Levy MA, Hoffman BM, Orme-Johnson WH (1986) ENDOR of the resting state of nitrogenase molybdenum-iron proteins from Azotobacter vinelandii: ^{1}H, ^{57}Fe, ^{45}Mo, ^{35}S studies. J Am Chem Soc 108:3487–3498
588. Syrtsova LA (1986) The structure and function of the nitrogenase active center. In: Shilov AE (ed) Fundamental research in homogeneous catalysis. Gordon and Breach, New York, pp 879–887
589. Likhtenshtein GI, Panteleeva NS, Skvortsevich EG, Syrtsova LA, Usenskaya AM (1980) On the role of adenosine triphosphatase in the functioning of nitrogenase. Mol Biol (Mosc) 14:147–156 (in Russian)
590. Likhtenshtein GI, Syrtsova LA, Tertyshnaya NI (1982) On the chemical mechanism of the ATPase reaction in energy transformation systems. Dokl Akad Nauk SSSR 262:1013–1016 (in Russian)

591. Kulikov AV, Syrtsova LA, Likhtenshtein GI, Popko EV, Druzhinin SYu, Usenskaya AM (1982) Preparation of the superreduced state of the MoFe protein of nitrogenase of Azotobacter by means of thermalized electrons. Dokl Akad Nauk SSSR 262:1177–1180 (in Russian)
592. Weston MF, Kotake S, Davis LC (1983) Interaction of nitrogenase with nucleotide analogs of ATP and ADP and the effect of metal ions on ADP inhibition. Arch Biochem Biophys 225:809–817
593. Hagen WR, Brasksma A (1985) The importance of quantitative Mössbauer spectroscopy of the MoFe proteins from Azotobacter vinelandii. Eur J Biochem 146:497–501
594. Syrtsova LA, Khramov AB, Usenskaya AM, Likhtenshtein GI, Moravskii AP, Shilov AE (1985) About proton involving in reaction of electron transfer in the nitrogenase. Dokl Akad Nauk SSSR 282:1445–1449 (in Russian)
595. Syrtsova LA, Druzhinin SYu, Usenskaya AM, Khramov AV, Moravskii AP, Likhtenshtein GI (1987) The kinetics of proton release in the process of electron transfer coupling with ATP hydrolysis by nitrogenase. Biofizika 32 (in press) (in Russian)
596. Robson RL (1984) Identification of possible adenine nucleotide-binding sites of nitrogenase Fe and MoFe proteins by amino-acid sequence comparison. FEBS Lett 173:394–398
597. Likhtenshtein GI (1985) The role of multielectron and synchronous processes in enzyme catalysis. Proc 16th FEBS Congr. Part A. VNU Science Press, pp 9–15
598. Likhtenshtein GI (1986) On mechanisms of redox reactions with participation of metalloenzymes and their models. In: Shilov AE (ed) Fundamental research in homogeneous catalysis. Gordon and Breach, New York, pp 325–334
599. Lontie R (ed) (1983) Copper proteins and copper enzymes. CRC, New York
600. Siegel H (ed) (1981) Metal ions in biological systems, vol 13, Dekker, New York
601. Malmström BG (1980) Metalloenzymes. In: Sigman DS, Brazier MA (eds) The evolution of protein structure and function. A symposium in honor of Professor Emil L Smith. Academic Press, New York, pp 87–96
602. Holmberg CG (1944) On the presence of a laccase-like enzyme in serum and its relation to the copper in the serum. Acta Physiol Scand 8:227–229
603. Frieden E, Hsieh HS (1976) Ceruloplasmin: the copper transport protein with essential oxidase activity. In: Meister A (ed) Advances in Enzymology, vol 44. Wiley, New York, pp 187–236
604. Zaitsev VN, Moshkov KA, Shavlovsky MM, Neifakh SA (1975) An X-ray diffraction study of human ceruloplasmin and its modified forms. Bioorgan Chem 1:1521–1527 (in Russian)
605. Miskowski V, Tang S-PW, Spiro TG, Shapiro E, Moss TH (1975) The copper coordination group in 'blue' copper proteins: evidence from resoncance Raman spectra. Biochemistry 14:1244–1250
606. Adman ET, Stenkamp RE, Sicker LC, Jensen LH (1978) A crystallographic model for azyrin at 3Å resolution. J Mol Biol 123:35–47
607. Aisen P, Morell AG (197) (1965) Physical and chemical studies on ceruloplasmin. III A stabilizing copper-copper interaction in ceruloplasmin. J Biol Chem 240:194
608. Deinum J, Vanngard T (1973) The stoichiometry of the paramagnetic copper and the oxidation-reduction potentials of type I copper in human ceruloplasmin. Biochim Biophys Acta 310:321–330
609. Nylen U, Petterson G (1972) The effect of diethylpyrocarbonate on some physical and chemical properties of ceruloplasmin. Eur J Biochem 27:578–584
610. Boden N, Holmes MC, Knowles PF (1974) Binding of water to 'types I and II' Cu^{2+} in proteins. Biochem Biophys Res Commun 57:845–848
611. Van Leeuwen FX, Wever R, van Gelder BF (1973) EPR study of nitric oxide-treated reduced ceruloplasmin. Biochim Biophys Acta 315:200–203
612. Szent-Gyorgyi A (1931) On the function of hexuronic acid in the respiration of the cabbage leaf. J Biol Chem 90:385–393
613. Blumberg WE, Peisach J (1966) The optical and magnetic properties of copper in Chenopodium album plastocyanin. Biochim Biophys Acta 126:269–273
614. Frank P, Pecht I (1986) Redox titration of type 2 copper-depleted Rhus laccase: reductive decoupling oxidative reconstitution of the type 3 site. J Phys Chem 90:3809–3814

615. Avigliano L, Davis JL, Groziani MT, Marchesini A, Mims WB, Mondovi B, Peisach J (1981) Electron spin echo spectroscopic studies of type 1 and type 2 copper in Rhus vernicifera laccase and in cucurbita pepo medullosa ascorbate oxidase. FEBS Lett 136:80–84
616. Allendorf MD, Spira DJ, Solomon EI (1985) Low-temperature magnetic circular dichroism studies of native laccase: spectroscopic evidence for exogenous ligand bridging at a trinuclear copper active site. Proc Natl Acad Sci USA 82:3063–3067
617. Dooley DM, Rawlings J, Dawson JH, Stephens PJ, Andreasson LE, Malmström BG, Gray HB (1979) Spectroscopic studies of type 1 and type 2 copper in Rhus vernicifera laccase and in cucurbita pepo medullosa and Polyporus versicolor laccase. Electronic structures of the copper sites. J Am Chem Soc 101:5038–5046
618. Ali Naki, Varfolomeev SD, Berezin IV (1981) Kinetic regularities of the reduction of molecular oxygen by a copper-containing oxidase (laccase). Dokl Akad Nauk SSSR 260:646–649 (in Russian)
619. Branden R, Deinum J, Coleman M (1978) A mass spectrometric investigation of the reaction between $^{18}O_2$ and reduced tree laccase. FEBS Lett 89:180–182
620. Lerch K (1981) Copper monooxygenases: Tyrosinase and dopamine β-monooxygenase. In: Sigel A (eds) Metal ions in biological systems. Copper proteins. Dekker, New York, 13:144–186
621. Kroneck PM, Jakob W (1985) Ascorbic oxidase: further investigations of its structure and catalytic properties. Rev Port Quim 27:47–48
622. Schoot A, Uiterkamp AJM, Mason HS (1973) Magnetic dipole-dipole coupled Cu(II) pairs in nitric oxide-treated tyrosinase: A structural relationship between the active sites of tyrosinase and hemocyanin. Proc Natl Acad Sci USA 70:993–996
623. Lerch K (1985) The binuclear copper center and the reaction mechanism of tyrosinase. Rev Port Quim 27:46–47
624. Barker SA, Shirley JA (1980) Glucose oxidase, glucose dehydrogenase, glucose isomerase, β-galactosidase and invertase. Microb Enzymes Bioconvers 37:117–226
625. Bordas J, Koch MHJ, Hartmann HJ, Weser U (1982) Tetrahedral copper-sulphur coordination in yeast Cu-thioneine: an EXAFS study. FEBS Lett 140:19–21
626. Tainer JA, Richardson JS, Richardson DC (1983) Structure and mechanism of copper, zinc-superoxide dismutase. Nature 306:284–287
627. Valentine JS, Pantoliano MW (1981) Superoxide dismutases. In: Spiro TG (ed) Copper proteins. Wiley, New York, pp 291–358
628. Free JA, Peisach J, Mims WB (1981) Superoxide dismutase: Examination of the metal binding sites by electron spin echo spectroscopy. J Biol Chem 256:1910–1914
629. Free JA, Briggs RG (1975) Studies on the reconstitution of bovine erythrocyte superoxide dismutase. V. Preparation and properties of derivatives in which both zinc and copper sites contain copper. Biochim Biophys Acta 400:439–450
630. Klug-Roth D, Fridovich I, Rabani J (1973) Pulse radiolytic investigations of superoxide catalyzed disproportionation mechanism for bovine superoxide dismutase. J Am Chem Soc 95:2786–2790
631. Van Bruggen EFJ (1980) Homocyanin: the mystery of blue blood. Trends Biochem Sci 5:185–188
632. Hol WGJ, Volbeda A (1985) The structure of the copper-containing oxygen-transporting hemocyanins from arthropods. Rev Port Quim 27:42–44
633. Brown JM, Powers L, Kincaid B, Larrabee JA, Spiro TG (1980) Structural studies of the hemocyanin active site. 1. Extended X-ray absorption fine structure (EXAFS) analysis. J Am Chem Soc 102:4210–4216
634. Freedman TB, Loehr JS, Loehr TM (1976) A resonance Raman study of the copper protein hemocyanin. New evidence for the structure of the oxygen-binding site. J Am Chem Soc 98:2809–2815
635. Colman PM, Freeman HC, Guss JM, Murata M, Norris VA, Ramshaw JAM, Venkatappa MP (1978) X-ray crystal structure analysis of plastocyanin at 2.7 Å resolution. Nature (Lond) 272:319–324
636. Wikström M (1986) Molecular energetics of the mitochondrial respiratory chains. Biological chemistry. Hoppe-Seyler. 17th FEBS Meeting, Berlin (west), August 24–29. de Gruyter, Berlin, p 66
637. Skulachev VP, Kozlov IA (1977) Protic phosphatases. Nauka, Moscow (in Russian)

638. Martonosi AM (ed) (1985) The enzymes of biological membranes. Plenum Press, New York
639. Green DE, Tzagoloff A (1966) The mitochondrial electron transfer chain. Arch Biochem Biophys 116:293–304
640. Munn EA (1974) The structure of mitochondria. Academic Press, New York, 465 pp
641. Beinert H, Albracht SPJ (1983) New insights, ideas and unanswered questions concerning iron-sulphur clusters in mitochondria. Biochim Biophys Acta 683:245–248
642. Luzikov VN (1980) Regulation of the formation of mitochondria. Molecular aspects. Nauka, Moscow (in Russian)
643. Wilson MT, Jensen P, Aasa R, Malmström BG, Vänngard T (1982) An investigation in EPR and optical spectroscopy of cytochrome oxidase during turnover. Biochem J 203:483–492
644. Mitchell P (1976) Possible molecular mechanisms of the proton-motive function of cytochrome systems. J Theor Biol 62:327–367
645. Williams RJP (1982) The nature of local chemical potential. FEBS Lett 150:1–3
646. Racker E (1976) A new look at mechanism in bioenergetics. Academic Press, New York
647. Kowal AT, Morningstar JE, Johnson MK, Ramsey RR, Singer TP (1986) Spectroscopic characterization of the number and type of iron-sulphur clusters in NADH: ubiquinonee oxidoreductase. J Biol Chem 261:9239–9245
648. Ragan CJ, Galante YM, Hatefi Y, Ohnishi T (1982) Resolution of mitochondrial NADH dehydrogenase and isolation of two iron-sulfur proteins. Biochemistry 21:590–594
649. Case GD, Leigh JS (1976) Intramitochondrial positions of cytochrome haem groups determined by dipolar interactions with paramagnetic cations. Biochem J 160:769–783
650. Boekema EJ, van Breemen JFL, Keegstra W, van Bruggen FJ, Albracht SPJ (1982) Structure of NADH: Q oxidoreductase from bovine heart mitochondria studied by electron microscopy. Biochim Biophys Acta 7 679:7–11
651. Maguire JJ, Johnson MK, Morningstar JE, Ackrell BA, Kearney EB (1985) EPR studies of mammalian succunate dehydrogenase. J Biol Chem 260:10909–10912
652. Ohnishi T, Harmon HJ, Warnung AJ (1985) EPR study of the spatial relationship of redox component of cytochrome *c* oxidase. Biochem Soc Trans 13:607–611
653. Salerno JC, Ohnishi T, Lim J, Widger WR, King TE (1977) Spin coupling between electron carriers in the dehydrogenase segments of the respiratory chain. Biochem Biophys Res Commun 75:618–624
654. Trumpower BL, Edwards CA (1979) Purification of a reconstructively active iron-sulfur protein (oxidation factor) from succinate, Cytochrome *c* reductase complex of bovine heart mitochondria. J Biol Chem 254:8697–8706
655. Van Jagow G, Schagger H, Engel WD, Machleidt J (1978) Beef heart complex III: isolation and characterization of cytochrome *b*. FEBS Lett 91:121–125
656. Capaldi RA, Darley-Usmar U, Fuller S, Millett F (1982) Structural and functional features of the interaction of cytochrome *c* with complex III and cytochrome *c* oxidase. FEBS Lett 138:1–7
657. Leonard K, Wingfield P, Arad T, Weiss H (1981) Three-dimensional structure of ubiquinol: cytochrome *c* reductase from Neurospora mitochondria determined by electron microscopy of membrane cyrystals. J Mol Biol 149:259–274
658. Witt SN, Blair DE, Chan SI (1986) Chemical and spectroscopic evidence for the formation of ferryl Fe_{a_3} intermediate during turnover of cytochrome *c* oxidase. J Biol Chem 261:8104–8107
659. Fuller SD, Capaldi RA, Henderson R (1979) Structure of cytochrome *c* oxidase in deoxycholate-derived two-dimensional crystals. J Mol Biol 134:305–327
660. Swanson MS, Quintanilha AT, Thomas DD (1980) Protein rotational mobility and lipid fludity of purified and reconstituted cytochrome *c* oxidase. J Biol Chem 255:7494–7502
661. Wikström M, Krab K, Saraste M (1981) Cytochrome oxidase. Academic Press, London
662. Powers L, Chance B, Ching Y, Muhoberac B, Weintraub ST (1982) Structural features of the copper-depleted cytochrome *c* oxidase from beef heart: iron EXAFS. FEBS Lett 138:245–248
663. Hill BC, Greenwood C (1984) The reaction of fully reduced cytochrome c oxidase with oxygen studied by flow-flash spectrometry at room temperature. Biochem J 218:913–921

664. Chance RB, Erecinska M, Chance EM (1973) The nature of the cytochrome chain. An experimental and theoretical study. In: King TE, Mason HS, Morrison M (eds) Oxidases and related redox systems, vol 2. University Park Press, Baltimore, pp 851–874
665. Khotchenkov VP, Drozdova NP, Krasnovsky AA (1987) Molecular organization of reaction centres of photosynthesizing bacteria. Biofizika 32:359–368
666. Loach PA (1980) Bacterial reaction center (RC) and photoacceptor complex (PRC) preparations. In: San-Pietro A (ed) Methods in enzymology, vol 69. Photosynthesis and nitrogen fixation, part C. Academic Press, New York, pp 155–172
667. Shuvalov VA, Krasnovskii AA (1981) Photochemical electron transfer in the reaction centers of photosynthesis. Biofizika 26:544–556 (in Russian)
668. Netzel TL, Rentzepis PM, Leigh J (1972) (1973) Picosecond kinetics of reaction centers containing bacteriochlorophyll. Science 128:238–241
669. Shuvalov VA, Parson WW (1981) Energies and kinetics of radical pairs involving bacteriochlorophyll and bacteriopheophytin in bacterial reaction centers. Proc Natl Acad Sci USA 78:957–961
670. Boso BB, Debrunner P, Okamura MY, Feher G (1981) Mossbauer spectroscopy studies of photosynthetic reaction centers from Rhodopseudomonas sphaeroides R-26. Biochim Biophys Acta 638:173–177
671. Eisenberger PM, Okamura MY, Feher G (1980) Investigation of the ferroquinone complex in reaction centers of R. sphaeroides R-26 by EXAFS. Fed Proc 39:1802
672. Kulikov AV, Bogatyrenko VR, Melnikov AV, Syrtsova LA, Likhtenshtein GI (1979) Determination of the distance between charges after their photoseparation in chromatophores from R. rubrum. Biofizika 24:178–181 (in Russian)
673. Tronrud DE, Schmid MF, Mattew BW (1986) Structure and X-ray amino-acid sequence of a bacteriochlorophyll protein from Prosthecochloris aestuaric refined at 1.9 Å resolution. J Mol Biol 188:443–454
674. Likhtenshtein GI, Kotelnikov AI, Kulikov AW, Syrtsova LA, Bogatyrenko VR, Melnikov AI, Frolov EN, Berg AI (1979) Some peculiarities of the electron transfer in redox enzymes. Int J Quantum Chem 16:419–435
675. Likhtenshtein GI, Kulikov AV, Kotelnikov AI, Bogatyrenko VR (1982) Structure and action mechanism of reaction centers of photosynthetic bacteria. Photobiochem Photobiophys 3:337–344
676. Michel H, Deisenhofer J (1986) X-ray diffraction studies on a crystalline bacterial photosynthetic reaction center. A progress report and conclusions on structure of photosystem II reaction center. In: Staehelin LA, Arntzen CJ (eds) Encyclopedia of plant physiology, vol 19, New Series. Springer, Berlin Heidelberg New York, pp 371–381
677. Wanden JT, Bolton JR (1973) Simultaneous quantitative comparison of the optical changes at 700 nm and EPR signals in system I of green plant photosynthesis. J Am Chem Soc 95:6435–6436
678. Ke B, Dolan E, Shuvalov VA, Klimov VV (1982) Early photochemical events in green-plant photosynthesis and EPR spectroscopic studies. In: Alfano RR (ed) Primary events in biology probed by ultrafast laser spectroscopy. Academic Press, New York, pp 55–77
679. Schloder E, Brettel K, With HT (1985) Relation between microsecond reduction kinetics of photooxidized chlorophyll a_{II} (P-680) and photosynthetic water oxidation. Biochim Biophys Acta 808:123–131
680. Shuvalov VA, Duysens LN (1986) Primary electron transfer reactions in modified reaction centers from Rhodopseudomonas spheroides. Proc Natl Acad Sci 83:1690–1694
681. Goodin DB, Yachandra VK, Britt RD, Sauer K, Klein MP (1984) The state of manganese in the photosynthetic apparatus. Light-induced change in X-ray absorption (K-edge) energies of Mn in photosynthetic membranes. Biochim Biophys Acta 767:209–216
682. De Groot A, Plijter JJ, Evelo R, Bobcock GT, Hoff AJ (1986) The influence of the oxidation state of the oxygen-evolving complex of photosystem II on the lattice relaxation time of signal II as determined by electron spin-echo spectroscopy. Biochim Biophys Acta 848:8–13
683. Evans MCW, Atkinson YE, Ford RC (1985) Redox characterization of the photosystem II electron acceptors. Evidence for two-electron carriers between pheophytin and Q. Biochim Biophys Acta 806:247–254

684. Govindjee, Kambara T, Coleman W (1985) The electron donor site of photosystem II: the oxygen-evolving complex. Photochem Photobiol 42:187–210
685. Dismukes GC (1986) The metal centers of the photosynthetic oxygen-evolving complex. Photochem Photobiol 43:99–115
686. Ikeuchi M,Yasa M, Inoue Y (1985) Simple and discrete isolation of an O_2-evolving PS II reaction center complex retaining Mn and the extrinsic 33 KD. FEBS Lett 185:316–322
687. Sato K, Ohno T, Katoh S (1985) An oxygen-evolving complex with a chloroplast structure from thermophilic cyanobacterium. FEBS Lett 180:320–330
688. Cheniae GM (1980) Manganese binding sites and presumed manganese proteins in chloroplasts. In: San Pietro A (ed) Methods in enzymology, vol 69. Academic Press, New York, pp 349–363
689. Wydrzynski T, Sauer K (1980) Periodic changes in the oxidation state of manganese in photosynthetic oxygen evolution upon illumination with flashes. Biochim Biophys Acta 589:56–70
690. Rutherford AW, Agalidis I, Reiss-Husson F (1985) Manganese-quinone interaction in the electron acceptor region of bacterial photosynthetic reaction center. FEBS Lett 182:151–157
691. Yocum CF, Yerkes CT, Blankenship RE, Sharp HA, Babocock GT (1981) Stoichiometry inhibitor sensitivity and organization of manganese association with photosynthetic oxygen evolution. Proc Natl Acad Sci USA 78:7507–7511
692. Kok B, Forbush B, McCloin M (1970) Cooperation of charges in photosynthetic O_2 evolution. I. A linear four-step mechanism. Photochem Photobiol 11:457–475
693. Semenov NN, Shilov AE, Likhtenshtein GI (1975) Multielectron oxidation-reduction processes in chemistry and biology. Dokl Akad Nauk SSSR 221:1374–1377 (in Russian)
694. Sandusky PO, Yocum CF (1986) The chloride requirement for photosynthetic oxygen evolution: factors affecting nucleophilic displacement of chloride from the oxygen-evolving complex. Biochim Biophys Acta 849:85–93
695. Domoder R, Klimov VV, Dismukes GC (1986) The effect of Cl^- depletion and X^- reconstitution on the oxygen evolution rate, the yield of the multiline manganese EPR signal and EPR signal II in the isolated photosystem II complex. Biochim Biophys Acta 848:378–391
696. Saygin Ö, Witt HT (1985) Evidence for the electrochromic identification of the change of charges in the four oxidation steps of the photoinduced water cleavage in photosynthesis. FEBS Lett 187:224–226
697. De Paula JC, Beck WF, Brudvig GW (1986) Magnetic properties of manganese in the photosynthetic O_2-evolving complex 2. Evidence for a manganese tetramer. J Am Chem Soc 108:4002–4009
698. Hoff AJ (1986) Magnetic interactions between photosynthetic reactions. Photochem Photobiol 43:727–745
699. Hansson Ö, Andreasson L, Vänngard T (1986) Oxygen from water is coordinated to manganese in the S_2 state of photosystem II. FEBS Lett 195:151–154
700. Dekker JF, van Gorkom HJ, Wensink J, Ouwehand L (1984) Absorbence difference spectra of the successive redox states of the oxygen-evolving apparatus of photosynthesis. Biochim Biophys Acta 767:1–9
701. Renger G, Weiss W (1986) Studies on the nature of the water-oxidizing enzyme system Y. Spectral characterization of the intermediatory redox state in the water-oxidizing enzyme system. Biochim Biophys Acta 850:184–196
702. Slabes AR, Evans MCW (1977) EPR signals in chloroplasts responding to illumination sequence of four flashes. Nature 270:169–171
703. Kulikov AV, Bogatyrenko VR, Likhtenshtein GI, Allakhverdiev SI, Klimov VV, Shuvalov VA, Krasnovsky AA (1983) Magnetic interaction of manganese with the pheophytin radical-anion in the reaction centers of photosystem 2. Biofizika 28:357–363 (in Russian)
704. Shafirovich VYa, Shilov AE (1986) Chemical models of photosynthesis. Reactions in organized systems. Mendeleev Chem J 31:542–550
705. Digman JD, Strobel HW (1977) NADPH-cytochrome P-450 reductase from rat liver:purification by affinity chromatography and characterization. Biochemistry 16:1116–1123

706. Ribbons DW (1975) Oxidation of C_1 compounds by particulate fractions from Methylococcus capsulatus: distribution and properties of methane-dependent reduced nicotinamide adenine denucleotide oxidase (methane hydrogenase). J Bacteriol 122:1351–1363
707. Woodland MP, Patil DS, Cammack R, Dalton H (1986) ESP studies of protein A of the soluble methane monooxygenase from Methylococcus capsulatus (Bath). Biochim Biophys Acta 873:237–242
708. Kawato S, Gat J, Cherry RJ, Winterhalter KH, Richter C (1982) Investigation of protein-protein interaction of cytochrome P-450 in phospholipid vesicles and liver microsome. J Biol Chem 257:7023–7029
709. Tamburini PP, MacFarquhar S, Schenkman JB (1986) Evidence of binary complex formations between cytochrome P-450, cytochrome b_5, and NADPH-cytochrome P-450 reductase of hepatic microsomes. Biochem Biophys Res Commun 134:519–526
710. Tamburini PP, White RE, Schenkman JB (1985) Chemical characterization of protein-protein interaction between cytochrome P-450 and cytochrome b_5. J Biol Chem 260:4007–4015
711. Rich PR, Tiede DM, Bonner WD (1979) Studies on the molecular organization of cytochromes P-450 and b_5 in the microsomal membrane. Biochim Biophys Acta 546:307–314
712. Udanova YI, Meckler VL, Fogel VR, Kulikov AV, Kotelnikov AI, Likhtenshtein GI, Berkovich MA, Karyakin AV, Archakov AI, Kaplun AP, Schvets VI (1986) Haem Location in haemoproteins by spin and triplet tools. Eur J Biochem 156:541–544
713. Blum H, Leigh JS, Salerno JC, Ohnishi T (1978) The orientation of bovin adrenal cortex cytochrome P-450 in submitochondrial particle multilayers. Arch Biochem Biophys 187:153–160
714. Subkhankulova TN, Lyakhovich VV, Resnikov BA, Elinova BI, Vainer LM (1986) Study of the interaction of lipophilic spin-labelled analog of the cytochrome P-450 substrate with liposomes and microsomes. Biomembrany 3:720–729 (in Russian)
715. Krainev AG, Weiner LM, Alferyev IS, Slynko NM (1985) Bifunctional compound study of the active-center location of cytochrome in a microsomal membrane ("float" molecules method). Biochim Biophys Acta 818:96–104
716. Tonge GM, Harrison DEF, Higgins IJ (1977) Purification and properties of the methane mono-oxygenase enzyme system from Methylosinum trichosporium OB3b. Biochem J 161:333–344
717. Green J, Dalton H (1986) Steady-state kinetic analysis of soluble methane mono-oxygenase from Methylococcus capsulatus (Bath). Biochem J 236:155–162
718. Gvozdev RI, Shushenacheva EV, Pylashenko-Novokhatny AI, Belova VS (1982) Investigation of enzymatic oxidation by methane monooxygenases of Methylococcus capsulatus, strain M, membrane structure. In: The 2nd International Specialists Meeting of the Combustion Institute on Oxidation. Technical University of Budapest, Budapest. Abstracts of papers, p 133
719. Gvozdev RI, Nikonova EL, Pilyashenko-Novokhatny AI, Shushenacheva EV, Grigoryan AN, Belova VS, Shirokova LA (1982) The role of CO-binding cytochrome *c* in enzymatic oxidation of methane by the bacterium Methylococcus capsulatus. Biochemistry 47:1118–1124
720. Shilov AE, Shteinman AA (1977) Activation of saturated hydrocarbons by metal complexes. Kinetika Kataliz 18:1129–1145 (in Russian)
721. Shilov AE (1983) Activation of saturated hydrocarbons by transition metal complexes. Reidel, Dordrecht
722. Aksenova NP, Gvrozdev RI (1987) Purification and properties of methane monooxidase from Methylococcus capsulatus (strain M). Biokhimiya (in press) (in Russian)
723. Tsuprun VL, Aksenova NP, Tagunova IV, Orlova EB, Grigoryan AN, Kiselev NA, Gvozdev RI (1987) Electron microscopy of methane monooxygenase from Methylococcus capsulatus. Dokl Akad Nauk SSSR 292:490–493 (in Russian)
724. Bill E, Trautwein AX, Winkler H, Bernhardt FH (1985) Nitrosyl-binding to the mononuclear non-heme iron of putidamonooxin: a model for the corresponding peroxo complex. Rev Port Quim 27:102–103

725. Belova VI, Gvozdev RI, Malashenko YuR, Sadkov AP, Yurchenko VV (1976) The isotope effect in the enzymatic oxidation of methane. Biokhimiya 41:1903–1904 (in Russian)
726. Lumry R, Rajender Sh (1970) Enthalpy-entropy compensation phenomenon in water solutions of proteins and small molecules: ubiquitous property of water. Biopolymers 9:1125–1227
727. Blumenfeld LA (1983) Physics in bioenergetic processes. Springer, Berlin Heidelberg New York
728. Lumry R, Gregory RB (1986) Free energy management in protein reactions: concepts, complications, and compensation. In: Welch GR (ed) The fluctuating enzymes. Wiley, New York, pp 3–185
729. Likhtenshtein GI (1966) Regularities in the enregy and entropy properties of enzymatic processes. Biofizika 11:24–32 (in Russian)
730. Gunsalus IC, Sligar SC (1978) Oxygen reduction by the P-450 monooxygenase systems. Advan Enzymol 47:1–44
731. Van Leenwen JW, Batler J, Swallow AJ (1981) A non-equilibrium state of deoxyhaemoglobin. T-dependence and O_2 binding. Biochim Biophys Acta 667:185–196
732. Lumry R, Eyring H (1954) Conformation changes of proteins. J Phys Chem 58: 110–120
733. Lumry R, Biltonen R (1969) Thermochemical and kinetical aspects of protein conformation. In: Tiamsheff SN, Fasman GD (eds) Structure and stability of Biological molecules, vol 2. biological macromolecules series. Dekker, New York, pp 146–306
734. Huang Y, Hara T, Sligar S, Coon MJ, Kumura T (1986) Thermodynamic properties of oxidation-reduction reactions of bacterial microsomal and mitochondrial cytochrome P-450. An entropy-enthalpy compensation effect. Biochemistry 25:1390–1394
735. Linderstrom-Lang KH, Schellman JA (1959) Protein structure and enzyme activity. In: Boyer PD, Lardy H, Myrback K (eds) The enzymes, vol 1. Academic Press, New York, pp 443–510
736. Koshland DE (1959) Mechanisms of transfer enzymes. In: Boyer PD, Lardy H, Myrback K (eds) The enzymes, vol 1. Academic Press, New York, pp 305–346
737. Englander SW, Kallenbach NR (1983) Hydrogen exchange and structural dynamics of proteins and nucleic acid. Q Rev Biophys 16:521–552
738. Lumry RW, Rosenberg A (1976) The water basis for mobile defects in proteins and the role of these defects in function. In: Alfsen A, Berteaud AJ (eds) L'eau et les sistemes biologiques, CNRS, Paris, pp53–62
739. Abaturov LV, Lebedev Yu, Nosova N (1983) The dynamical structure of proteins. Mol Biol (Mosc) 17:543–558 (in Russian)
740. Kossiakov AA (1982) Protein dynamics investigated by the neutron diffraction hydrogen exchange technique. Nature 296:713–721
741. Likhtenshtein GI, Avilova TV (1973) The dynamical structure of proteins and enzyme catalysis. Uspekhi Biol Nauk 75:26–45 (in Russian)
742. Morishima J, Yizuka T (1974) NMR studies of hemoproteins. Restricted relation of a heme side chain methyl group in some ferric myoglobin complexes. J Am Chem Soc 96:7365–7367
743. Oster O, Neireiter AW, Clouse AO, Gurd FRN (1975) Deuterium NMR of deuterium-labeled diacetyl-deuterohemin incorporated into sperm whale myoglobin. J Biol Chem 250:7990–7996
744. Moore GR, Huang ZX, Eley CG, Barker HA, Williams G, Robinson MN, Williams RJP (1982) Electron transfer in biology. The function of cytochrome *c*. Faraday Disc Chem Soc 74:311–329
745. Wutrich K, Wagner G (1975) NMR investigation of the dynamics of the aromatic amino acid residues in the basic pancreatic trypsin inhibitor. FEBS Lett 50:265–268
746. Wilbur DJ, Norton RS, Cluse AO, Addlman R, Allerhand A (1976) Determination of rotational correlation times of proteins in solutions from ^{13}C spin lattice relaxation measurements. Effect of magnetic field strength and anisotropic rotation. J Am Chem Soc 98:8250–8254
747. Aksenov SI (1977) Investigation of the conformational mobility of proteins in aqueous media by the NMR technique. In: Burshtein EA (ed) Equilibrium dynamics of the native structure of protein. Institute of Biophysics, Pushchino, pp 42–59 (in Russian)

748. Likhtenshtein GI (1986) Water and dynamics of proteins and membranes. Stud Biophys 111:89–100
749. Lakowicz JR (1983) Principles of fluorescence spectroscopy. Plenum Press, New York
750. Permyakov ZEA, Burshtein EA (1977) Relaxation processes in frozen aqueous solutions. Effects of water isotopic composition and chromophore localization on the relaxation freezing. Stud Biophys 64:163–172
751. Lakovicz JR, Weber G (1973) Quenching of protein fluorescence of oxygen. Biochemistry 12:4171–4179
752. Frolov EN, Likhtenshtein GI, Khurgin YuI, Belonogova OV (1973) The mobility of the spin probe in the hydrophobic pocket of α-chymotrypsin. Izv Akad Nauk SSSR (Ser Khim) 1:132 (in Russian)
753. Frolov EN, Kharakhonycheva NV, Likhtenshtein GI (1977) Investigation of the dynamical structure of human serum albumin by the spin probe technique. Mol Biol (Mosc) 6:886–893 (in Russian)
754. Likhtenshtein GI, Troshkina TV, Akhmedov YuD, Shuvalov VF (1969) On the possibility of the determination of the energy parameters of local conformational transitions in proteins by the paramagnetic label technique. Mol Biol (Mosc) 3:413–420 (in Russian)
755. Likhtenshtein GI, Kotelnikov AI (1983) Study of the functional intramolecular mobility of proteins by the physical probe technique. Mol Biol (Mosc) 17:505–518 (in Russian)
756. Likhtenshtein GI, Kotelnikov AI, Kulikov AV (1982) New approaches to studies of water-protein and water-membrane interactions by physical labeling. Stud Biophys 91:23–27
757. Berg AI, Kononenko AF, Noks PP, Khrymova IN, Frolov EN, Rubin AB, Likhtenshtein GI, Uspenskaya N, Khideg K (1979) The conformational mobility and functional dependence of photosynthetic reaction centers from Rhodospirillum rubrum. Mol Biol (Mosc) 13:469–477 (in Russian)
758. Noks PP, Kononenko AA, Rubin AB, Likhtenshtein GI, Kochetkov VV, Kol'tover VK, Kotelnikov AI, Fogel VR, Grishanova NP (1982) Intramolecular dynamics and electron transfer in photosynthetic reaction centers. Investigations carried out by EPR and luminescence techniques. The 1st USSR Congress. Abstracts of papers, vol 1, No 346 (in Russian)
759. Likhtenshtein GI (1983) The role of intramolecular dynamics of proteins in enzyme catalysis. Uspekhi Biol Khim 23:3–35 (in Russian)
760. Careri G, Fassella P, Gratton E (1975) Statistical time events in enzymes. Time estimation. Crit Rev Biochem 3:141–164
761. Artemiuk PJ, Blake CCF, Grace DEP, Catley SJ, Phillips DS, Hernberg MJE (1979) Crystallographic studies of the dynamic properties of lysozyme. Nature 280:563–566
762. Frauenfelder H, Petsko GA, Tchernoglou D (1979) T-dependent X-ray diffraction as a probe of protein structural dynamics. Nature 280:558–563
763. Shaitan KV, Rubin AB (1980) Conformational mobility and the theory of the Mössbauer effect in biosystems. Mol Biol (Mosc) 14:1323–1334 (in Russian)
764. Goldansky VI, Kononenko AA, Rubin AB, Frolov EN, Chernavsky DS (1981) On the relationship between electron tunnelling and conformational mobility and degrees of freedom of biomolecules. Dokl Akad Nauk SSSR 257:491–494 (in Russian)
765. Käiväräinen AI (1985) Solvent-dependent flexibility of proteins and principles of their function. Reidel, Dordrecht
766. Lifshitz IM, Grosberg AYu, Khokhlov AR (1978) Some problems of the statistical physics of polymer chains with volume interaction. Rev Modern Phys 50:683–713
767. Sarvazyan AP, Hemmes P (1979) Relaxation cantributions to protein compressibility from ultrasonic data. Biopolymers 18:3015–3024
768. Shepard JW (1975) Non-elastic scattering of laser beam by synthetic and biological polymers. Rep Progr Phys 38:565–620
769. Lumry R (1979) Dynamical aspects of a small molecule protein interaction. In: Briabanti A (ed) Bioenergetics and thermodynamics. NATO Advanced Study. Parma, Italy, pp 435–454
770. Krinichny VI, Grinberg OYa, Yudanova EI, Lubashevskaya EB, Antsiferova LI, Likhtenshtein GI, Lebedev YaS (1987) Study of spin-labeled lysozyme by 2-mm ESP spectroscopy. Biofizika 32:215–220 (in Russian)

771. Krinichny VI, Grinberg OYa, Yusupov IKh, Marupov RM, Bobodzhanov PKh, Likhtenshtein GI, Lebedev YaS (1986) Two-millimeter band ESR study of spin-labeled cotton fiber. Biofizika 31:482–485 (in Russian)
772. Likhtenshtein GI (1963) On the possible explanation of the compensation effect in some processes. Kinetika i Kataliz 4:35–38 (in Russian)
773. Karush F (1950) Heterogeneity of the binding site of bovine serum albumin. J Am Chem Soc 72:2705–2713
774. Sergeev PV, Magai IV, Lexina VP, Stepanyants AU, Likhtenshtein GI (1976) Investigation of the interaction of steroids with human serum albumin by the spin echo-paramagnetic probe technique. Mol Biol (Mosc) 10:1166–1174
775. Demchenko AP (1982) On the nanosecond mobility in proteins. Edge excitation fluorescence red shift of protein-bound 2(p-toluidinylnaphthalene)6-sulphonate. Biophys Chem 15:101–109
776. Krupyanskii YE, Parak F, Gaubman EE, Wagner FM, Goldanskii VI, Mössbauer R, Suzdalev IP, Vogel H (1980) Investigation of the dynamics of metmyoglobin by Rayleigh scattering of Mössbauer radiation. J Phys 41:489–490
777. McCammon JA, Lee CY, Northrup SH (1983) Side chain rotational isomerization in proteins: a mechanism involving gating and transient packing. J Am Chem Soc 105:2232–2237
778. Bauminger ER, Cohen SG, Nowik I, Offer S, Yariv J (1983) Dynamics of heme iron in crystals of metmyoglobin and deoxymyoglobin. Proc Natl Acad Sci USA 80:736–740
779. Mayo KH, Kucheida D, Parak F, Chien JC (1983) Structural Dynamics of human deoxyhemogobin and hemochrome investigated by Mössbauer spectroscopy. Proc Natl Acad Sci USA 80:5294–5296
780. Krinichny VI, Grinberg OYa, Bogatyrenko VR, Likhtenshtein GI, Levedev YaS (1985) Study of the microenvironment effect on magnetic resonance parameters of spin-labeled human serum albumin in a 2 mm range. Biofizika 30:216–219 (in Russian)
781. Peterson-Kennedy SE, McGourty JL, Hoffman BM (1984) Temperature dependence of long-range electron transfer in [Zn, Fe^{III}] hybrid hemoglobin. J Am Chem Soc 106:5010–5012
782. Austin RH, Beeson KW, Ensenstein L, Frauenfelder H, Gunsalus IC (1975) Dynamics of ligand binding to myoglobin. Biochemistry 14:5355–5362
783. Johnson ME (1978) Vibrational motion of an immobilized spin label hemoglobin. Biochemistry 17:1223–1228
784. Levedev YaS, Burstein AI (1976) Elementary kinetics in molecular solids. Free volume model of the diffusion-limited processes. Chem Phys 12:259–267
785. Agmon N, Hopfield JJ (1983) CO binding to hem proteins; a model for barrier height distributions and slow conformational changes. J Chem Phys 79:2042–2053
786. Calef DF, Wolynes PG (1983) Classical solvent dynamics and electron transfer. II Molecular aspects. J Chem Phys 78:470–482
787. Privalov PL, Khechinashvili NN (1974) A thermodynamic approach to the problem of stabilization of globular protein structure: a colorimetric study. J Mol Biol 86:665–684
788. Mrevlishvili GM (1979) Low-temperature colorimetry of biomolecules. Uspekhi Fiz 120: 273–312 (in Russian)
789. Goldanskii VI, Krupynaskii YuF, Flerov BN (1983) Glass-like model of protein. Dokl Akad Nauk SSSR 272:978–981 (in Russian)
790. Likhtenshtein GI (1985) Amorphous-crystalline model of intramolecular dynamics of globular proteins. Biofizika 30: 27–30 (in Russian)
791. Khurgin YuI, Chernavsky DS, Shnol' SE (1967) The protein molecule as a mechanical system. In: Frank CM (ed) Vibrational processes in biological and chemical systems. Nauka, Moscow, pp 130–138 (in Russian)
792. Gavish B, Werber MM (1979) Viscosity dependent structure fluctuations in enzyme catalysis. Biochemistry 18:1269–1275
793. Goldanskii VI, Krupynaskii YuF, Fedotov EN (1983) The role of conformational substrates in protein reactivity. Mol Biol (Mosc) 17:532–542 (in Russian)
794. Lippmaa ET, Olivson AI, Jarvet YuI (1983) ^{13}C NMR spectroscopy study of carbanhydrase. J Mol Biol (Mosc) 17:484–491 (in Russian)

795. Calhann DB, Vanderkooi JM, Woodrow GV, Englander SW (1983) Penetration of dioxygen into proteins studied by quenching of phosphorescence. Biochemistry 22:1526–1532
796. Lipari G, Szabo A, Levy RM (1982) Protein dynamics and NMR relaxation: comparison of simulations with experiment. Nature 300:197–198
797. Salemme FR (1982) Cooperative motion and hydrogen exchange stability in protein β-sheets. Nature 299: 754–756
798. Likhtenshtein GI (1976) Water and protein dynamics. In: Alfsen A, Bertrand AJ (eds) L'eau et les systems biologiques. CNRS Paris: 45–53
799. Churg AK, Weiss RM, Warshel A (1983) On the cytochrome *c*: correlating geometry changes upon oxidation with activation energy of electron transfer. J Phys Chem 87:1683–1694
800. Churg AK, Warshel A (1985) Modeling the activation energy and dynamics of electron transfer in proteins. In: Clementi E (ed) Structure and motion: membranes, nucleic acids and proteins. Academic Press, New York, pp 97–105
801. Redi MH, Gerstman BS, Hopfield JI (1981) Hemoglobin-carbon monoxide binding rate and low T magnetic-optical detection of spin tunneling. Biophys J 35:471–484
802. Gatteschi D, Kahn O, Willett RD (eds) (1986) Magneto-structural correlation in exchange coupled systems. Reidel, Dordrecht
803. Goldanskii VI (1979) Facts and hypotheses of molecular chemical tunnelling. Nature 279:109–115
804. Berdnikov VM, Bogadchikov GA (1979) On the calculation of the probability of the outer-sphere electron transfer in dilute aqueous solutions. Zh Fiz Khim 53:273–280 (in Russian)
805. Likhtenshtein GI (1987) Clusters in biological systems. Mendeleev Zh Khim 32:140–161 (in Russian)
806. Owen Y (1962) Spin resonance of ion pairs in crystal lattices. J Appl Phys Suppl 33:355–357
807. Kryshtalik LI, Kharkatz UI (1985) Reorganization energies of the medium in the charge transfer reaction in the protein globule. J Theor Biol 112:221–227
808. Zauhar RJ, Morgan RS (1985) A new method of computing of macromolecular electric potential. J Mol Biol 186:815–820
809. Ritsman M, Bosshard HR (1986) Sulfite oxidase from chicken liver. The role of imidazole and carboxyl groups for the reaction with cytochrome *c*. Eur J Biochem 150:493–497
810. Churg AK, Warshel A (1986) Control of redox potential of cytochrome *c* and microscopic dielectric effect in proteins. Biochemistry 25:1675–1681
811. Mitovska MI, Dancheva KI, Atanasov BP (1984) Charge interaction of cytochrome *c* with cytochrome *c* oxidase. Int J Biochem 16:1059–1064
812. Marcus RA, Sutin N (1985) Electron transfer in chemistry and biology. Biochim Biophys Acta 811:265–322
813. Kotelnikov AI, Fogel VR, Bystryak IM, Likhtenshtein GI (1986) Factors determining electron transfer in proteins. In: Shilov AE (ed) Fundamental research in homogeneous catalysis. Gordon and Breach, New York, pp 773–784
814. Vogel VR, Rubtsova ET, Kotelnikov AI, Likhtenshtein GI (1986) Investigation of protein diffusion collisions by triplet labels. Biofizika 31:152–153 (in Russian)
815. Fogel VR, Likhtenshtein GI, Kotelnikov AI (1985) An estimation of the electron transfer parameters in cytochrome *c*. Biofizika 30:394–399 (in Russian)
816. Petrov EG (1980) Mechanism of electron transfer through proteins. In: Petrov EG (ed) Progress in quantum chemistry and quantum biology. Proc of the Int Conf, part 1. Naukova Dimka, Kiev (in Russian)
817. Closs GL, Calcaterra LT, Green NJ, Rentfield KW, Miller JR (1986) Distance, stereoelectronic effects and the Marcus inverted region in organic intramolecular electron transfer in organic radical anions. J Phys Chem 90: 3673–3683
818. Moore GR (1983) Control of redox properties of cytochrome *c* by specific electrostatic interactions. FEBS Lett 161:171–175
819. Sider P, Cave RJ, Marcus RA (1984) A model for orientation of effects in electron transfer reactions. J Chem Phys 81:5613–5625
820. Wherland S, Grey HB (1977) Electron transfer mechanism employed by metalloproteins. In: Addison AW, Cullen WR, Dolphin D (eds) Biological aspects of inorganic chemistry. Wiley Interscience, New York, pp 289–325

821. Moore GR, Williams RJP (1976) Electron transfer proteins. Coord Chem Rev 18:125–197
822. Zinth W, Knapp EW, Fischer SF, Kaiser W, Deisenhofer J, Michel H (1985) Correlation of structural and spectroscopic properties of a photosynthesis reaction centre. Chem Phys Lett 119:1–4
823. Gupta RK, Koenig SH, Redfield AG (1972) On the electron transfer between cytochrome molecules as observed by NMR. J Magn Res 7:66–73
824. Reynolds JG, Coule CL, Holm RH (1980) Electron exchange kinetics of $Fe_4S_4(SR)_4{}^{-2}/Fe_4S_4(SR_4)^{-3}$ systems. Estimates of the intrinsic self-exchange rate constant of the 4 Fe sites in oxidezed and reduced ferredoxins. J Am Chem Soc 102:4350–4354
825. Mizzahi JA, Meyer TE, Gusanovitch A (1980) Kinetics of oxidation and reduction of high-potential iron-sulfur proteins with nonphysiological reaction. Biochemistry 21:4727–4733
826. Goldberg M, Pecht J (1976) Kinetics and equilibria of the electron transfer between azurin and hexacyanoiron (II/III) couple. Biochemistry 15:4197–4205
827. Mizzahi JA, Gusanovitch MA (1980) Electron transfer reactions of high-potential iron-sulfur proteins and *c*-type cytochromes. Biochemistry 21:4733–4747
828. Wilson NT, Silverstrini MC, Morpurgo L, Brunori M (1979) Electron transfer kinetics between stellacyanin and cytochrome *c*. J Inorg Biochem 11:95–100
829. Marcus RA, Sutin N (1975) Electron-transfer reactions with unusual activation parameters. A treatment of reactions accompanied by large entropy decreases. Inorg Chem 14:213–223
830. Wherland S, Pecht J (1978) Protein-protein electron transfer. A Marcus theory analysis of reaction between *c*-type cytochrome and blue copper proteins. Biochemistry 17:2585–2590
831. Salemme FR (1976) Computer modelling of the cytochrome c-cytochrome b_5 interaction. J Mol Biol 102:563–569
832. Poulos TL, Kraut J (1980) Computer modelling of the cytochrome *c* oxidase-cytochrome c interaction. J Biol Chem 255:10322–10330
833. Davydov RM, Magonov SN, Arutyunayan AM, Sharonov YuA (1978) The absorption and magnetic circular dichroism spectra of heme-containing proteins in nonequilibrium states. IV cytochrome *c* and its derivatives. Mol Biol (Mosc) 12:1341–1347 (in Russian)
834. Larsson S (1982) Some theoretical remarks concerning electron pathways. Faraday Disc Chem Soc 74:390–392
835. Newton MD (1982) Medium effect in electron transfer reactions. Faraday Disc Chem Soc 74:110–111
836. Gray HB (1985) Electron transfer at fixed distances in metalloproteins. Rev Port Quim 27:8–9
837. Brunschwig BS, de Laive PJ, English M, Goldberg M, Gray HB, Mayo SL, Sutin N (1985) Kinetics and mechanisms of electron transfer between blue copper proteins and electronically excited chromium and ruthrenium polypyridine complexes. Inorg Chem 24:3743–3749
838. Gunner MR, Robertson DE, Dutton PL (1986) Kinetic studies on the reaction center protein from Rhodopseudomonas spheroides: the temperature and free energy dependence of electron transfer between quinones in the Q site and the oxidezed bacteriochlorophyll dimer. J Phys Chem 90: 3783–3795
839. Debus RJ, Feher G, Okamua MY (1986) Iron-depleted reaction center from Rh. spheroides R-26. Characterization and reconstitution with Fe^{2+}, Mn^{2+}, Ni^{2+}, Cu^{2+}, Zn^{2+}. Biochemistry 25:2276–2287
840. Phillips WD, McDonald, Stombaugh NA, Orme-Johnson WH (1974) PMR and magnetic susceptibility characterization of ferredoxin I from Bacillus polymyxa. Proc Natl Acad Sci USA 71:140–143
841. Chamorovsky SK, Kononenko AA, Petrov EG, Pottosin II, Rubin AB (1986) Effects of dehydration and low temperature on the oxidation of high-potential cytochrome *c* by photosynthetic reaction center in Ectothiorhodospira shaposhnikovii. Biochim Biophys Acta 848:402–410
842. De Vault B, Parks JH, Chance B (1967) Electron tunnelling in cytochromes. Nature 215:642–665
843. Dogonadze RR, Zakharia MG (1984) Theory of low-temperature processes of electron transfer. Biofizika 29:548–552 (in Russian)

844. Bystryak IM, Likhtenshtein GI, Kotelnikov AI, Hankovskii O, Hideg K (1986) Influence of solvent molecular dynamics on photochemical reduction of nitroxide radicals. Zh Fiz Khim 60:2796–2802 (in Russian)
845. Berezin IV, Varfalomeev SD (1979) Biokinetics. Nauka, Moscow (in Russian)
846. Likhtenshtein GI, Syrtsova LA, Samuilov VD, Frolov EN, Borisov AYu, Bogatyrenko VR (1975) Study on primary steps of electron transfer in photosynthesis. In: Abstracts of the papers presented at the XII International botanic Congress. Nauka, Leningrad, p 429 (in Russian)
847. Shin IV, Kotelnikov AI, Likhtenshtein GI (1982) Kinetics of electron phototransfer in the eosin-casein complex. Biofizika 27:208–211 (in Russian)
848. Dzhabiev TS, Shilov AE (1980) Photosynthesis and its chemical models. Zh Vses Khim Ova Im DI Mendeleeva 5:503–514 (in Russian)
849. Ford WE, Otvos JW, Corvin M (1979) Photosensitized electron transfer across lipid vesicle walls. Proc Natl Acad Sci USA 76:3590–3593
850. Lomyr' SV, Tsvetkov IM, Parmon VN, Zamaraev KI (1982) Electron phototransfer across the membranes of lipid vesicle walls. Khim Phys Fiz 3:405–408 (in Russian)
851. Krasnovsky AA, Semenova AN, Nikandrov VV (1982) Modelling of photosystem I by means of chlorophyll-containing liposomes. Dokl Akad Nauk SSSR 262:469–472 (in Russian)
852. Semenov NN, Shilov AE, Likhtenshtein GI (1975) Polynuclear processes in chemistry and biology. Dokl Akad Nauk SSSR 221:1374–1377 (in Russian)
853. Lyanagi T, Yamazaki I, Anan KF (1985) One-electron oxidation-reduction properties of ascorbic acid. Biochim Biophys Acta 806:255–261
854. Krasnov KS, Timoshinin VS, Danilova TG, Khondoshko SV (1968) The molecular constants of inorganic compounds. A handbook. Khimiya, Leningrad (in Russian)
855. Parai-Koshits MA, Atovmyan LO (1974) Crystallochemistry of coordination membranse compounds. Nauka, Moscow (in Russian)
856. Blumenfeld LA (1978) The physical aspects of energy transduction in biosystems. Q Rev Biophys 11:251–308
857. Racker E (1976) A new look at mechanism in bioenergetics. Academic Press, New York
858. Dame JB, Scorborough GA (1981) Identification of the phosphorylated intermediate of the Neurospora plasma membrane as β-aspartyl phosphate. J Biol Chem 256:10724–10730
859. Panteleeva NS (1975) Myosin (exchange and phosphorylation). Leningrad University Press, Leningrad (in Russian)
860. De Jonge PC, Westerhoff HV (1982) The proton per electron stoichiometry of site 1 of oxidative phosphorylation at high proton-motive force is close to 1.5. Biochem J 204:515–523
861. Proteau G, Wrigglesworth JM, Nicholls P (1983) Proton-motive functions of cyt *c* oxidase in reconstituted vesicles. Biochem J 210:199–205
862. Marvin DA (1983) Proton current and proton motion in membranes. FEBS Lett 156:1–5
863. Mitchell P, Moyle J (1983) Alternative hypotheses of proton ejection in cytochrome *c* oxidase vesicles. FEBS Lett 151:167–178
864. Likhtenshtein GI, Shilov AE (1976) On the chemical mechanism of coupling of ATP hydrolysis reactions. Dokl Akad Nauk SSSR 229:1127–1130 (in Russian)
865. Lenaz G, Esposti MD, Timoneda J, Barber T (1984) Studies on the mechanism of protein translocation in the mitochondrial bc_1 complex. In: Third International Conference on Water and Ions in Biological Systems. Bucharest, Rumania, 2–6 Oct. Abstracts, pp 79–80
866. Papa S (1984) Proton-motive activity of energy-transfer systems of mitochondria. In: Third International Conference on Water and Ions in Biological Systems. Bucharest, Rumania, 2–6 Oct. Abstracts, pp 80–81
867. Malmström BG (1985) Cytochrome *c* oxidase as a proton pump. A transition state mechanism. Biochim Biophys Acta 811:1–12
868. Porras AG, Palmer G (1982) The room temperature potentiometry of the flavin, molybdenum and iron-sulfur centers. J Biol Chem 257:11617–11626
869. Likhtenshtein GI, Syrtsova LA, Tyrtyshnaya NI (1982) About the chemical mechanism of the ATPase reaction in energy-transformation systems. Dokl Akad Nauk SSSR 262:1013–1016 (in Russian)

870. Nicholls DG (1982) Bioenergetics and introduction to the chemiosmotic theory. Academic Press, New York
871. Ida S, Mikami B (1986) Spinach ferredoxin-nitrite reductase: a purification procedure and characterization of chemical properties. Biochim Biophys Acta 871:167–176
872. Sutch K, Yamamoto K, Wakabayashi (1986) Electron microscopic visualization of the ATPase site of myosin by photoaffinity labeling with a biotynylated photoreactive ADP analog. Proc Natl Acad Sci USA 83:221–226
873. Hol WGJ, Volbeda A (1985) The structure of the copper-containing oxygen-transporting hemocyanins from Arthropods. Rev Port Quim 27:42–44
874. Ohnishi T, Harmon HJ, Warnung AJ (1985) EPR Studies on the spatial relationship of redox components of cytochrome oxidase. Biochem Soc Trans 13:607–611
875. Groves JT, McMurry TJ (1985) Synthetic analogs of oxidized heme proteins. Preparation and characterization of iron (IV) porphyrins. Rev Port Quim 27:102–103
876. Fukuyama K, Hase T, Matsumoto S, Tsukihara, Katsube Y, Tanaka N, Kakudo M, Wada K, Matsubara H (1980) Structure of S. platensis [2Fe.2S] ferredoxin and evolution of chloroplast-type ferredoxins. Nature 286:522–524
877. Kabsch AW, Sander C (1983) How good are products of protein secondary structure? FEBS Lett 155:179–186
878. Ptitsyn OB, Finkelstein V (1983) Theory of protein secondary structure and algorithm of its prediction. Biopolymers 22:15–23
879. Bixon M, Jortner J (1986) Coupling of protein modes to electron transfer in bacterial photosynthesis. J Phys Chem 90:3795–3800
880. McCarthy MB, White RE (1983) Functional differences between peroxide compound I and the cytochrome P-450 reactive oxygen intermediate. J Biol Chem 258:9153–9158
881. Witt HT (1983) Biophysical aspects of the molecular machine of photosynthesis. Stud Biophys 94:161–165
882. Fukuyama K, Hase T, Matsumoto S, Tsukihara T, Katsube Y, Tanaka N, Kakudo M, Wada K, Matsubara H (1980) Structure of S. platensis 2Fe-2S ferredoxin and evolution of chloroplast-type ferredoxins. Nature 286:522–524
883. Brzezinski P, Malmström BG (1986) Electron transport driven proton pumps display nonhyperbolic kinetics. Simulation of the steady-state kinetics of cytochrome *c* oxidase. Proc Natl Acad Sci USA 83:4282–4286
884. Shilov AE (1978) N_2 fixation in solution in presence of transition metal complexes. In: Hardy RWF, Bottomley F, Burns RC (eds) A treatise on dinitrogen fixation. Wiley Interscience, New York, pp 37–97
885. Likhtenshtein GI (1982) Kinetic and structural features of polynuclear metalloenzymes. In: Likhtenshtein GI (ed) Redox enzymes and their models, part 1. Nauka, Moscow, pp 3–28 (in Russian)
886. Schwarz D, Pirrwitz J, Rückpaul K (1982) Rotational diffusion of cytochrome P-450 in the microsomal membrane-evidence for a clusterlike organization from saturation transfer EPR spectroscopy. Arch Biochem Biophys 216:322–327
887. Sevdimaliev PM, Vosnjak (1986) Magnetic field-sensitive stage of the charge recombination process in photosystem II reaction center of higher plants. A two-radical-pair mechanism. Stud Biophys 115:69–80
888. Moura JJG, Teixeira M, Moura I, Xavier AV, le Gall J (1985) Desulfobvibrio gigas hydrogenase. Catalytic cycle and activation process. Rev Port Quim 27:63–66
889. Purmal AP (1967) Redox catalysis by transition ions. In: Braunstein AE, Yakovlev VA (eds) Mechanisms of respiration, photosynthesis and dinitrogen fixation. Nauka, Moscow, pp 167–172 (in Russian)
890. Travina OA, Travin SO, Purmal AP (1986) Ascorbic acid catalytical oxidation by kation $Fe_2CrO(AcO^-)_6(H_2O)$. Khim Fizika 5:1123–1129 (in Russian)
891. Mayo SL, Ellis WR, Crutchley RJ, Gray HB (1986) Long-Range electron transfer in heme proteins. Science 233:948–952
892. Barynin VV, Vagin AA, Melik-Adamjan VP, Vainstein BK (1986) Structure of T-catalase. Dokl Acad Nauk 228:877–880
893. Spiro DJ, Man Sung CO, Solomon EI, Hodgson KO (1983) EXAFS investigation of the binuclear cupric site in met T2D Rhus Laccase and its azide bound derivative. Biochem Biophys Res Commun 112:746–753

Subject Index

MIX
Papier aus verantwortungsvollen Quellen
Paper from responsible sources
FSC® C105338

If you have any concerns about our products,
you can contact us on
ProductSafety@springernature.com

In case Publisher is established outside the EU,
the EU authorized representative is:
Springer Nature Customer Service Center GmbH
Europaplatz 3, 69115 Heidelberg, Germany

Printed by Libri Plureos GmbH
in Hamburg, Germany